Nanoparticles in the Water Cycle

Fritz H. Frimmel · Reinhard Niessner
Editors

Nanoparticles in the Water Cycle

Properties, Analysis and Environmental Relevance

Springer

Editors
Prof. Fritz H. Frimmel
Karlsruhe Institute of Technology (KIT)
Engler-Bunte-Institut
Bereich Wasserchemie
Engler-Bunte-Ring 1
76131 Karlsruhe
Germany
fritz.frimmel@kit.edu

Prof. Dr. Reinhard Niessner
TU München
Institut für Wasserchemie
und Chemische Balneologie
Marchioninistr. 17
81337 München
Germany
reinhard.niessner@ch.tum.de

ISBN 978-3-642-10317-9 e-ISBN 978-3-642-10318-6
DOI 10.1007/978-3-642-10318-6
Springer Heidelberg Dordrecht London New York

Library of Congress Control Number: 2010922457

Cover design: WMXDesign GmbH, Heidelberg

Printed on acid-free paper

Springer is part of Springer Science+Business Media (www.springer.com)

Foreword

The Gottlieb Daimler and Karl Benz Foundation explores as models new paths of interdisciplinary research, accompanied by a continuous dialogue between theory and practice. In the projects for clarifying the "Interplay between Humanity, Technology and the Environment" sponsored by the foundation, the search for solutions to societally relevant questions is combined with new research methods which are integrated in networks.

Interdisciplinary studies in the "Ladenburg Collegia" stand in the programme's centre of interest and have investigated, among other subjects, "Group Interaction in High-Risk Environments", the "Zwischenstadt" (City in Between) or the relationship between the "Internal Clock and Shift-Work".

In the "Ladenburg Discourses", the foundation discusses new research projects, together with scientists and experts from practice. In just such a discourse, the authors of this volume have concerned themselves with the subject of "Engineered Nanoparticles – An Issue for the Next Decades". Conferences – like the "Berlin Colloquium" – and lectures open to the public additionally bring current scientific topics to the attention of a broader public.

Encouragement of young scientists and aid to developing nations are further points of the foundation's concentration: a fellowship programme finances the stay for doctoral students from all subject areas at research centres abroad, respectively, in Germany. Seminars in Vietnam, North Korea, and Myanmar bring German scientists together with colleagues from these countries.

Ladenburg, Germany

Gottlieb Daimler
and Karl Benz Foundation

Preface

Colloids have been addressed as neglected phase. And indeed the borderline between purely dissolved molecules and particulates was for a long time not too well defined. The non-availability of powerful characterisation methods and a limited field of colloid-based daily life products may be reasons for that. This has dramatically changed during the last two decades. Engineered NanoParticles (ENP) with promising properties were tailored and produced in technical scale. They found their way as advanced materials in broad application, e. g. in surface treatment, catalysis, paints, medicine and personal care products. The impact of ENP-based products on our life style has even led to the expression of the "Nano Age".

Unfortunately the fascination of the great possibilities of the nanoworld has not reached so far as to care about the life cycles of most of its products. That is to say, the fate of ENP after use and their function in the environment are still broadly unknown. Only air pollution by nanoparticles has found some attention, and the availability of quantification methods has even led to the enforcement of legal limits. Concerning aquatic systems, the situation is less clear.

Nanoparticles in the water cycle have been identified as timely subject for discussing the state of art, the possibilities and the needs for their sound assessment. A selected group of scientists has come together to interdisciplinarily address this issue. The result is given in this volume. The first section deals with basic aspects of nanoparticles and their function in air, water, soil and water treatment. The second part contains examples for the application of advanced instrumentation for quantification and characterisation of nanoparticles with special respect to aqueous samples. Bioeffects of nanoparticles in water are elaborated in the third section, and finally standardisation activities for the assessment of nanoparticles in the aquatic environment are presented. We thank all the authors and the reviewers for their fine work and enthusiasm they put into this project.

The content of this book could not have been assembled without the generous support by the *Gottlieb Daimler and Karl Benz Stiftung* which also hosted the two workshops for the topic and funded the editing of the book. The great personal interest of Prof. Dr. zu Putlitz and Prof. Dr. Dietrich acting as chairmen of the foundation as well as Dr. Klein, Frau Hallenberger and Herr Schmitt from the foundation's administration made the days in Ladenburg a wonderful experience of science and enjoyment for the participants of the workshops.

We also want to thank Markus Delay and Luis Tercero for the excellent work they put into author contacts and manuscript handling.

Karlsruhe, Germany
München, Germany,
February, 2010

Fritz H. Frimmel
Reinhard Niessner

Contents

Contributors

Matthias Ballauff Institute for Physical Chemistry, University of Bayreuth, D-95440 Bayreuth, Germany, matthias.ballauff@uni-bayreuth.de

Thomas Baumann Institute of Hydrochemistry, Technische Universität München, 81377 München, Germany, thomas.baumann@ch.tum.de

John Coates Department of Plant and Microbial Biology, University of California, 271 Koshland Hall, Berkeley, CA 94720, USA

Markus Delay Engler-Bunte-Institut, Chair of Water Chemistry, Karlsruhe Institute of Technology (KIT), Engler-Bunte-Ring 1, D-76131 Karlsruhe, Germany, markus.delay@kit.edu

Karl Fent School of Life Sciences, University of Applied Sciences Northwestern Switzerland, CH-4132 Muttenz, Switzerland; Department of Environmental Sciences, Swiss Federal Institute of Technology (ETH Zürich), CH-8092 Zürich, Switzerland, karl.fent@bluewin.ch

Fritz H. Frimmel Engler-Bunte-Institut, Chair of Water Chemistry, Karlsruhe Institute of Technology (KIT), Engler-Bunte-Ring 1, D-76131 Karlsruhe, Germany, fritz.frimmel@kit.edu

Sophie-Charlotte Gleber Institute for X-ray Physics, University of Göttingen, Friedrich-Hund-Platz 1, 37077 Göttingen, Germany

Hartmut Gliemann Institute of Functional Interfaces, Karlsruhe Institute of Technology (KIT), D-76131 Karlsruhe, Germany, hartmut.gliemann@kit.edu

Birgit C. Gordalla Engler-Bunte-Institut, Chair for Water Chemistry, Karlsruhe Institute of Technology (KIT), Engler-Bunte-Ring1, D-76131 Karlsruhe, Germany, birgit.gordalla@kit.edu

Anastasios D. Karathanasis Department of Plant and Soil Sciences, University of Kentucky, N-122 K Ag, Science-North, Lexington, KY 40546, USA akaratha@uky.edu

George Metreveli Engler-Bunte-Institut, Chair of Water Chemistry, Karlsruhe Institute of Technology (KIT), Engler-Bunte-Ring 1, D-76131 Karlsruhe, Germany

Jürgen Niemeyer Institute for Applied Biotechnology in the Tropics at the Georg-August-University Göttingen, 37079 Göttingen, Germany

Reinhard Niessner Institute of Hydrochemistry, Technische Universität München, Marchioninistrasse 17, D-81377 München, Germany, reinhard.niessner@ch.tum.de

Carsten Prasse Federal Institute of Hydrology (BfG), Am Mainzer Tor 1 D-56068 Koblenz, Germany

Jens Rieger BASF SE, Polymer Physics, 67056 Ludwigshafen, Germany

Thomas Schimmel Institute of Applied Physics and Institute of Nanotechnology, Karlsruhe Institute of Technology (KIT), D-76128 Karlsruhe, Germany, thomas.schimmel@kit.edu

Heiko Schwegmann Engler Bunte Institut: Chair of Water Chemistry, Karlsruhe Institute of Technology (KIT), Engler-Bunte-Ring 1, D-76131 Karlsruhe, Germany, heiko.schwegmann@kit.edu

Julia Sedlmair Institute for X-ray Physics, University of Göttingen, Friedrich-Hund-Platz 1, 37077 Göttingen, Germany

Luis A. Tercero Espinoza Engler-Bunte-Institut, Chair of Water Chemistry, Karlsruhe Institute of Technology (KIT), Engler-Bunte-Ring 1, D-76131 Karlsruhe, Germany

Thomas Ternes Federal Institute of Hydrology (BfG), Am Mainzer Tor 1 D-56068 Koblenz, Germany, Ternes@bafg.de

Jürgen Thieme NSLS-II Project, Brookhaven National Laboratory, Upton, NY 11973, USA, jthieme@bnl.gov

Harald Zänker Institute of Radiochemistry, Forschungszentrum Dresden-Rossendorf, D-01314 Dresden, Germany, h.zaenker@fzd.de

Chapter 1
Introducing the "Nano-world"

Fritz H. Frimmel and Markus Delay

1.1 Setting the Scene

"Nano" means 10^{-9}, consequently, the diameter of nanoparticles (NP) lies in the range of 10^{-7} to 10^{-9} m. The atomic dimension lies around 10^{-10} m which is equal to the old unit 1 Å.

Regarding the world of small particles, colloids have a long tradition in chemistry and physics and also in chemical engineering. Originally, glue or lime were the first representatives of that specific state of matter which is characterized by the particularly fine distribution of very small particles. Another famous example from the old days is the ruby glass which contains colloidal gold clusters of 20–100 nm size.

In the meantime, the classical colloids have got brothers and sisters. They are represented by the new matter called nanoparticles and also include engineered ones. This booming field of production and application is of great influence on our daily life, in industry, biology, farming, food production, and medicine.

There is no question, colloidal and nano systems have been present in nature from the beginning. The structure and function of shells is based on small $CaCO_3$ particles of about 200 nm size combined with proteins. A principle which is used in new synthetic materials. Or the fascinating color of butterfly wings: It is the result of light scattering and interferences on nano scale matter. This has been imitated in innovative coatings in which the nanoparticle size determines the color.

The application of modern engineered nanoparticles (ENP) includes colors, self-cleaning surfaces and scratch-resistant coatings, sunscreens and UV blocking clothes, fibers and fascicles in fabrics, high-performance insulation, and the broad field of food technology where dispersibility and particle size distribution are important not only for the optical appearance of many products but also for their taste (Paschen, 2004; Wiesner et al., 2006).

F.H. Frimmel (✉)
Engler-Bunte-Institut, Chair of Water Chemistry, Karlsruhe Institute of Technology (KIT), Engler-Bunte-Ring 1, D-76131 Karlsruhe, Germany
e-mail: fritz.frimmel@kit.edu

F.H. Frimmel, R. Niessner (eds.), *Nanoparticles in the Water Cycle*,
DOI 10.1007/978-3-642-10318-6_1, © Springer-Verlag Berlin Heidelberg 2010

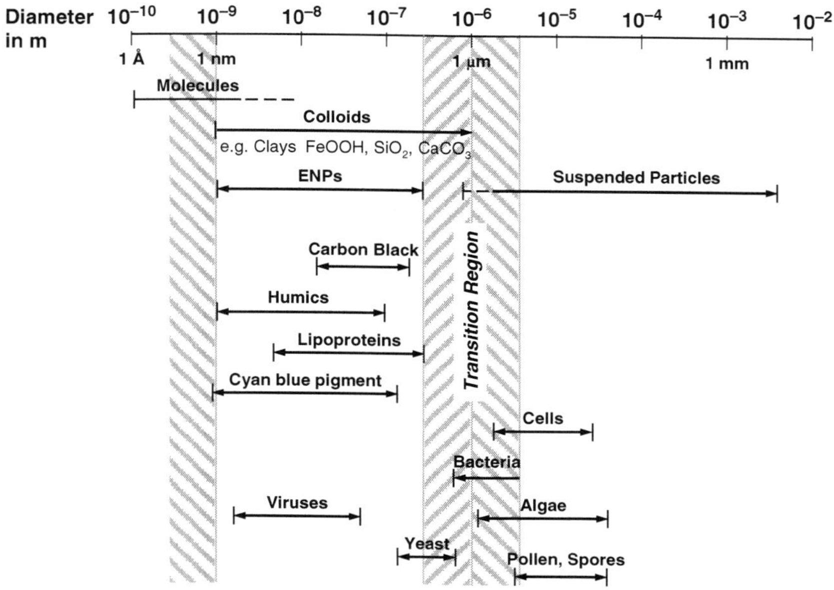

Fig. 1.1 Size ranges for ENP and colloids in aqueous systems (modified from Lead and Wilkinson, 2007; Nowack and Bucheli, 2007)

It is interesting to look for the key aspects responsible for the special properties of both colloids and ENP in aqueous matrices.

According to ASTM Standard E2456-06 (2006), nanoparticles are ultrafine particles with lengths in two or three dimensions greater than 1 nm and smaller than about 100 nm. Figure 1.1 shows some examples for colloids and nanoparticles in between the range of molecular distribution on the lower side and larger particles as aggregates, flocs cells, etc., on the higher side.

1.2 Surface Area and Surface Properties

As a consequence of small particles size, the mass-specific surface of NP is quite high. This can easily be demonstrated with a theoretical experiment in which a cube of given volume (V) and mass is cut into smaller cubes with bisecting sides. Repeating this again and again leads to a fast increase of the surface (A) of all produced cubes at constant total volume and mass. Increasing surface also means more molecules or atoms of the matter of the initial cube are in the surface compared to inside (Fig. 1.2).

The energy necessary to create a new unit area of surface (A) is given according to Eq. (1.1) (Everett, 1988):

$$\Delta G = \Delta W = 2\delta^0 A \tag{1.1}$$

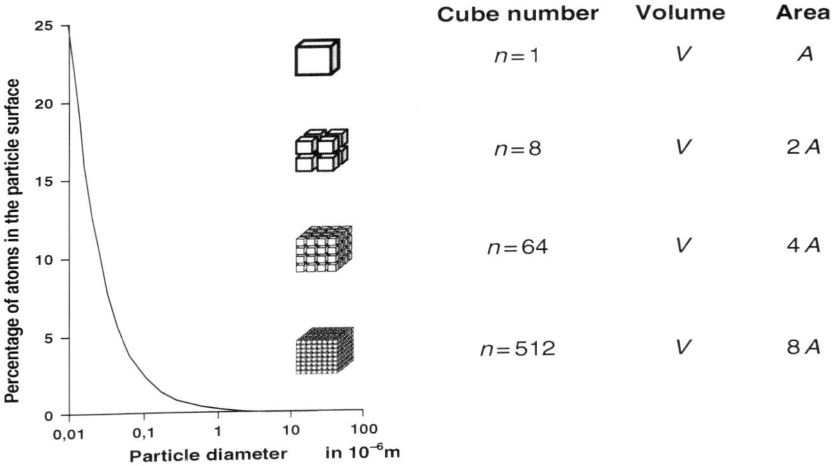

Fig. 1.2 Mass-specific surface areas of divided cubes (modified from Everett, 1988)

with

ΔG: increase of free energy,
ΔW: work needed to separate the pieces reversibly against the forces of attraction,
δ^0: proportionality factor (surface or interfacial tension).

The surface tension δ^0 of the free surface is reduced by $\Delta\delta^0$ on immersion in a fluid medium, e.g., in water compared to the situation in vacuum.

These interactions are crucial for the stability of the disperse systems and depend on the special chemical composition of the surfaces. The interactive forces are a result of the LONDON–VAN DER WAALS molecular forces and BORN repulsion. Experiments and the calculated interactive forces between particles suggest that the energy of interaction falls off much less with their distance in comparison to single molecules (Brezesinski and Mögel, 1993).

1.3 Surface Charge

Figure 1.3 shows some examples for simplified NP surfaces of different chemical composition. It is obvious that surfaces in which oxygen atoms are involved mostly have negative charges and that the surface charges are strongly dependent on the pH value of the liquid phase.

The surface charge of the NPs is of great influence on their water solubility and on the stability of their suspensions. The resulting attractive (+–) and repulsive (++ or – –) forces are dependent on the intervening medium, its pH value, and the

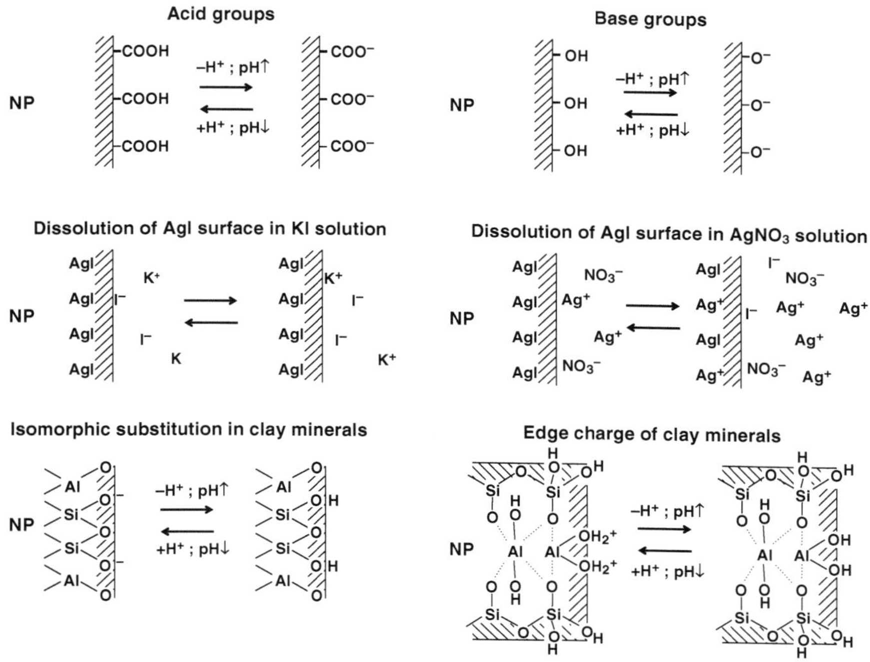

Fig. 1.3 Surface charges and their generation (modified from Everett, 1988)

kind and concentration of its electrolytes. In addition, the concentration of the NP forming agents, the presence of polymers, and the temperature influence the stability of suspensions. Figure 1.4 shows the pH dependence on the formation of colloidal metal hydroxides and the relevant hydrolyzed species. The lowest solubility of the hydroxides is at the pH value of their electro-neutral formulation.

The effect of electrolytes on the agglomeration of charged NPs can be explained with the help of the model of diffuse double layer (Fig. 1.5).

The negative surface charge of NP attracts the cations of the electrolyte more than its anions. This leads to a partial neutralization of the surface and the resulting potential decreases exponentially from the NP's surface into the bulk solution (GOUY-CHAPMAN layer) with direct influence on the stability of the suspension. It is reasonable to assume that increasing concentrations of electrolyte and increasing cation charge both result in a steeper decrease of the potential, i.e., in a compression of the GOUY–CHAPMAN layer. The consequence is a stimulation of aggregation and floc formation.

There is the empirical SCHULZE–HARDY rule (Eq. 1.2) which allows the estimation of salt concentrations of differently charged cations necessary to reach a similar destabilization effect of a NP system (Crittenden, 2005):

$$c(Me^{+})\colon c(Me^{2+})\colon c(Me^{3+}) = \left(\frac{1}{1}\right)^{6} : \left(\frac{1}{2}\right)^{6} : \left(\frac{1}{3}\right)^{6} \tag{1.2}$$

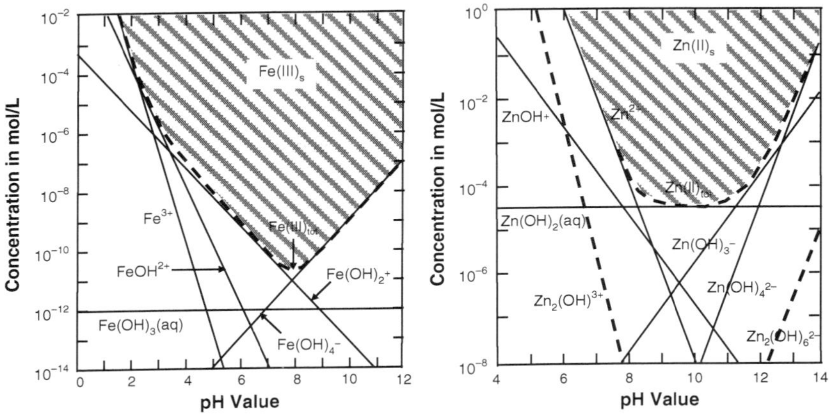

Fig. 1.4 pH related Fe and Zn species in aqueous solutions (modified from Stumm and Morgan, 1996)

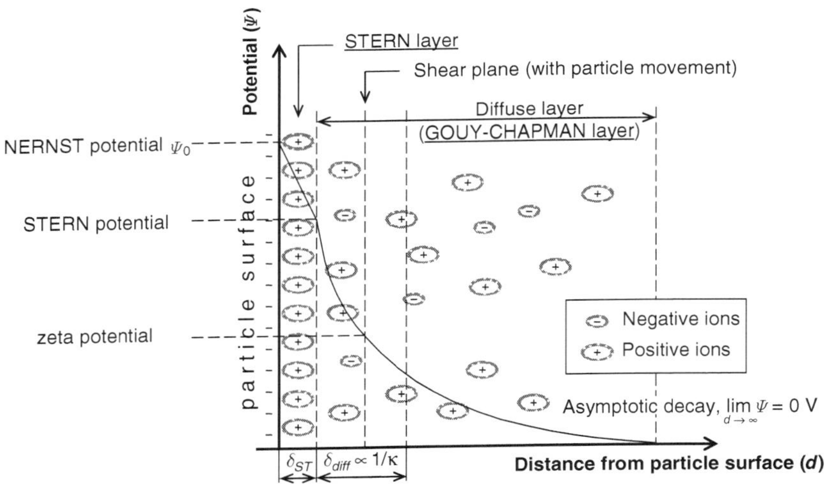

Fig. 1.5 Model of the diffuse double layer (modified from Stumm and Morgan, 1996)

In water treatment, realistic concentrations for the removal of geogenic colloids by floc formation are 20–150, 0.5–3, and 0.01–0.1 mmol/L for mono-, di-, and trivalent cations, respectively.

1.4 Water Cycle

There are several hot spots in the water cycle for NPs and their interaction with other water constituents. Table 1.1 gives an overview of the different domains of aquatic systems and examples for the NPs to be expected.

It is clear that the broad application of ENP is closely related to the water cycle by the fact that we use water for drinking, cleaning, production, and even for leisure purposes. Tracing the NP there is not easy due to the manifold other water constituents. Interactions are to be expected including coating, mixed agglomerations and flocculation, sorption onto surfaces and cell walls, uptake in biofilms, and biotransformation. It is a great challenge to apply and develop the analytical tools for seeing the fate of the ENP in this complex environment. To learn about their distribution, reactions, and functions will hopefully bring us to a better understanding of the downside of their application and by this open the doors for an even better tailoring and selection of the products and the best kind of their use.

There is the ubiquitous geogenic refractory organic matter called humic substances, a kind of higher molecular polyelectrolyte and, by this, polyfunctional member of the NP-family. There are many mineral-derived NP of silicate, carbonate, or metal oxide structure. There is the large zoo of xenobiotics which can adsorb, react, or interact in a complex way (Frimmel et al., 2007).

Table 1.1 Occurrence of nanoparticles and colloids in the water cycle

Domain	Nanoparticles and colloids	Topic of interest
Atmosphere, precipitation	Aerosols	Cloud formation, transport
Infiltration, groundwater	Airborne NP	Soil interaction
	Dry deposition Geogenic NP	Crop interaction Fate in aquifers
Rivers/lakes	Airborne NP Geogenic NP NP from waste water (treated)	Transport Sedimentation Bio-uptake Bio-transformation Photo-transformation
(Waste) Water treatment	Geogenic NP	Sedimentation
	ENP	Bio-uptake
	Technical NP (to be eliminated, to be used)	Elimination processes Advanced treatment methods

1.5 Analytical Tools and Characterization Systems

Good analytical methods are meant to be powerful in identification and quantification. Their direct application for the characterization of analytes in complex matrices, however, is often limited and lead to data with low accuracy. This also applies to many environmental samples which on the other hand should be analyzed without major changes to avoid contamination and artifact formation. Needless to say that low concentrations which are quite normal for aqueous samples bear another challenge for the method development and the operator.

In Table 1.2, some important NP parameters and suitable methods for their determination in aqueous samples are given (see also Hofmann et al., 2003).

As often in analytical chemistry the accuracy of data and their meaning can be improved by combination of powerful separation methodology with sensitive identification tools. Practicability increases in cases with direct on-line coupling of the two approaches.

(Asymmetrical) Flow field-flow fractionation ((A)F^4) has turned out to be a highly promising method for aqueous NP characterization and particle separation especially when coupled with highly sensitive detection methods like inductively

Table 1.2 Methods to characterize NP in aqueous samples

NP parameter	Method	Direct application to water samples
Hydrodynamic radius Size distribution Concentration	Field-flow fractionation (FFF) – flow field – sedimentation – thermal field – electric field	Yes
	Multi-angle laser light scattering (MALLS)	Yes
Electrophoretic mobility Zeta-potential Point of zero charge (PZC)	Photon correlation spectroscopy (PCS) Electrophoresis Sedimentation potential	Yes
Chemical composition	Scanning electron microscope (SEM)	No
	Energy-dispersive X-ray spectroscopy (EDX)	No
	X-ray absorption spectroscopy (XAS)	No/Yes
Size distribution and elemental composition	Asymmetrical flow field-flow fractionation or size exclusion chromatography coupled with inductively coupled plasma mass spectrometry (SEC/ICP-MS or AF^4/ICP-MS)	Yes

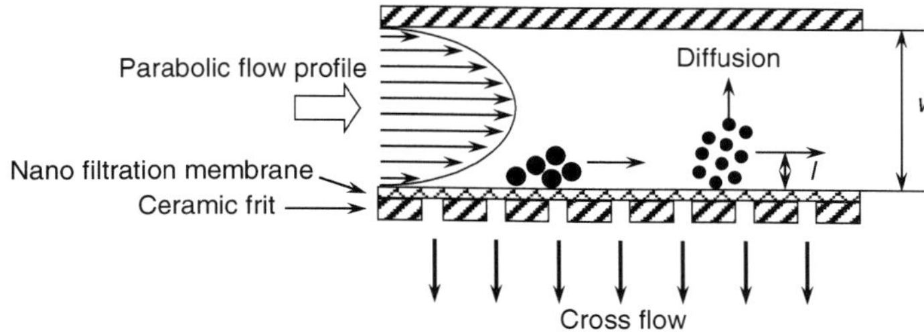

Separation is based on the hydrodynamic diameter of the analyte particles.

Retention parameter λ_{AFFFF}

$$\lambda_{AFFFF} = \frac{l}{w} = \frac{kTV^0}{3\pi\eta\dot{V}_c d w^2}$$

k	Boltzmann's constant
T	Absolute temperature
V^0	Geometric volume of the channel
η	Dynamic viscosity of the fluid
$\dot{V}_c$	Volumetric rate of cross flow
d	Hydrodynamic diameter of the particle
w	Channel thickness

Fig. 1.6 Asymmetrical flow field-flow fractionation (AF^4)

coupled plasma mass spectrometry (ICP-MS). The (A)F^4 method is based on non-turbulent flow with parabolic profile in a thin layer cell which often is followed by a cross flow to the main flow direction through a membrane at the cell bottom (Schimpf et al., 2000). The principle of the method is given in Fig. 1.6.

In case of the sedimentation FFF, a gravity field (G) is used instead of the cross flow. By this, the retention time of particles of diameter d is proportional to d^3 and this is proportional to $1/G$.

The applicability of the two methods is given in Fig. 1.7.

Flow FFF is normally more powerful than sedimentation FFF, the application of which can be pushed to lower particles sizes by increasing the density difference between NP and liquid phase. This can be a limitation for unknown suspensions on the one hand, but can lead to specific information about mixtures on the other hand.

on-line detection after high-resolution separation with the so-called hyphenated systems is of special value for the characterization of complex mixtures. This also includes aqueous samples of NP. It is of great advantage if those systems run in standardized mode. There should be a sample injection port possibly fed by an automated sample provider, the separation unit can be an analytical column or fractionation device or optional even a packed column with real-life porous material like sand. For precise results the *on-line* detection system has to include an internal standard addition device to calibrate the data obtained. It is essential to run and control the whole system by a data system. A scheme of such a setup is shown in Fig. 1.8.

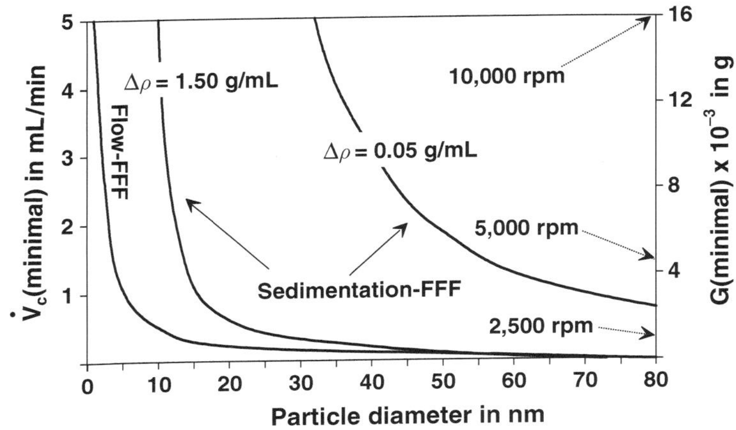

Fig. 1.7 Cross fields needed for sedimentation and flow FFF to just retard particles depending on their diameter ($\lambda = 0.1$)

An example for the characterization of several monodisperse NP standard suspensions (polystyrene particles) with different diameter is given in Fig. 1.9.

Calibration of unknown NP has to be done cautiously for often authentic calibration samples are not available or cannot be assumed due to unknown or altered functional structure of the samples.

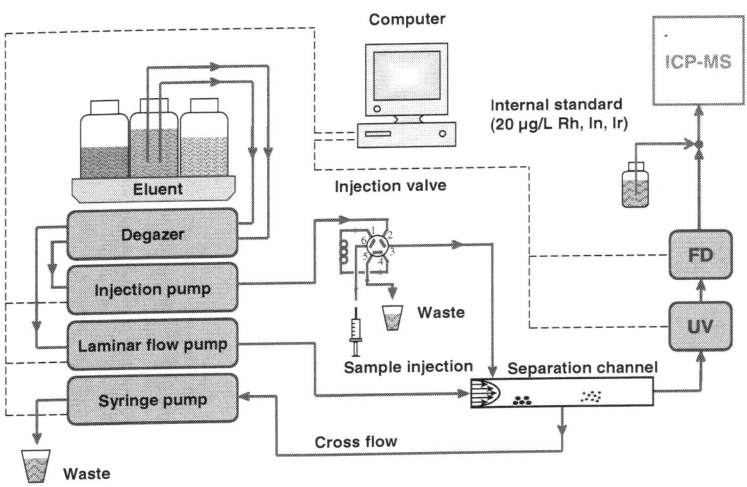

Fig. 1.8 Coupling of a separation unit (here: AF^4) with UV and fluorescence detection (FD) and ICP-MS

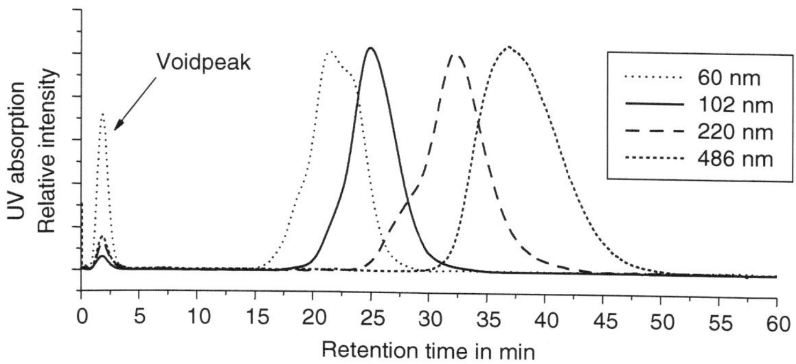

Fig. 1.9 AF^4 fractograms of monodisperse particle standards (polystyrene particles)

1.6 Conclusions

Colloids and ENP have great similarities but they can also have specifically different properties. Whereas most geogenic colloids show typically a broad particle size distribution, ENP often have a narrow distribution and are defined to be monodisperse. In general, however, the basic properties can be understood for both species on the same theoretical basis. From this it can be concluded that also the characterization methods which can be applied are the same in both cases.

Whereas plenty of information has become available for the properties of ENP and their production and application, relatively little is known about their fate after use. It is evitable that ENP have found and will find their way into the environment. This asks for investigations on the amount, the distribution, the interactions, and reactions with abiotic and biotic matter.

Many powerful methods have become available for characterization of ENP and their behavior. Unfortunately a direct application to samples with complex matrices like water, wastewater, sediments, and soil does not lead to meaningful results. Therefore further development is needed to open the door for tracing ENP in the dominant phases of their distribution in the environment including living cells. Hyphenated methods are promising to separate and identify the different nano-sized species. In addition, the question on their function has to be brought into the focus of our work. Do the environmentally aged ENP still have the properties of their "clean" phase of production? With which partners did they affiliate? And which products came out of that? How much did all this change the properties of the matter? Are there any bioeffects?

These questions have to be addressed adequately, for they supply the key for a better understanding of the whole "nano-world," for tailoring our products and for using and applying them in the right, i.e., sustainable way.

References

ASTM Standard E 2456-06 (2006) Terminology for Nanotechnology. West Conshohocken, USA.

Brezesinski G, Mögel HJ (1993) Grenzflächen und Kolloide: Physikalisch-chemische Grundlagen. Spektrum Akademischer Verlag, Heidelberg.

Crittenden JC (2005) Water Treatment: Principles and Design, 2nd ed. Wiley, New York.

Everett DH (1988) Basic Principles of Colloid Science. Royal Society of Chemistry, London.

Frimmel FH, von der Kammer F, Flemming HC (eds.) (2007) Colloidal Transport in Porous Media, Springer-Verlag, Berlin, Heidelberg.

Hofmann T, Baumann T, Bundschuh T, von der Kammer F, Leis A, Schmitt D, Schäfer T, Thieme J, Totsche K-U, Zänker H (2003) Aquatische Kolloide II: Eine Übersichtsarbeit zur Probennahme, Probenaufbereitung und Charakterisierung. Grundwasser 8: 213–223.

Lead JR, Wilkinson KJ (2007) Environmental colloids and particles: current knowledge and future developments. In: Wilkinson KJ and Lead JR (eds.) Environmental Colloids and Particles Behaviour, Separation and Characterisation. IUPAC Series on Analytical and Physical Chemistry of Environmental Systems. John Wiley and Sons, Chichester, pp. 1–16.

Nowack B, Bucheli TD (2007) Occurrence, behavior and effects of nanoparticles in the environment. Environ Pollut 150: 5–22.

Paschen H (2004) Nanotechnologie: Forschung, Entwicklung, Anwendung. Springer, Berlin.

Schimpf ME, Caldwell K, Giddings, JC (eds.) (2000) Field-Flow Fractionation Handbook. John Wiley and Sons, New York.

Stumm W, Morgan JJ (1996) Aquatic Chemistry: Chemical Equilibria and Rates in Natural Waters, 3rd ed. John Wiley & Sons, New York.

Wiesner M, Lowry G, Alvarez P, Dionysiou D, Biswas P (2006) Assessing the risks of manufactured nanomaterials. Environ Sci Technol 40: 4336–4345.

Chapter 2
Nanoparticles Acting as Condensation Nuclei – Water Droplet Formation and Incorporation

Reinhard Niessner

Obviously the situation is more complex than expressed by the statement "without particles no precipitation", but through assessment of the water cycle and identification of its bottlenecks we will see that nanoparticles definitely play a major role.

Accepting the fact that water can only be recycled and not newly synthesized by nature, one has to deal with the water cycle in more detail. Figure 2.1 depicts a simple scenario for this.

To begin the discussion one should note that water vapour is already present in the atmosphere to a large extent. Depending on variables such as ambient temperature, the area of free water surfaces, wetted soil surfaces, and plant exhalation, billions of tons of water in the gaseous state is translocated and dispersed globally by meteorological factors. This water vapour can be inferred to be the real source of fresh water through various mechanisms which act to change the water vapour from the gaseous state to a condensed one; the link being air-dispersed particles, i.e., aerosols.

Cloud or fog formation is not possible without the existence of submicron-sized particles. This had already been recognized in previous centuries, and Paul-Jean Coulier (1875) was the first to develop an experimental setup to visualize aerosol particles by condensing water on them. Many obscure ancient technologies including the use of large cannons in Styria (Austria) or the still-present technology of cloud seeding (to prevent hailstorm formation) by spreading artificially generated silver iodide aerosols into the cloud base, had the aim of modifying the weather, namely precipitation events.

In order to understand the role of nano-engineered particles within the natural water cycle we need to have greater knowledge of the transition of water vapour to the condensed state.

R. Niessner (✉)
Institute of Hydrochemistry, Technische Universität München, Marchioninistrasse 17, D-81377, München, Germany
e-mail: reinhard.niessner@ch.tum.de

F.H. Frimmel, R. Niessner (eds.), *Nanoparticles in the Water Cycle*,
DOI 10.1007/978-3-642-10318-6_2, © Springer-Verlag Berlin Heidelberg 2010

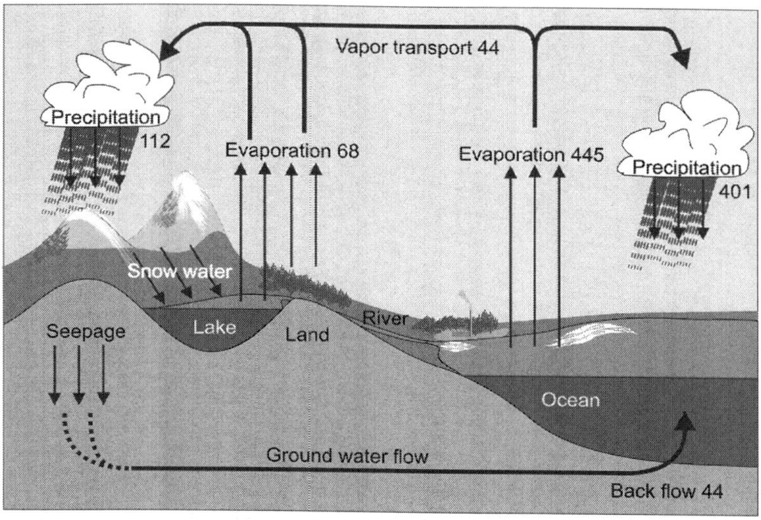

Fig. 2.1 The total water cycle and its bottlenecks. The numbers show the transported water volumes for 1,000 km^3 per year

2.1 Condensation of Water Vapour onto Aerosol Particles

Many reviews are available giving some insight into the different peculiarities of the condensation process within a cloud (or fog). The most comprehensive standard textbook on this was written by Pruppacher and Klett (1997).

For the evaluation of the role and impact of nano-engineered particles some requirements have to be fulfilled:

The ambient atmosphere often becomes slightly water supersaturated. This typically occurs as air masses become elevated when crossing higher orographic obstacles, for example accessing a mountain from the direction of the sea. Other mechanisms include convective cooling during changes in daylight. A supersaturated air mass exceeding 100% relative humidity will hold a few percent more water in the gaseous state for a period of time than expected in the equilibrium state.

Usually there are two main routes for the return to equilibrium conditions. The first is dilution by dry air masses – in this case no condensation of water will be observed. The second mechanism is condensation on ultrafine water-soluble particles. Simply put, water vapour forms a brine at the particle surface (e.g. NaCl). Because of solvolysis the water vapour pressure above such droplet surfaces becomes depleted (according to Raoult's law). As described by Fick's law water diffusion remains active in the direction of the "sink" (the brine), as long as there is a vapour pressure gradient between the particle acting as the condensation nuclei and the surrounding supersaturated air mass. This process comes to a halt after some time because of increasing dilution of the brine at the particle surface.

Once the concentration of ionizable surface material is in the trace regime a notable vapour pressure gradient no longer exists to stimulate further water vapour condensation. Because of the release of condensation heat during diffusion to and from a condensing nucleus the whole process might become very complex. Qualitatively, well dissolvable particle cores will serve as very efficient condensation nuclei. In essence, the particles will become incorporated into the water cycle and have a chance to become precipitated.

For the assessment of the role of dispersed nano-engineered particles within the water cycle, we have to identify those which can be involved in such water condensation processes. As already mentioned above, dissolvable material would fit perfectly. In contrast, hydrophobic particles as individuals can only act as condensation nuclei when a certain water supersaturation is present. The well-known Kelvin equation describes such a situation of condensing water vapour without any dissolution process. Originally set only for pure water droplets this equation at least qualitatively gives an estimate for the ruling principle:

$$X_K = \frac{4\sigma M}{\rho_l RT \ln S_{eq}}$$

.

X_K is the size of a water droplet, which is in equilibrium with the saturation ratio S_{eq} of condensable water vapour. Since the surface tension and density of water is constant, as is molecular weight M and temperature T, the existence of a water droplet is only dependent on the water supersaturation S_{eq}, at a first approximation. This would mean that nanoparticles would never be incorporated into condensing water, since the required water supersaturation for acting as condensation nuclei would be tremendously high, by far exceeding the available 1 or 2% within a cloud or fog situation. The only route to lowering the required supersaturation is described by Roault's law, where dissolvable compounds increase the surface tension (as surface tension is directly proportional to the number of ionized molecules). Hence, any change of particle surfaces, either by a chemical reaction or agglomeration with ionizable material, will lead to an actively participating condensation nucleus. If there is a change of wettability of such a particle system, for example by agglomeration with other small particles during residence time in air, lower supersaturation is needed to trigger condensation. This is due to dissolution and diameter increase.

2.2 Experimental Verification I: Water Uptake

Besides the considerable theoretical framework published in the past there has also been convincing experimental work. To assess the condensation properties of particles and hence their incorporation into the water cycle a test system mimicking the cloud formation process would be necessary. Fortunately, such experiments have been known for decades. What is needed? According to the statements made above beside surface properties mainly the particle diameter will govern the partitioning behaviour of particles within a condensing system.

Jean-Paul Coulier (1875) was the first to observe a droplet growth change, when changing the surrounding humidity.

Today, such experiments are performed by a so-called Tandem Differential Mobility Classifier experiment. The particles of interest are dispersed by an atomizing system creating a polydisperse aerosol system. This means it is an assembly of differently sized particles. This aerosol is now sorted by a first mobility classifier. The working principle of such a classifier is based on the particle's electrical mobility, which means the aerosol must be electrically charged. This is achieved by coagulation with charge carriers previous to the sorting process. Like in an electronic band pass filter, only those charged particles will pass the electric "gate" whose electric mobility (charge and size) fits the gate's dimension perfectly. In essence, the first classifier forms a monodisperse fraction of aerosol from any polydisperse fraction. This aerosol is in the next step subjected to different water humidities. Depending on its water-attracting properties it takes up water vapour and changes its size. This becomes measurable when such an aerosol is fed into a second classifier. To pass this "gate", after becoming swollen by water uptake, a wider gate dimension needs to be set. For NaCl aerosols the observed behaviour agrees well with calculations based on Köhler's theory of hygroscopic growth of water-soluble aerosol particles. Applying this experimental approach to artificially produced PbS nanoparticles of the same size leads to a different behaviour. When exposed to increased relative humidity up to 83% shrinkage was observed. In the case of NaCl recrystallization from the originally chain-like condenstion nuclei via droplet-like brine solutions is the explanation, whereas in the case of PbS, insolubility prohibits this. Only a slight restructuration has been observed (Krämer et al., 2000).

So, for water-insoluble material used as nanoparticulate matter a different water uptake behaviour is expected.

2.3 Experimental Verification II: Inhibition of Water Condensation

When nanoparticles, like insoluble carbon nanotubes or flame-processed particles become released to the atmosphere the question arises, whether they can become initiators of water condensation, or not. If not, the residence time for such nanoparticles will become considerably prolonged as they are small and not subjected to the typical removal mechanisms of diffusion, sedimentation and interception. The only removal mechanism acting would be dry deposition, triggered by turbulent diffusion of air packages. Diffusion, deposition or sedimentation are not very effective in a size range of 300–500 nm, the size range of many nano-engineered particle systems.

The consequences connected with a prolonged residence time are numerous, including increased interaction with radiation and higher risk of inhalation.

Hence, experimental verification of the condensation properties of ultrafine particles is highly desirable. In the 1980s this question had already been raised, since airborne cloud measurements have led to the observation of the "cloud interstitial aerosol". These aerosol particles consist of material that was not activated during cloud or fog formation during the actually existing water vapour supersaturation. They remained as small particles within the cloud or fog system. The chemical composition of these particles was thought to be similar to "fresh hydrophobic particles" (such as soot). The first experiments to understand this stem from van der Hage (1984) and Liu et al. (1984).

How does one figure out the different surface properties by means of a condensation process? Niessner et al. (1990) presented an instrumental setup, which at least provides semi-quantitative information on the influence of different surfaces of nanoparticles as condensation nuclei. They made use of a multistep condensation nuclei counter (CNC) as the detector. The features of these experiments are such that they were carried out with quasi-monodisperse aerosols and hence the only variable parameters were surface composition (wettability, dissolution properties etc.) and water vapour supersaturation. Particle number concentration and diameter were also kept constant for all experiments when comparing different nanoparticle systems. In this manner variation in the condensational behaviour of the particles can be related exclusively to differences in surface structure and surface composition.

In the following the characterization of differently coated nanoparticle systems (e.g. H_2SO_4 as an extremely hygroscopic particle core with various coating materials) is discussed briefly. From a sulphuric acid aerosol source a monodisperse fraction is selected by a differential mobility analyzer (DMA). This aerosol fraction, now monodisperse and of known size and number concentration, becomes mixed with vapours of hydrophobic coating materials, for example paraffines, terpenes etc. The coating thickness was independently measured by estimation of the diffusion coefficient of the coated and uncoated aerosol.

For a measurement the differently coated aerosols (monodisperse) are injected into the multistep condensation nuclei counter. In the CNC (Fig. 2.2) the aerosol is

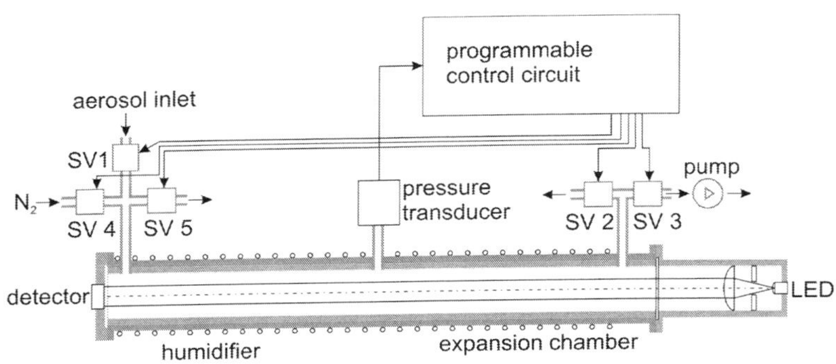

Fig. 2.2 Sectional drawing of the multistep condensation nuclei counter

fed into the expansion chamber via solenoid valve SV1. After closing SV1 and SV3, clean, particle-free nitrogen is introduced through SV4. By this means the chamber pressure, p_0, is increased up to a preselected pressure value. After some seconds of humidification and equilibration a rapid adiabatic expansion (t_{exp} <15 ms) occurs through opening of the two valves SV2 and SV5, resulting in a certain supersaturation corresponding to p/p_0. In the case where the introduced particles acted as condensation nuclei at the employed p/p_0 ratio, they will have taken up water vapour, and the intensity of the transmitted light through the expansion chamber due to light scattering and absorption by the droplets formed. Measurement of the light absorption before and after the expansion led to a CNC-indicated particle–droplet transition. The system automatically runs through 16 stages of increasing p/p_0 to produce an activation spectrum of the nanoparticles under investigation.

Figure 2.3 depicts directly the influence of a hydrophobic surface in such activation spectra comparing H_2SO_4 droplets (diameter: 20 nm) with and without a 5-nm thick n-hexadecanol layer. Onset of droplet growth is seen at the 11th stage, in comparison to the 7th stage (for uncoated, hygroscopic H_2SO_4 droplets).

Activation spectra for various nano-engineered particles, but all of the same size (~20 nm) are shown in Fig. 2.4.

Water-soluble particles like H_2SO_4 or NaCl initiated droplet growth at rather low water supersaturation, much lower than calculated by the Kelvin equation theoretically. This is because of the earlier-mentioned effect of creating a salt dissociation within the growing water phase. In the case of non-soluble Al_2O_3 (pyrosynthesized from heating up $AlCl_3$ in the aerosol phase) and pure carbon agglomerates

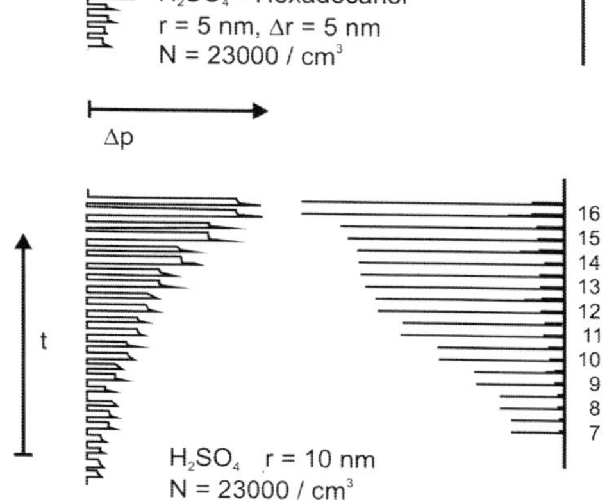

Fig. 2.3 Applied pressure difference and light intensity plots for pure and hydrophobically coated sulfuric acid droplets

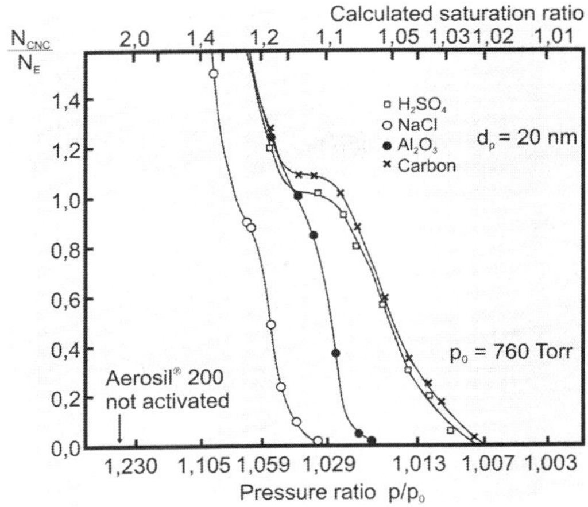

Fig. 2.4 Activation spectra of sulfuric acid droplets, NaCl, Al_2O_3, carbon, and Aerosil 200 nanoparticles (diameter: 20 nm)

Table 2.1 Behaviour of different oxidic particles towards condensation of water vapour

Pigment type	Chemical composition	Specific surface area (m^2/g)	Manufacturer	Counting efficiency (%)
Aerosil 200	SiO_2	206	Degussa	0
Aerosil 380	SiO_2	375	Degussa	1
Aerosil MOX 170	SiO_2 + 1% Al_2O_3	170	Degussa	1.2
Aerosil COK 84	SiO_2 + 16% Al_2O_3	170	Degussa	2.5
Rutil PKP 09045	TiO_2		Bayer	4.2
Titanium dioxide PKP 09040	TiO_2		Bayer	8.3
Titanium dioxide P 25	TiO_2		Degussa	2.7
Sicotrans Orange L 2515 D 80825	Fe_2O_3		BASF	1.8
Carbonyl iron oxide	Fe_2O_3	80	BASF	1
Carbonyl iron oxide	Fe_2O_3	25	BASF	1.7
Anatas PKP 09044	TiO_2		Bayer	4.3
Aluminium oxide C	Al_2O_3		Degussa	1.2

Counting efficiency = N_{CNC}/N_E (N_{CNC}, particle number concentration as indicated by the CNC, N_E, particle number concentration as measured with an aerosol electrometer in front of the inlet of the CNC; see Fig. 2.1).

a tremendous shift in needed water supersaturation is seen. Such supersaturation will never exist in cloud or fog formation, where supersaturations rarely exceed 1 or 2%. More than 10% would be needed to incorporate these particle systems into the vertical water cycle.

Extremely interesting behaviour can be reported for one of the currently most used nano-engineered particulates: Aerosil (Mathias and Wannemacher, 1988). Aerosil is a perfectly spherical particle consisting of SiO_2 (synthesized by high-temperature pyrolysis of $SiCl_4$ in hydrogen/oxygen flames). For Aerosil 200, no water vapour condensation was observed even at a saturation ratio of 230%, a value never obtained under natural conditions.

Several nano-engineered materials have been tested for their activation properties. The results are compiled in Table 2.1. All particle systems represent a considerable part of the nanomaterials widely used today. All these particle systems will not become incorporated into a water condensing system. Only after agglomeration with water-attracting or dissolvable particles may a direct incorporation into the water cycle occur.

2.4 Particle Collection During Precipitation

What remains to be discussed is the assessment of nanoparticle collection by impaction with already existing water droplets, for example during rain fall. This situation will be a common one, especially when other ambient (the majority) particles seeded cloud or fog formation. Water (rain) droplets can uptake during their fall non-activated particles.

This has been known for several decades from studies dealing with radioactive fall-out. Falling droplets may interact with particles by interception, diffusion or impaction when in the very near neighbourhood. All mechanisms together give rise to the so-called "Greenfield gap" (Kellogg, Rapp & Greenfield, 1957). Typically, around a particle size of 300–700 nm there is a minimum in the collision efficiency for such uptake. This is understandable due to the various size-dependencies of the respective mechanisms. This means that nano-engineered particles suspended in the ambient atmosphere will again have a large chance to remain as a (aero)sol in the air.

Beside this, the wettability of a contacted nanometre particle decides whether it becomes really incorporated into an existing droplet after a successful collision. Hydrophobic particles become squeezed out and do not enter a water droplet. This was shown experimentally by Mikhailov et al. (2001).

2.5 Conclusions

An attempt was made herein to describe the fate of air-dispersed nanoparticles and their incorporation route into the water cycle.

Cloud or fog formation can be seen as the origin of new water formation. This is meant physically, since only a change of state from gaseous water vapour to the condensed water phase occurs. A slight water supersaturation will serve as a water pool as long as particles are available with seeding properties. Also agglomeration of suspended dry nanoparticles with already-formed water droplets will not be a very successful route for incorporation. Wettability properties of nanoparticles govern the attachment process.

It could be demonstrated experimentally that most of the presently used nanomaterials, because of their inherent insolubility and hydrophobicity, do not activate water condensation under natural conditions. Depending on the wettability, or agglomeration with wettable or hygroscopic nanoparticles, they will have an unknown prolonged residence time in the ambient atmosphere. The consequences will be unknown impact on the radiation balance and occupational health.

References

Coulier J-P (1875) Note sur une nouvelle propriété de l′ air. Journal de Pharmacie et de Chimie Ser. IV 22: 165.

Kellogg WW, Rapp RR, Greenfield SM (1957) Close-in fallout. J Meteorol 14: 1–8.

Krämer L, Pöschl U, Niessner R (2000) Microstructural rearrangement of NaCl condensation aerosol particles on interaction with water vapor. J Aerosol Sci 31: 673–685.

Liu BYH, Pui DYH, McKenzie R, Agarwal J, Pohl F, Preining O, Reischl G, Szymanski W, Wagner P (1984) Measurements of Kelvin-equivalent size distributions of well-defined aerosols with particle diameters >13 nm. Aerosol Sci Technol 3: 107–115.

Mathias J, Wannemacher G (1988) Basic characteristics and applications of aerosil. 30. The chemistry and physics of the aerosil surface. J Colloid Interface Sci 125: 61–68.

Mikhailov E, Vlasenko S, Krämer L, Niessner R (2001) Interaction of soot aerosol particles with water droplets: influence of surface hydrophilicity. J Aerosol Sci 32: 697–711.

Niessner R, Daeumer B, Klockow D (1990) Investigation of surface properties of ultrafine particles by application of a multistep condensation nucleus counter. Aerosol Sci Technol 12: 953–963.

Pruppacher H, Klett J (1997) Microphysics of Clouds and Precipitation. Kluwer Academic Publishers, Dordrecht.

van der Hage J (1984) Drop formation on insoluble particles. J Colloid Interface Sci 101: 10–18.

Chapter 3
Nanoparticles in Groundwater – Occurrence and Applications

Thomas Baumann

3.1 Nanoparticles in Groundwater

3.1.1 Occurrence and Stability of Nanoparticles

Natural nanoparticles have been addressed in a review by Wiggington et al. (Wigginton et al., 2007). Originating from geochemical and biotic processes, natural nanoparticles are abundant in soils and groundwater. Typical concentrations are in the lower mg/L range. In the vicinity of waste disposals and other groundwater contaminations the concentration levels typically increase (Baumann et al., 2006, 1998). Looking at a typical grain size distribution of a sand and gravel aquifer, up to 5 % of the mineral matrix is in the size range of nanoparticles. However, in contrast to synthetic nanoparticles, natural nanoparticles represent some degree (if not total) of equilibration with their environment. The ingredients that make up the nanoparticles are present in higher amounts; therefore, abiotic growth of nanoparticles to larger colloids as well as biotic growth is likely. During growth, nanoparticles may co-precipitate and adsorb metal contaminants like arsenic, uranium, or technetium.

Humic acid, present in the topsoil in larger quantities, has a stabilizing effect on most nanoparticles. For instance citrate-anion-stabilized gold nanoparticles were found to show lower aggregation in the presence of Suwanee River humic acid (Diegoli et al., 2008). The stability of fullerene C_{60} was improved in the presence of humic acid due to steric stabilization, but otherwise exposed aggregation behavior according to DLVO theory. At $CaCl_2$ concentrations of 40 mM, bridging of fullerene nanoparticles and humic acid was observed that led to high aggregation rates (Chen and Elimelech, 2007). Similarly, the stability of multiple-walled carbon nanotubes was found to increase in the presence of natural organic matter (Hyung et al., 2007). Likewise, aggregates of nC_{60} were found to disaggregate in the presence of Suwanee River humic and fulvic acids as a result of changes in the particle structure (Xie et al., 2008).

T. Baumann (✉)
Institute of Hydrochemistry, Technische Universität München, 81377 München, Germany
e-mail: thomas.baumann@ch.tum.de

F.H. Frimmel, R. Niessner (eds.), *Nanoparticles in the Water Cycle*,
DOI 10.1007/978-3-642-10318-6_3,

The immission of silver nanoparticles, which are used in fabrics for their antimicrobial effects, has been assessed for river waters (Blaser et al., 2008). A total of 20–130 t/a of silver was expected to reach natural waters, with a predicted accumulation effect of up to 300 μg/L in river waters and up to 12 mg/kg in river sediments. Due to its strong tendency to bind to sulfur in freshwater systems, silver is most likely to form either silver–sulfur clusters or silver–sulfur complexes with organic matter.

Based on literature data we are likely facing increased emissions of nanoparticles. While the properties of synthetic nanoparticles are usually well described for their respective application, the effect of synthetic nanoparticles on wildlife is still questionable.

3.1.2 Transport Properties

The transport of single nanoparticles in groundwater is governed by colloid filtration theory (CFT), which has been successfully applied on the microscale (Baumann and Werth, 2005). Upscaling is aided by the correlation equation developed by Tufenkji and Elimelech, which describes a broad range of field observations pretty well (Tufenkji and Elimelech, 2004). Due to their small size, the main driving force for collisions of nanoparticles with the matrix is diffusion (Fig. 3.1). Interception and gravitation can safely be ignored when assessing the transport of single nanoparticles. Gravitational effects have to be considered for aggregates of nanoparticles which might be very dense (i.e., metal and metal oxide nanoparticles). CFT predicts an exponential decrease of the nanoparticle concentration with increasing distance from the source. However, some field data (see data compiled by Pang (2009), Bradford and Toride (2007)) show a hyperexponential or bimodal decrease of the particle concentration. This is because the parameters controlling CFT are varying in the spatial and time domain. Physical and chemical heterogeneities along the flow

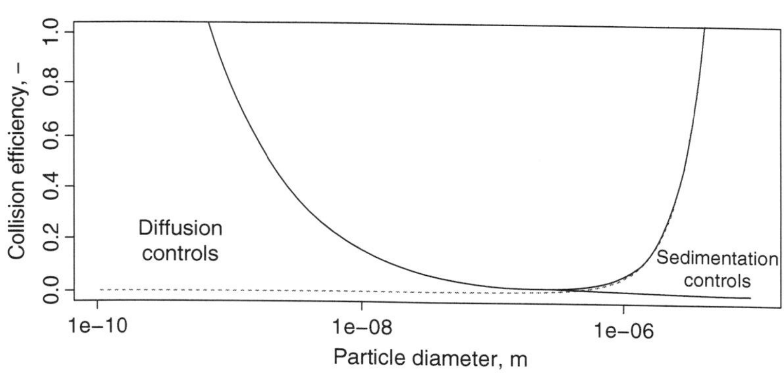

Fig. 3.1 Collision efficiency for nanoparticles

path contribute to a variation of the retention rates. Also, concentration-dependent effects like aggregation, clogging, and filter ripening or blocking close to the source have to be considered. Stochastic simulations (Bradford and Toride, 2007) can help to quantify the probability density function (pdf) for particle attachment and detachment. However, the pdf is of little help in explaining the underlying processes.

The spreading of nanoparticles can be described with an advection–dispersion transport equation and adding an attachment/detachment term. Colloid dispersion is generally smaller compared to a solute tracer. Size and charge exclusion cause faster breakthrough and less dispersion on the microscale. Attachment and aggregation lead to a bias of the nanoparticle breakthrough in favor of nanoparticles moving in larger pores.

The diffusion of nanoparticles in a fluid with viscosity η at temperature T can be calculated with Stokes–Einstein equation (3.1, (Kholodenko and Douglas, 1995)) and scales linearly with the nanoparticle radius R:

$$D = \frac{kT}{6\pi \eta R} \tag{3.1}$$

The Stokes number St describes the behavior of suspended particles in a fluid flowing around an obstacle (Eq. 3.2). For nanoparticles in sandy aquifers St $\ll$ 1 inertial moments can be neglected and particles follow the streamlines of the fluid around the obstacle:

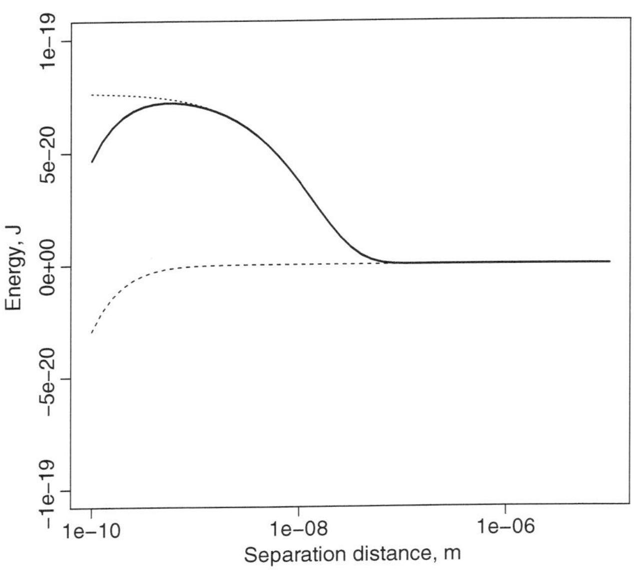

Fig. 3.2 Schematic of attractive and repulsive forces for a 20 nm nanoparticle and a quartz surface

$$\mathrm{St} = \frac{\upsilon (2R)^2 \rho_p}{18 l \eta} \tag{3.2}$$

Here, υ is the velocity of the fluid, ρ_p is the density of the particle, l is the characteristic length of the obstacle, and η is the viscosity of the fluid.

On the other hand, the relative importance of nanoparticle–nanoparticle and nanoparticle–surface interactions increases, thus affecting the filtration negatively (Fig. 3.2). While the interaction energy scales linearly with R, the forces driving the nanoparticle to the surface scale with the area (R^2) and the mass/volume (R^3) of the nanoparticle. Therefore attachment rates for nanoparticles are likely to be small. Taking a closer look at the interaction forces between nanoparticle and collector, retention of nanoparticles in the secondary energy minimum adjacent to the repulsive barrier seems to become an important factor (Johnson et al., 2007). This retainment, however, is different from an attachment in the primary energy minimum at the collector, as the interaction forces are smaller and the retainment might be non-permanent.

Straining of nanoparticles in pore throats which are too small to be passed by the nanoparticle is characterized through d_p/d_c, where d_p is the diameter of the nanoparticle and d_c is the diameter of the grain or collector and starts at a ratio of 0.154 (Herzig et al., 1970). This ratio decreases to 0.05 when grain-to-grain contacts are taken into account. At higher ionic strength, the critical ratio is further reduced to values around 0.0016 (Shen et al., 2008). That way, for nanoparticles with a diameter of 20 nm, straining is unlikely for sediments which are coarser than silty sands. At low ionic strength, d_p/d_c was found in the 0.027 range. This suggests that straining, which is an important factor for bigger particles, is negligible for nanoparticles in aquifer sediments.

Filtration theory predicts increasing collision rates with decreasing particle size, but, as has been shown above, the attachment efficiency is likely decreasing. The surface modification of synthetic nanoparticles can affect the nanoparticle–surface and nanoparticle–nanoparticle interactions by a factor of 100 (Saleh et al., 2008; Pelley and Tufenkji, 2008). This behavior of synthetic nanoparticles is in agreement with the theoretical framework. However, as of today, experimental data on the transport properties are sparse. Thus, a prediction of the transport of synthetic nanoparticles based on observations of natural colloids will be risky at least.

The transport of fullerene $n\mathrm{C}_{60}$ was also consistent with theory, but dependent on the preparation method of the $n\mathrm{C}_{60}$ aggregates (THF, TTA). Tannic acid, comparable to fulvic and humic acids, increased the stability and therefore the mobility when injected together with $n\mathrm{C}_{60}$ aggregates. In contrast a decrease of the mobility was found when injected together with alginate, a polysaccharide-like natural organic matter (Espinasse et al., 2007). Addition of Ca should further increase the attachment efficiency, as it promotes the formation of polysaccharide gels.

The transport of nanoparticles may significantly alter in the presence of other colloids. Pang et al. (2008) have shown that MS2, a bacteriophage with a diameter of 26 nm, attaches to colloids, causing the removal rates to decrease compared to MS2 without colloids. The reason is that MS2 is less negatively charged than soil colloids. One can imagine that surface-modified nanoparticles might exhibit the same behavior. Fluorescent polystyrene beads with casein covalently bound to the surface were used to mimic the behavior of MS2, as well as other viruses and bacteria (Pang et al., 2009). This example also illustrates the wide variety of nanoparticle–matrix interactions.

Colloid transport in the unsaturated zone has been reviewed recently (Bradford and Torkzaban, 2008). In the unsaturated zone, flow velocities are highly dynamic imposing physical stress on attached colloids (Bergendahl and Grasso, 2000). Film flow can occur at low saturation ratios (Crist et al., 2005, 2004; Wan and Tokunaga, 1997) and can cause straining of colloids. Capillary forces during wetting and drying cycles can force nanoparticle attachment and detachment, although this process is size dependent (Shang et al., 2007). The capillary forces of colloids retained in films or at the air–water–solid interface are several orders of magnitude higher than DLVO forces (Gao et al., 2008).

Assessment of the fate of nanoparticles in the unsaturated zone has to carefully take into account the surface functionality. Hydrophobic particles will likely be found at the air–water interface, whereas hydrophilic particles will float in the water film. One has also to be aware of a change of the ionic strength during a wetting/drying cycle, especially in arid or semi-arid environments, significantly affecting the interaction distances between particles and collectors or other particles.

3.1.3 Nanoparticles in Remediation Applications

Synthetic nanoparticles offer highly interesting options for groundwater remediation, especially if the contamination is not easily accessible. Zerovalent iron (ZVI, Fe^0) is an excellent electron donor and reducing agent, with a variety of environmental applications at hand. ZVI nanoparticles have been proposed to reduce and immobilize chromate (Cr(VI)) (Xu and Zhao, 2007; Li et al., 2008). Here X-ray spectroscopic measurements indicate that Cr(VI) is incorporated into the Fe-oxyhydroxide shell of the ZVI nanoparticle. This Fe–Cr-hydroxide shell then acts as a sink for further Cr(VI). Reaction kinetics is fast and the shell is considered stable (Li et al., 2008). In a similar way arsenic can be removed from groundwater, either by injection of ZVI nanoparticles (with surfactant-modified surface to avoid clogging) in an in situ setup or by coating sand with ZVI nanoparticles (Kanel et al., 2006, 2007) in a reactive barrier setup. Other applications of ZVI nanoparticles include the degradation of atrazine (Satapanajaru et al., 2008), PCB (Choi et al., 2008), TCE (Liu et al., 2005) or the promotion of microbial reduction of nitrate (Shin and Cha, 2008).

While a complete survey of possible applications for synthetic nanoparticles is beyond the scope of this contribution, it is worthwhile to highlight some of the common findings and processes. Generally ZVI nanoparticles exhibit a high reaction rate, but the contaminant-specific reaction rate will decrease in the presence of competing solutes like nitrate and non-competing solutes like phosphate, hydrogen carbonate, or chloride (Liu et al., 2007). From existing data it seems evident that synthetic nanoparticles have to be stabilized to avoid aggregation and blocking of the pore space. This applies in particular to cases where nanoparticles are administered in high concentrations as with in situ applications. The long-term fate of the surface modification remains to be questioned, as polymers might swell, disintegrate, and deposited to the aquifer matrix in the presence of organic surfaces (Baumann and Niessner, 2006). Surfactants might dissolve under environmental conditions if other detergents like humic substances are present.

Another prerequisite for successful environmental applications is that the synthetic nanoparticles reach the source zone and match the stoichiometry of the source. Currently there are only few pilot-scale studies showing the feasibility and potential of the concepts (Elliot and Zhang, 2001; Quinn et al., 2005). On the other hand, although there are very detailed studies highlighting the processes on the scale of a single particle, studies imposing spatial constrictions also are scarce.

In the following micromodel study, we will look at the application of ZVI nanoparticles to remediate a spill of trichloroethylene (TCE). The treatment of groundwater contaminated with chlorinated solvents through reactive barriers filled with ZVI is a state-of-the-art technique (Schäfer et al., 2003). TCE dechlorination requires 3 moles of ZVI per mole of TCE, that is, 1 g of Fe^0 will dechlorinate 1.3 g TCE. In order to maintain the reactivity and stability in groundwater ZVI nanoparticles have to be modified, e.g., with anionic polyelectrolytes (Phenrat et al., 2008). The polyelectrolyte layer increases the hydrodynamic diameter of the nanoparticles (original diameter 5–40 nm) to 100 nm and up, depending on the equilibrium times. Of course this also affects transport and attachment. Other approaches include polymerization, e.g., with a tri-block polymer (Liu et al., 2005), or using bimetallic synthetic nanoparticles (Elliot and Zhang, 2001).

However, the case of a reactive barrier dechlorination applies to the downstream groundwater only; the source zone is left untouched. In cases where the groundwater flow is slow and the source zone is not accessible, deployment of ZVI nanoparticles might be an alternative. As nanoparticles are highly reactive due to their small size a rapid decrease of the concentration levels is expected.

In order to reach the contamination area, the nanoparticles have to be mobile in porous media. Permanent attachment to uncontaminated matrix should be minimized while the sticking efficiency to the interface of chlorinated solvent pools has to be high. In addition, the reaction products and "used nanoparticles" have to leave the contaminated area cleanly to avoid blocking. Kanel et al. (2008) demonstrated in a sandbox experiment that ZVI nanoparticles without protective coating to reduce oxidation are not mobile in an aquifer system made of silica beads (diameter 1.1 mm), whereas nanoparticles modified with polyacrylic acid were moving without much retardation, but exhibited a significant density effect.

Dechlorination can take place in aqueous solution, at the interface between water and solvent, and in the solvent itself (Fig. 3.3). For the latter two scenarios, the particles have to attach to the interface and possibly penetrate it. This requires a hydrophobic coating.

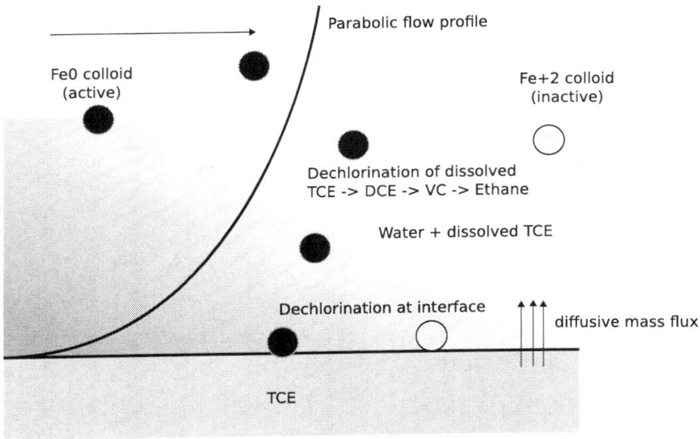

Fig. 3.3 Schematic of TCE dechlorination with synthetic nanoparticles

Micromodel experiments with polymer nanoparticles have shown that most of the nanoparticles are mobile in the aqueous phase. The ZVI colloids were stable and showed only little tendency for aggregation compared to bare iron colloids, which quickly aggregated. The residence times of the ZVI colloids in close proximity to the TCE interface were on the order of seconds. Nanoparticles attaching to the interface were rare and kept sticking to the interface. No penetration of the interface was observed. Overall attachment rates were small compared to attachment rates of polystyrol latex colloids (Fig. 3.4).

The implications of this experiment are twofold: First, the dechlorination takes place mostly in the aqueous phase. This results in a quick drop of the TCE concentration. The TCE pool, however, is hardly affected. The release of TCE is controlled by diffusion. While the presence of ZVI nanoparticles might increase the diffusive mass flux due to the immediate dechlorination of TCE in the aqueous phase, the

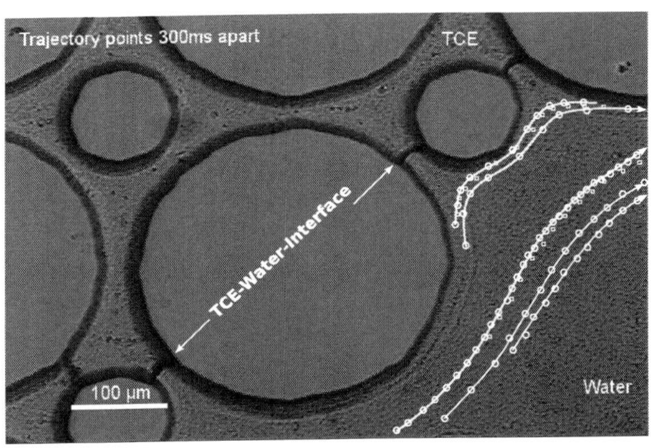

Fig. 3.4 Trajectories of ZVI colloids in a micromodel experiment

reaction products have to be removed immediately, because otherwise the diffusion will be limited. Second, there seems to be a problem with the delivery of large quantities of synthetic nanoparticles at heavily contaminated sites because of blocking effects and steric interactions.

In response to the limited accessibility of contamination pools to nanoparticles due to flow restrictions, nanoparticles with sorbing and degrading properties are actively developed. Choi et al. (2008) presented a system of activated carbon impregnated with Fe/Pd nanoparticles for the dechlorination of PCB. The size of the activated carbon particles was about 1–3 mm and the Fe/Pd nanoparticles were adsorbed in the mesopores of the activated carbon. The size of these particles is clearly tailored toward applications in water treatment systems, funnel-and-gate systems, or reactive capping systems.

3.2 Nanoparticles at the Interface

Nanoparticles might be generated in situ as a result of weathering processes, intentionally applied to the land surface, e.g., in agricultural applications, or deposited to the soil surface through atmospheric emission. Either way, wet deposition and dry deposition have to be distinguished as they affect the initial transport conditions.

In the topsoil a variety of abiotic and biotic processes might alter the surface properties and the stability of the nanoparticles. Humic acid and other organic acids present in the soil surface might stabilize the nanoparticles, while bacteria and biofilms may increase removal rates or adversely be affected by nanoparticle toxicity. Transport through the topsoil is expected to be slow, but preferential flow paths and cracks in the soil may enhance mobility (see Chap. 4 by A.D. Karathanasis, this book).

Transport of nanoparticles in the unsaturated zone of an aquifer is governed by attachment/detachment processes reflecting the wetting/drying cycles (Shang et al., 2007). Transport of bacteriophages MS2 and ϕX174 was found to increase in unsaturated sand columns with increasing concentrations of metal oxides. The effect of phosphorous, calcium, and organic carbon concentrations was found to be negligible (Chu et al., 2003). This underlines the importance of reactions in topsoil for the transport properties of nanosized particles.

Both the negligible occurrence of pathogens and viruses and the generally low concentrations of colloids even in shallow groundwater indicate that the soil is an effective filter for particles in the size range of colloids and nanoparticles. However, bypass flow can occur during winter months after the soil has been frozen and during summer months when the soil is dried out. Kühnhardt has shown at the "Munich Well," a field site which provides access to the unsaturated zone of the Munich Gravel Plain (Fig. 3.5), that after the soil has been frozen, there is significant transport of polycyclic aromatic hydrocarbons through the topsoil down to 6 m into the vadose zone with the snow melt (Fig. 3.6). In addition wormholes, roots, and other preferential flow paths may facilitate the passage through the soil. It seems very likely that nanoparticles are following the same route.

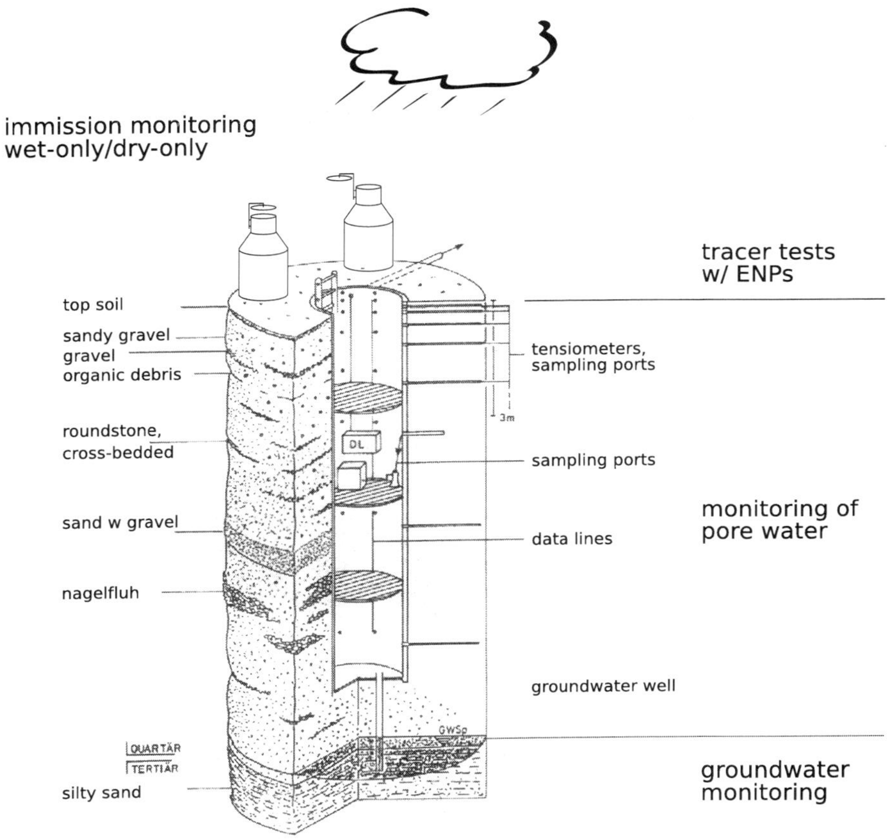

Fig. 3.5 Setup of the field laboratory

Therefore, the immission situation of groundwater with regard to nanoparticles applied to land surfaces depends on a variety of factors, with only few factors of these being explored at present.

3.3 Sampling of Nanoparticles

Colloids and nanoparticles in ground- and surface water are sensitive to any change of the environmental conditions. Sampling and characterization of colloids have been laid out in detail (McCarthy and Degueldre, 1993). These rules will also apply to nanoparticles. In particular it seems necessary to prevent aggregation of nanoparticles with other colloids as well as disaggregation of nanoparticle–colloid aggregates to assess the exposure routes.

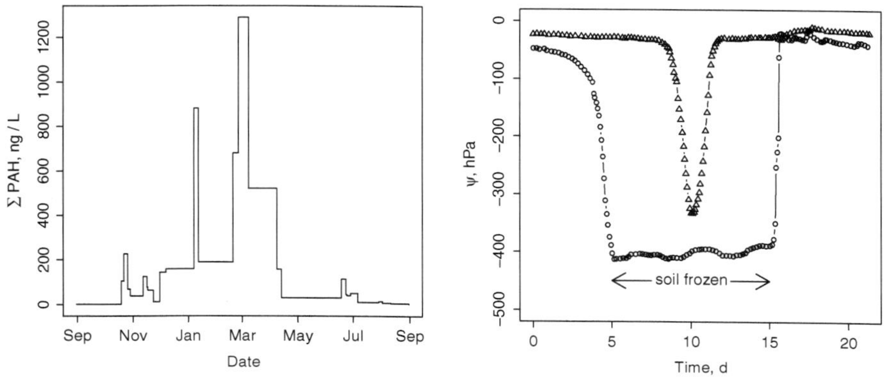

Fig. 3.6 PAH concentrations (sum of fluoranthene, benzo(b)fluoranthene, benzo(k)fluoranthene, benzo(a)pyrene, benzo(ghi)perylene, indeno(123-cd)pyrene) in seeping water (*left*) and capillary pressure for the 0.2 and 0.5 m sensors (*right*). The capillary pressure surrounds the January PAH peak. Data from Kühnhardt (1994)

An assessment of potential sampling errors might distinguish between technical issues, hydraulic disturbances, and hydrochemical changes. Technical artifacts are often seen with newly installed sampling wells, where leftovers of the drilling fluid contaminate the sample. Taking the sample too early can also be considered a technical issue. It has been shown that steady – state conditions with respect to the colloid concentrations significantly lag behind the hydraulic and hydrochemical equilibrium (Hofmann et al., 2003). Contaminated sampling materials also fall into this category.

The second category of sampling artifacts relates to changes in the hydraulic conditions during sampling transfer and storage. From numerous studies it is evident that an increase of the flow rate causes a detachment and mobilization of colloids of all sizes, leading to higher concentrations and eventually to a biased particle size distribution. The second type of artifacts, like very low flow conditions, may cause a settlement of bigger particles in the well, eventually dragging nanoparticles with them. Again, a biased particle size distribution will be the result. Obviously, under no flow conditions the particles are more likely to get into contact by diffusion as compared to porous media where diffusion is restricted to the pore space. An increase of the temperature during transport and storage enhances this effect.

Hydrochemical artifacts during sampling and characterization also include changes in the redox potential and the pH value, changes in the gas composition due to exposure to ambient air or due to a pressure drop causing degassing, and change of the ionic strength, e.g., during filtration or field flow fractionation. The latter might also introduce a change of the dominating cation from calcium to sodium.

To overcome these artifacts inline sampling and characterization technique have been proposed (Nilsson et al., 2007). The pressure conditions of the aquifer system were maintained in an argon atmosphere, while filters for ultrafiltration are prepared in a portable glove box.

Seeping water samples in the unsaturated zone are often collected using suction cups. To maintain a good hydraulic contact to the aquifer, suction cups are embedded into fine sand. This setup likely causes a biased particle size distribution due to filtration in the sand pack and the suction cup. The negative pressure gradient in the suction cup will also cause a degassing of CO_2, thus affecting the carbonate equilibrium. Open pipe sampling devices and lysimeters avoid these artifacts, but have a higher technical footprint.

As a rule of thumb and as long as in situ characterization techniques are not available or not feasible, changes of the environmental conditions during sampling and the time between sampling and characterization have to be kept to a minimum.

Acknowledgments Financial support by the Gottlieb Daimler- und Karl Benz-Stiftung, Ladenburg, Germany, is gratefully acknowledged.

References

Baumann T, Fruhstorfer P, Klein T, Niessner R (2006) Water Res 40: 2776.
Baumann T, Fruhstorfer P, Niessner R (1998) Grundwasser 3: 3.
Baumann T, Niessner R (2006) Water Resour Res 42: W12S04. DOI 10.1029/2006WR004893.
Baumann T, Werth CJ (2005) Coll Surf A 265: 2.
Bergendahl J, Grasso D (2000) Chem Eng Sci 55: 1523.
Blaser SA, Scheringer M, MacLeod M, Hungerbuehler K (2008) Sci Tot Environ 390(2–3): 396. DOI 10.1016/j.scitotenv.2007.10.010.
Bradford SA, Toride N (2007) J Environ Qual 36: 1346. DOI 10.2134/jeq2007.0004.
Bradford SA, Torkzaban S (2008) Vadose Zone J 7: 667. DOI 10.2136/vzj2007.0092.
Chen KL, Elimelech M (2007) J Coll Interf Sci 309(1): 126. DOI 10.1016/j.jcis.2007.01.074.
Choi H, Al-Abed SR, Agarwal S, Dionysiou DD (2008) Chem Mater 20(11): 3649. DOI 10.1021/cm8003613.
Chu Y, Jin Y, Baumann T, Yates MV (2003) J Environ Qual 32: 2017.
Crist JT, McCarthy JF, Zevi Y, Baveye P, Throop JA, Steenhuis TS (2004) Vadose Zone J 3: 444.
Crist JT, Zevi Y, McCarthy JF, Throop JA, Steenhuis TS (2005) Vadose Zone J 4: 184.
Diegoli S, Manciulea AL, Begum S, Jones IP, Lead JR, Preece JA (2008) Sci Tot Environ 402(1): 51. DOI 10.1016/j.scitotenv.2008.04.023.
Elliot DW, Zhang WX (2001) Environ Sci Technol 35: 4922.
Espinasse B, Hotze EM, Wiesner MR (2007) Environ Sci Technol 41(21): 7396. DOI 10.1021/es0708767.
Gao B, Steenhuis TS, Zevi Y, Morales VL, Nieber JL, Richards BK, McCarthy JF, Parlange JY (2008) Water Resour Res 44: W04504. DOI 10.1029/2006WR005332.
Herzig J, Leclerc DM, Legoff P (1970) Ind Eng Chem 62(8): DOI: 10.1021/ie50725a003.
Hofmann T, Baumann T, Bundschuh T, van der Kammer F, Leis A, Schmitt D, Schäfer T, Thieme J, Totsche KU, Zänker H (2003) Grundwasser 8: 213.
Hyung H, Fortner JD, Hughes JB, Kim JH (2007) Environ Sci Technol 41: 179. DOI 10.1021/es061817g.
Johnson WP, Tong M, Li X (2007) Water Resour Res 43: W12S13. DOI 10.1029/2006WR005770.
Kanel SR, Goswami RR, Clement TP, Barnett MO, Zhao D (2008) Environ Sci Technol 42(3): 896. DOI 10.1021/es071774j.
Kanel S, Greneche J, Choi H (2006) Environ Sci Technol 40(6): 2045. DOI 10.1021/es0520924.
Kanel SR, Nepal D, Manning B, Choi H (2007) J Nanopart Res 9(5): 725. DOI 10.1007/s11051-007-9225-7.
Kholodenko AL, Douglas JF (1995) Phys Rev E 51(2): 1081. DOI 10.1103/PhysRevE.51.1081.

Kühnhardt M, Untersuchungen zur Dispersion von polyzyklischen aromatischen Kohlenwasserstoffen (PAHs) in der ungesättigten und gesättigten Zone eines fluvioglazialen Schotters. Dissertation, Techn. Univ. München (1994)
Li XQ, Cao JS, Zhang WX (2008) Industr Eng Chem Res47(7): 2131. DOI 10.1021/ie061655x—ISSN 0888-5885.
Liu Y, Majetich SA, Tilton RD, Sholl DS, Lowry GV (2005) Environ Sci Technol 39: 1338.
Liu Y, Phenrat T, Lowry GV (2007) Environ Sci Technol 41: 7881.
McCarthy JF, Degueldre C (1993) Sampling and characterization of groundwater colloids for studying their role in the subsurface transport of contaminants.. In: J Buffle, HP van Leeuwen (ed.) Environmental Particles, Vol. 2. Lewis Publisher, Boca Raton, pp. 247–315.
Nilsson AC, Hedqvist I, Degueldre C (2008) Anal Bioanal Chem 391(4): 1327. DOI 10.1007/s00216-007-1782-1. 14th European Conference on Analytical Chemistry, Antwerp, BELGIUM, SEP 09-14, 2007.
Pang L (2009) J Environ Qual 38: 1531.
Pang L, Nowostawska U, Ryan J, Williamson W, Walshe G, Hunter K, Flintoft. M Body doubles – modifying the surface charge of pathogen-sized microspheres to study pathogen transport in groundwater. 1st International Conference on Microbial Transp. and Survival in Porous Media, 10–13 May 2009, Niagara-on-the-Lake (2009)
Pang L, Walshe G, Nowostawska U, Hunter K, Flury M, Close M, Sinton L, Leonard M (2008) Geophys Res Abstr 10: EGU2008.
Pelley AJ, Tufenkji N (2008) J Coll Interface Sci 321(1): 74. DOI 10.1016/j.jcis.2008.01.046.
Phenrat T, Saleh N, Sirk K, Kim HJ, Tilton RD, Lowry GV (2008) J Nanopart Res 10(5): 795. DOI 10.1007/s11051-007-9315-6.
Quinn J, Geiger C, Brooks K, Koon C, O'Hara S, Krug T, Major D, Yoon WS, Gavaskar A, Holdsworth T (2005) Environ Sci Technol 39: 1309. DOI 10.1021/es0490018.
Saleh N, Kim HJ, Phenrat T, Matyjaszewski K, Tilton RD, Lowry GV (2008) Environ Sci Technol 42(9): 3349. DOI 10.1021/es071936b—ISSN 0013–936X.
Satapanajaru T, Anurakpongsatorn P, Pengthamkeerati P, Boparai H (2008) Water Air Soil Poll 192(1–4): 349. DOI 10.1007/s11270-008-9661-8.
Schäfer D, Köber R, Dahmke A (2003) J Cont Hydrol 65: 183.
Shang J, Flury M, Chen G, Zhuang J (2007) Water Resour Res 44: W06411 DOI 10.1029/2007WR006516.
Shen C, Huang Y, Li B, Jin Y (2008) Water Resour Res 44: W05419. DOI 10.1029/2007WR006580.
Shin KH, Cha DK (2008) Chemosphere 72(2): 257. DOI 10.1016/j.chemosphere.2008.01.043.
Tufenkji N, Elimelech M (2004) Environ Sci Technol 38: 529.
Wan JM, Tokunaga TK (1997) Environ Sci Technol 31: 2413.
Wigginton NS, Haus KL, Hochella MF Jr. (2007) J Environ Monit 9(12): 1306. DOI 10.1039/b712709j.
Xie B, Xu Z, Guo W, Li Q (2008) Environ Sci Technol 42(8): 2853. DOI 10.1021/es702231g.
Xu Y, Zhao D (2007) Water Res 41(10): 2101. DOI 10.1016/j.watres.2007.02.037.

Chapter 4
Composition and Transport Behavior of Soil Nanocolloids in Natural Porous Media

Anastasios D. Karathanasis

Natural nanoparticles and nanocolloids are ubiquitous in soil environments playing important roles in many environmental processes, including soil genesis, nutrient cycling, dispersion/flocculation, sorption, precipitation, dissolution, contaminant transport, biogeochemical transformations, bioavailability, and various remediation practices (Christian et al., 2008). They are generally defined as mineral or organic particulate or aggregate material with at least one dimension of less than 100 nm. Therefore, nanoparticles and nanocolloids constitute a sub-fraction of colloidal material whose upper size range extends to 1,000 nm. Soil nanomaterials usually exhibit heterogeneous and complex structure, composition, and surface chemistry that make their behavior very difficult to predict. Due to their small size, they usually have exceptionally high surface area-to-mass ratio and exhibit specially enhanced surface properties, which may be variable due to the heterogeneous composition of the particle (Ryan and Elimelech, 1996). For example, the core of the particle may be comprised of SiO_2 or $Al(OH)_3.SiO_2$, but the surface (shell) may be coated with $Fe(OH)_3$ or organic functional groups. Since the surface of the nanoparticle is the most interactive with the environment, understanding its composition is critical for predicting its functional behavior. In fact, manipulated functionalization of this "shell" layer is the focus of many engineered nanoparticles that are being used for various environmental applications.

Soil nanoparticles can be of natural origin, commonly formed as mineral weathering byproducts, new mineral precipitates from supersaturated solutions, and biogenic products of microbial activity (Wigginton et al., 2007) or of anthropogenic origin, released in the soil environment as byproducts of technological processes (Christian et al., 2008). Under certain environmental conditions different microorganisms may induce the formation of mineral nanoparticles through redox transformation or surface nucleation processes. The morphology of soil nanoparticles varies considerably with their origin, composition and environmental setting, while their fate, behavior, and mobility are closely related to their intrinsic

A.D. Karathanasis (✉)
Department of Plant and Soil Sciences, University of Kentucky, N-122K Ag, Science-North, Lexington, KY 40546, USA
e-mail: akaratha@uky.edu

F.H. Frimmel, R. Niessner (eds.), *Nanoparticles in the Water Cycle*,
DOI 10.1007/978-3-642-10318-6_4,

properties. It is well established that environmental nanoparticles can travel very long distances since their average displacement through diffusion processes is proportional to the inverse square root of the radius of the particle. Their transport potential, however, is also dictated by the colloidal stability (potential of the particle to remain dispersed) imposed by the dispersion media and its ionic environment. While the small size of the nanoparticles implies considerable stabilization potential, their high surface energy may induce multiple interparticle collisions and significant agglomeration, particularly under elevated ionic concentrations (Bradford et al., 2007). This is because the increased ionic strength can shield the charge of two approaching particles and reduce the effectiveness of the Coulomb repulsion, especially in the presence of polyvalent ions, which could lead into rapid coagulation and flocculation. Conversely, the stability of soil nanoparticles may be enhanced in association with organic colloids or thin-film coatings of adjacent mineral surfaces (Lead and Wilkinson, 2007).

The stability, mobility, and reactivity behavior of soil nanoparticles is of particular interest in environmental pollution and remediation processes. Not only could their presence in large quantities cause a significant deterioration of water quality, but their association with and co-transport of organic and inorganic pollutants may adversely affect the integrity of sensitive ecosystems and pose a great contamination risk for groundwater supplies (Kaplan et al., 1993; Ouyang et al., 1996). Unfortunately, very little information exists about the characteristics and the behavior of natural soil nanoparticles in the environment. Thus, the understanding of their environmental impact could significantly benefit from existing research on environmental colloids (Lead and Wilkinson, 2007). The high surface area-to-mass ratio of soil nanoparticles is expected to induce significant association with organic or inorganic contaminants through sorption, coprecipitation, or organic complexation processes (Lyven et al., 2003). Engineered carbon nanotube particles have already been used for sorption of a variety of organic compounds and metals from water (Liang et al., 2006; Zhou et al., 2006). Their sorption capacity is enhanced with increasing pH above the point of zero charge or by surface coatings of natural organic matter (NOM) (Liang et al., 2006). Recent technological advances in fractionation techniques, involving field-flow fractionation (FFF) and atomic force microscopy, have allowed improved separation and characterization of natural environmental nanoparticle fractions (Baalousha and Lead, 2007). Ultrafiltration methods combined with high-resolution techniques involving FFF-ICP-MS, TEM, EDS, and NanoSims have become especially important in determining the elemental association and distribution of contaminants among and within nanoparticle fractions (Ranville et al., 2005; Lead and Wilkinson, 2007). Recent studies with riverbed and floodplain sediments from Clark Fork River in Montana, USA, involving HR-TEM-EDS identified several metal-bearing nanoparticles, including Mn-oxides, Fe-oxides, and sulfides, with particular preference to Zn, Cu, As, and Pb (Wigginton et al., 2007).

The objective of this chapter is to provide information obtained through mesocosm laboratory experiments on soil nanocolloid composition and behavior in soil environments, their interaction with selected metals and herbicides, and their potential to facilitate contaminant transport in soil porous media.

4.1 Materials and Methods

Water dispersible mineral colloids were fractionated from upper Bt horizons of three soils in Kentucky (Table 4.1), with smectitic, mixed, and kaolinitic mineralogy, according to USDA criteria for mineralogical family classes. Similarly, water dispersible biosolid colloids were fractionated from one agricultural (poultry manure, PMBC) and two municipal (lime stabilized, LSBC, and aerobically digested, ADBC) biosolid wastes. The colloid fractionation was accomplished by mixing 10 g of material with 200 mL of de-ionized water, shaking overnight, and centrifuging at 130×*g* for 3.5 min. Mineral colloid suspensions of 300 mg/L spiked with 10 mg/L of Cu, Zn, and Pb-chloride salts and 2 mg/L of atrazine and metolachlor herbicides were infused into undisturbed soil monoliths consisting of A and Bt horizons of a fine-silty, mixed, mesic Typic Argiudoll at a rate of 2.2 cm/h, sustaining unsaturated flow conditions for a continuous 7 pore volume leaching cycle. Biosolid colloid suspensions of 200 mg/L spiked with 10 mg/L of the same metal chloride salts were also infused into undisturbed soil monoliths of a fine, mixed, mesic Typic Argiudoll and a sandy, mixed, mesic Typic Udifluvent at a rate of 0.7 cm/h, sustaining unsaturated flow conditions for a continuous 16 pore volume leaching cycle. Leaching cycles with a conservative tracer ($CaCl_2$), de-ionized water, and metal solutions without colloids were used as controls. Leaching outputs were monitored continuously with respect to eluent volume and colloid, metal, and herbicide concentrations. All leaching experiments were duplicated with an acceptable standard error level of <15%.

Eluent colloid concentrations were determined gravimetrically and turbidimetrically by measuring the optical density of the colloid suspensions at 540 nm with a Bio-Tec instruments microplate reader. Metal concentrations were measured via inductively coupled plasma spectrometry. Herbicide levels were determined using liquid chromatography (US-EPA, 1994). Physicochemical colloid properties were determined according to procedures of the National Soil Survey Laboratory Manual (NRCS, 1996). Mineralogical compositions were assessed with X-ray diffraction (XRD) and thermal analyses (Karathanasis and Hajek, 1982). Electrophoretic mobility was determined with a Delsa 440 Doppler Light Scattering Analyzer and the mean colloid diameter via a Beckman Coulter N5 Submicron Particle Size Analyzer.

4.2 Results and Discussion

4.2.1 Soil Nanocolloid Characteristics

4.2.1.1 Mineral Nanocolloids

The water dispersible mineral colloids (Table 4.1) showed a mean colloid diameter range of 220–1,050 nm (Seta and Karathanasis, 1997a). Infusion of these colloid suspensions into undisturbed soil monoliths of a fine-silty, mixed, mesic Typic Argiudoll produced eluted colloids with a mean diameter range of 50–120 nm,

Table 4.1 Colloid characteristics

Colloid	pH	NOM (%)	SM	HISV	MICA	KLN	Q	FEOX	$CaCO_3$	$Ca(OH)_2$	MCD (nm)	SA (m^2/g)	CEC ($cmol_c/kg$)	EM (μm cm/V s)
Smectitic	6.2													
(INF)		0.8	60		20	15	4	1			220	390	63.4	−1.8
(EL)		0.9	85		5	4	1	5			50	560	91.8	−2.3
Mixed	6.3													
(INF)		0.8		46	10	35	8	1			700	190	47.5	−1.4
(EL)		0.8		58	5	23	2	12			95	330	62.3	−1.7
Kaolinitic	5.2													
(INF)		0.4		20	10	55	12	3			1050	114	29.0	−0.8
(EL)		0.3		31	6	44	3	16			120	180	42.0	−1.0
LSBC	11.3													
(INF)		20.1	10						50	10	410	361	80.4	−2.4
(EL)		43.5	25						28		76	415	112.7	−3.8
ADBC	7.2													
(INF)		60.6	15			3		2			460	506	101.0	−3.9
(EL)		69.4	20					6			13.4	707	116.3	−4.3
PMBC	5.5													
(INF)		47.9			15	15	5		5		380	578	93.2	−4.2
(EL)		75.6			10	10					83	783	125.4	−5.0

INF – infused colloids; EL – eluted colloids; LSBC – lime-stabilized biosolid colloids; ADBC – aerobically digested colloids; PMBC – poultry manure biosolid colloids; NOM – natural organic matter; SM – smectite; HISV – hydroxy-interlayered smectite/vermiculite; KLN – kaolinite; Q – quartz; FEOX – iron oxyhydroxide; $CaCO_3$ – calcium carbonate; $Ca(OH)_2$ – calcium hydroxide; MCD – mean colloid diameter; SA – surface area; CEC – cation exchange capacity; EM – electrophoretic mobility.

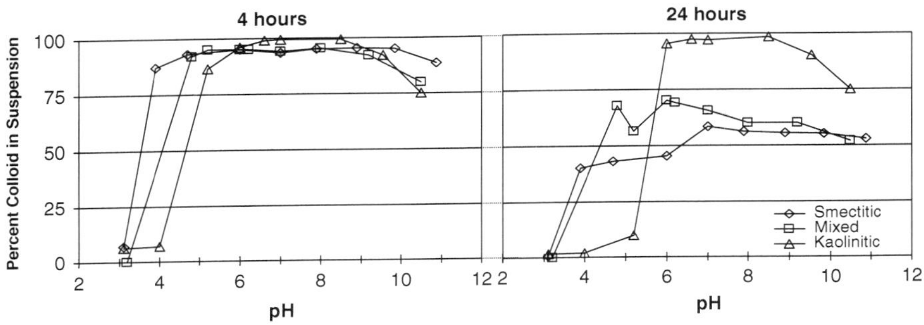

Fig. 4.1 Mean settling rates of smectitic, mixed, and kaolinitic nanocolloids eluted from Argiudoll soil monoliths

suggesting that the majority of the eluted particles were within the nanosize range. The mineralogical composition of the eluted nanocolloids was slightly different than the infused colloids, involving consistent increases in 2:1 minerals (smectite, mica, HIV), as well as Fe-oxyhydroxides, and decreases in 1:1 minerals and quartz. These compositional changes were accompanied by significant increases in surface reactivity, including surface charge (24–31%), surface area (30–42%), and electrophoretic mobility (18–22%) (Table 4.1).

Settling rate experiments with the eluted nanocolloids suggested considerable stability above pH 4 for the smectitic, pH 5 for the mixed, and pH 6 for the kaolinitic particles, ranging from 50 to 75% over a 24 h period (Fig. 4.1). The instability below these respective pH ranges is associated with the proximity of each nanocolloid to the zero point of charge (ZPC) level of their dominant mineral, which is lower for smectite and higher for kaolinite. As the pH of the medium approaches the nanocolloid ZPC, edge-to-face bonding and Fe–Al bonding to particle surfaces increase, causing a drop in the net surface potential to the point that coagulation occurs. Above the ZPC, bonding is limited, thus increasing nanocolloid dispersion and stability. All nanocolloids experienced some decrease in stability over time, likely due to increased collision frequency rather than ionic strength effects, which were maintained at low levels and relatively constant in all eluents (EC < 0.2 μS/cm). Surprisingly, the kaolinitic colloids maintained higher stability at upper pH ranges than the smectitic or the mixed, in spite of their larger diameter and lower NOM content, which at high concentrations may promote dispersion by steric hindrance (Kretzschmar et al., 1995). Apparently, greater abundance of Fe-oxyhydroxide coatings on the surface of these particles induced increased stability above their $pH_{(zpc)}$ level.

4.2.1.2 Bio-nanocolloids

The infused water dispersible biosolid colloids had a mean colloid diameter range of 380–460 nm (Karathansis and Johnson, 2006). Eluted colloids from the undisturbed

Argiudoll and Udifluvent soil monoliths were within or near the nanosize range, with mean diameter of 76–134 nm. With the exception of NOM content, which increased significantly in the eluted colloids (13–51%), only minor differences were detected in the composition of the infused vs. the eluted colloids. However, the eluted nanocolloids showed drastic increases in surface area (13–28%), surface charge (13–29%), and electrophoretic mobility (16–37%) (Table 4.1), suggesting a much more interactive behavior.

Settling rate experiments of the eluted bio-nanocolloids indicated considerable stability over a wide range of pH conditions (4–8) in a 24 h period, following the sequence LSBC > PMBC > ADBC (Fig. 4.2). More than 85% of the LSBC nanocolloids remained in suspension after 24 h, suggesting a remarkable potential mobility over great distances, even at pH 4, following considerable carbonate dissolution and an EC increase to 68 μS/cm. The eluted ADBC were considerably less stable (50–65%) than the LSBC in spite of their higher NOM content. It is likely that the large size and the presence of Fe-oxyhydroxides in these colloids may have

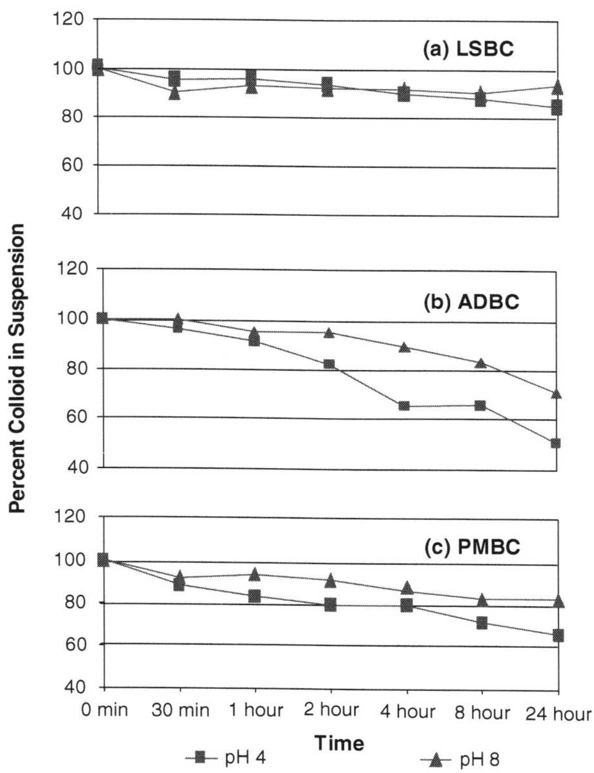

Fig. 4.2 Mean settling rates of LSBC, ADBC, and PMBC nanocolloids eluted from Argiudoll soil monoliths

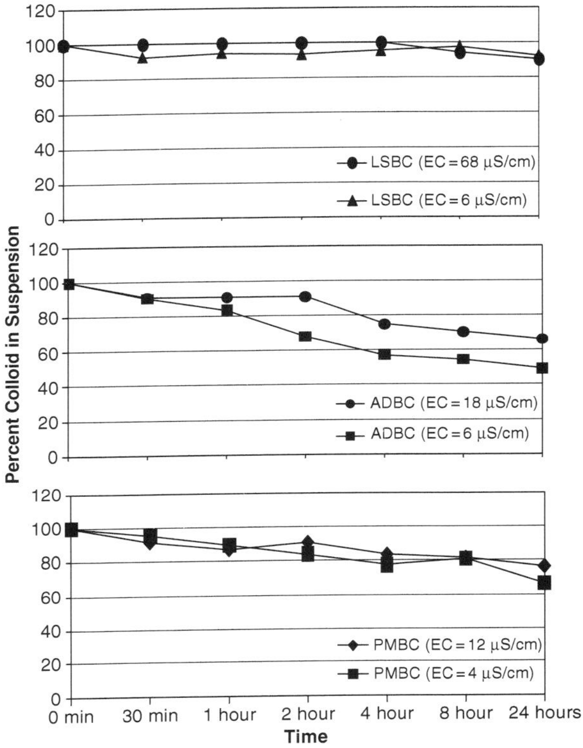

Fig. 4.3 Mean settling rates of LSBC, ADBC, and PMBC nanocolloids eluted from Argiudoll soil monoliths at different EC levels

inhibited their stability, particularly at low pH levels, where a charge reversal on Fe-oxyhydroxide surfaces and a charge reduction on the organic particle surfaces can lead to further destabilization. The PMBC nanocolloids showed an intermediate stability (70–80%) range, consistent with their size, but inconsistent with their highest NOM content. Apparently, the prevalence of different types of organic functional groups in these colloids may have had a variable effect on their surface properties, not proportional to their NOM content (Jekel, 1986). Noticeably, the stability range of the colloids within the nano-range (LSBC and PMBC) remained consistently high (70–90%) and was least affected by more than 3-fold increases in their inherent EC environments (Fig. 4.3). Apparently, their small size and surface properties of their organic constituents overcame the destabilizing effects of the increased ionic strength (Roy and Dzombak, 1997).

4.2.1.3 Other Colloids

Additional experiments involving leaching of undisturbed monoliths from reclaimed mine soils with de-ionized water at various rates produced eluted colloids with a mean diameter range of 314–1,127 nm, which is considerably larger than the nanosize range (Miller, 2008). The size of the eluted colloids increased even more when the monoliths were amended with biosolid wastes. It is likely that the large macroporosity of these monoliths allowed large-size colloids to migrate through the matrix without significant straining.

4.2.2 Nanocolloid Transport Behavior

4.2.2.1 Mineral Nanocolloids

Breakthrough curves (BTC) of mineral colloid suspensions eluted through the soil monoliths of the fine-silty, mixed, mesic Typic Argiudoll showed an initial rapid elution ($C/C_0 = 0.75$), suggesting significant preferential flow (Fig. 4.4) (Seta and Karathanasis, 1997a). In spite of the fact that size exclusion appeared to be a dominant mechanism controlling the transport of the smectitic and the mixed colloids, as evidenced by the initial breakthrough prior to the conservative tracer, significant straining of large-size colloids occurred and reduced the mean diameter of the eluted colloids 4- to 7-fold (Table 4.1) (Gang and Flury, 2005). This is particularly evident with the kaolinitic colloids that showed an initial breakthrough after the conservative tracer and experienced the greatest reduction (60%) in recovery and size (9-fold) as a result of extensive sorption and filtration processes. After an initial breakthrough of approximately 0.4 pore volumes, colloid elution gradually leveled off and approached a steady state at about 3 pore volumes, indicating considerable colloid filtration and attachment within the matrix that restricted colloid mobility (Kretzschmar et al., 1995). Switching the input solution to de-ionized water after 5 pore volumes produced a steep drop in colloid elution, indicating irreversible sorption of the colloids by the monolith soil matrix (Fig. 4.4).

After 5 pore volumes of leaching, eluent colloid recovery ranged from 90% in the smectitic to 35% in the kaolinitic colloids. The high transportability of the smectitic colloid is attributed to the integrated effects of their lowest size and the highest

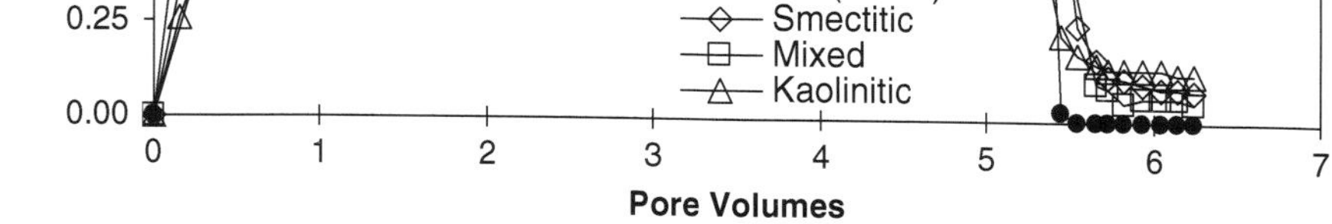

Fig. 4.4 Mean breakthrough curves for smectitic, mixed, and kaolinitic nanocolloids eluted from Argiudoll soil monoliths

surface area, surface charge, and electrophoretic mobility, while the low transportability of the kaolinitic colloid is consistent with their largest size and reduced surface reactivity (Table 4.1). The increased amounts of Fe-oxyhydroxides in the kaolinitic colloid may have also contributed to lower transportability since its eluent pH was near the ZPC of these minerals. The similar mineralogical composition of the eluted colloids compared to the infused colloids suggested very little preferential filtration or sorption of specific mineral types by the monolith matrix during the leaching process. Eluent ionic strength was too low (<0.2 μS/cm) to have any significant impact on colloid stability and transportability in these experiments.

4.2.2.2 Bio-nanocolloids

Breakthrough curves of biosolid colloids eluted through the Argiudoll and the Udifluvent soil monoliths showed somewhat more irregular patterns than the mineral colloids (Fig. 4.5) (Karathanasis and Johnson, 2006). Breakthrough was generally slower than the conservative tracer, indicating considerable interaction through sorption and physical entrapment processes in their flow path through the soil monolith matrix (Jacobsen et al., 1997). In contrast to the mineral colloids that reached an elution steady state after 3 pore volumes, biosolid colloid

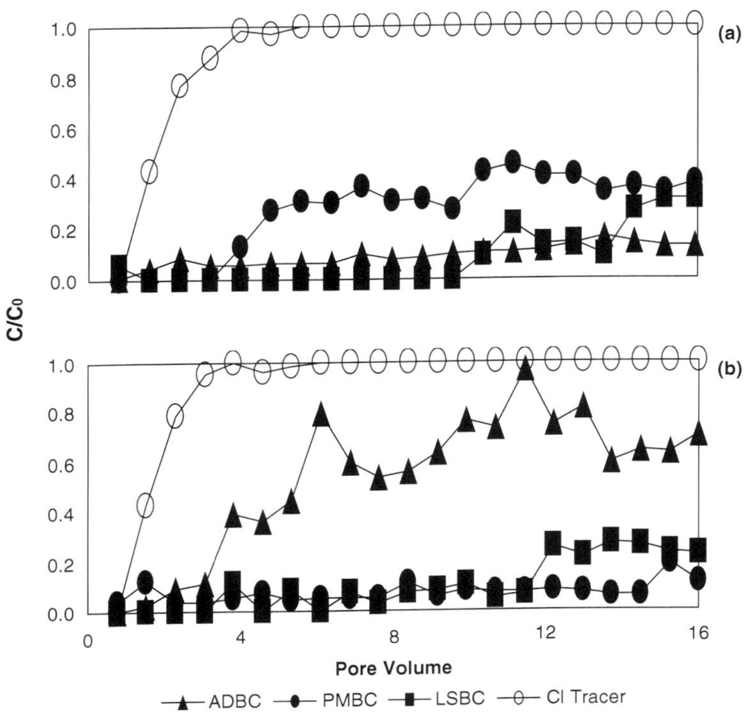

Fig. 4.5 Mean breakthrough curves of LSBC, ADBC, and PMBC nanocolloids eluted from Argiudoll (**a**) and Udifluvent (**b**) soil monoliths

breakthrough showed several maxima and minima throughout the leaching cycle. This elution pattern is consistent with a cluster type of transport resulting from attachment/detachment cycles that coincide with macropore blockage and flushing incidents.

The PMBC showed the highest transportability through the Argiudoll monolith reaching a maximum of 0.45 C/C_0, at 12 pore volumes before tapering off to about 0.3 C/C_0 at the end of the leaching cycle. These results are consistent with the small size and the surface properties of the colloid. In contrast, the elution of PMBC through the Udifluvent monolith was the lowest of all the colloids, barely reaching 0.1 C/C_0 throughout the leaching cycle. In spite of its coarser texture, this soil had lower hydraulic conductivity and macroporosity than the Argiudoll, which apparently hindered colloid mobility. However, a more crucial factor contributing to the reduced transportability of the PMBC might be the lower pH of the Udifluvent compared to the Argiudoll (4.0 vs. 6.1), which could have caused a significant surface charge reduction of the organic constituents and induce coagulation and flocculation (Gang and Flury, 2005).

The ADBC showed the highest transportability through the Udifluvent monolith, reaching maxima of 0.8 C/C_0 and 1.0 C/C_0, at 6 and 11 pore volumes, respectively, before tapering off to 0.6 C/C_0 at the end of the leaching cycle. The modest filtration experienced by these colloids through the Udifluvent compared to the PMBC, in spite of their larger size, is attributed to their higher pH (7.2 vs. 5.5), which apparently buffered considerably the low pH effect of the soil matrix and overcame coagulation and flocculation setbacks. In contrast, the elution of ADBC through the Argiudoll was low, remaining below 0.2 C/C_0 throughout the leaching cycle. The reduced transportability of these colloids through the Argiudoll matrix cannot be attributed to pH or EC effects and most likely are dictated by physical processes.

Surprisingly, the LSBC, although having the highest overall stability, showed very low transportability through both soils during the first 10 pore volumes before peaking to 0.4 and 0.25 C/C_0 in the Argiudoll and Udifluvent, respectively, at the end of the leaching cycle. The limited transportability of these colloids is probably the result of their high eluent EC caused by carbonate dissolution in early stages of leaching, which resulted in significant colloid flocculation and filtration within the monolith matrix. The above findings underline the complex nature of colloid transport behavior through natural porous media, which is the integrated effect of colloid size, surface properties, and interactions with the soil matrix. These interactions are not always predictable with simple tests and experiments of colloid characterization alone and require thorough assessments of colloid behavior under natural environmental settings.

4.2.3 Facilitated Transport of Pollutants

4.2.3.1 Metals

Figures 4.6 and 4.7 show BTCs of Cu, Zn, and Pb eluted with the mineral colloids through the Argiudoll monoliths (Karathanasis, 1999, 2000, 2003). In the absence of

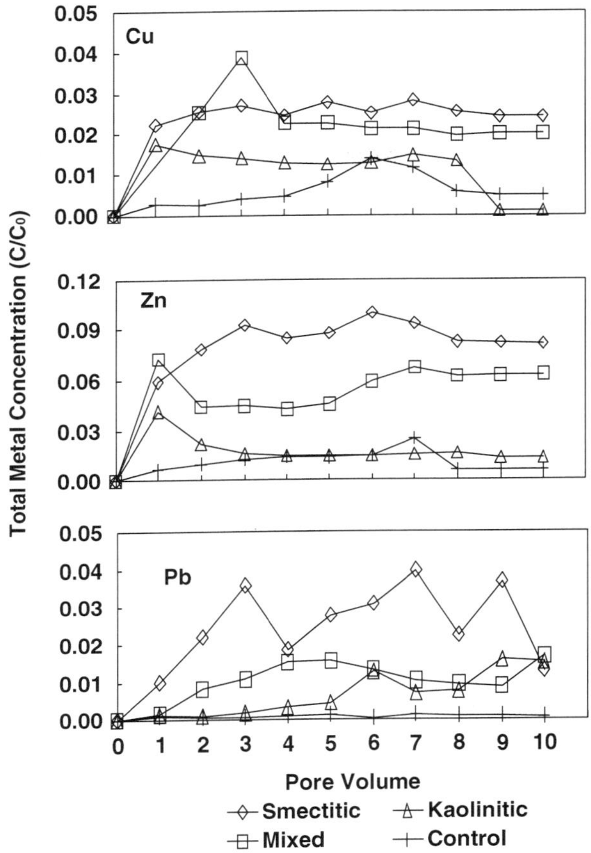

Fig. 4.6 Mean total metal loads (Cu, Zn, Pb) eluted with smectitic, mixed, and kaolinitic nanocolloids from Argiudoll soil monoliths

colloids (controls), practically none of the metals exhibited any meaningful breakthrough, suggesting nearly complete sorption by the soil matrix. The presence of colloids considerably enhanced total metal elution and in most cases even soluble metal elution, thus providing strong evidence for colloid-mediated transport. Most BTCs showed considerable asymmetry attributed to partial clogging and flushing cycles and/or chemical interactions between solutes, colloids, and soil matrix. These interactions are anticipated to be manifested through physical and chemical processes associated with multiple colloid attachment/detachment stages and variable affinities of metals for colloid and soil surfaces. Generally, total metal elution was higher than soluble metal elution. Considering that the difference between total metal and soluble metal load represents the colloid-bound fraction and given the strong correlation between total metal and colloid elution, it appears that the colloids were acting as carriers of the majority of the metal load. As was the case with the colloid elution, the metal load-carrying efficiently followed the sequence smectitic > mixed > kaolinitic, indicating a strong relationship with colloid surface charge

Fig. 4.7 Mean soluble metal loads (Cu, Zn, Pb) eluted with smectitic, mixed, and kaolinitic nanocolloids from Argiudoll soil monoliths

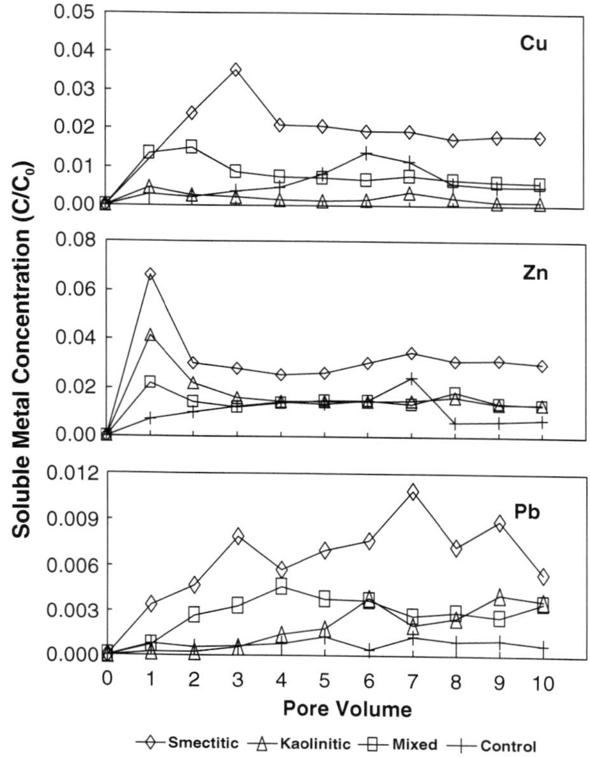

properties. Therefore, this provides compelling evidence that the primary mechanism for the enhanced metal transport was metal chemisorption to reactive colloid surfaces, especially in cases where metal affinity for colloid sites was greater than that for soil matrix sites. However, competitive metal sorption between colloid and soil matrix may also occur during the leaching cycle, in spite of metal affinities, in the process of establishing local equilibrium between the two solid phases.

Metal transport increases were also metal specific, following the sequence Zn > Pb > Cu for total metal elution and Zn > Cu > Pb for soluble metal elution. Overall, between 30 and 90% of Cu was transported in the soluble fraction, while >60% of Zn and Pb were transported in the colloid-sorbed fraction. This is generally consistent with the metal affinities of the different colloids in conjunction with the NOM content and colloid size differences. Average increases of total Cu transported in the presence of colloids were 3-fold for kaolinitic, 5-fold for mixed, and 6-fold for smectitic colloids compared to the controls (absence of colloids). The respective average increases for Zn transport were 1.5-fold for kaolinitic, 6-fold for mixed, and 9-fold for smectitic colloids. Average increases for total Pb were the highest, ranging from 7-fold for kaolinitic up to 30-fold for smectitic colloids. Average soluble metal elution increases were not as dramatic for Cu and Zn (up to 3-fold), but more substantial for Pb (up to 11-fold), with the maxima being associated with either smectitic or mixed colloids. The similar soluble Cu load transported by all colloids regardless of mineralogy is attributed to the strong affinity of this metal to form organic complexes (Camobreco et al., 1996; Zhou and Wong, 2001). This mechanism may also be partially responsible for the additional soluble metal loads of Zn and Pb recovered in the presence of colloids. Furthermore, exclusion of soluble metal species from soil matrix sites blocked by colloids and elution of metal ions associated with the diffuse layer of colloid particles may have increased the soluble metal load. These findings clearly demonstrate the role of colloid mineralogical composition on their ability to induce and mediate the transport of heavy metals in subsurface soil environments. The observed metal transport patterns appeared to be influenced primarily by inherent and/or accessory mineralogical and physicochemical properties of the colloids, such as surface charge, surface area, electrophoretic mobility, and mean colloid diameter, and less by coincidental factors, such as NOM, pH, Fe-Al hydroxides, and ionic strengths, normally encountered in soil environments.

Figures 4.8, 4.9, and 4.10 show BTCs for Cu, Zn, and Pb eluted through the Argiudoll and the Udifluvent soil monoliths in the presence and absence (controls) of bio-nanocolloids (Karathanasis et al., 2005). Metal elution was essentially zero

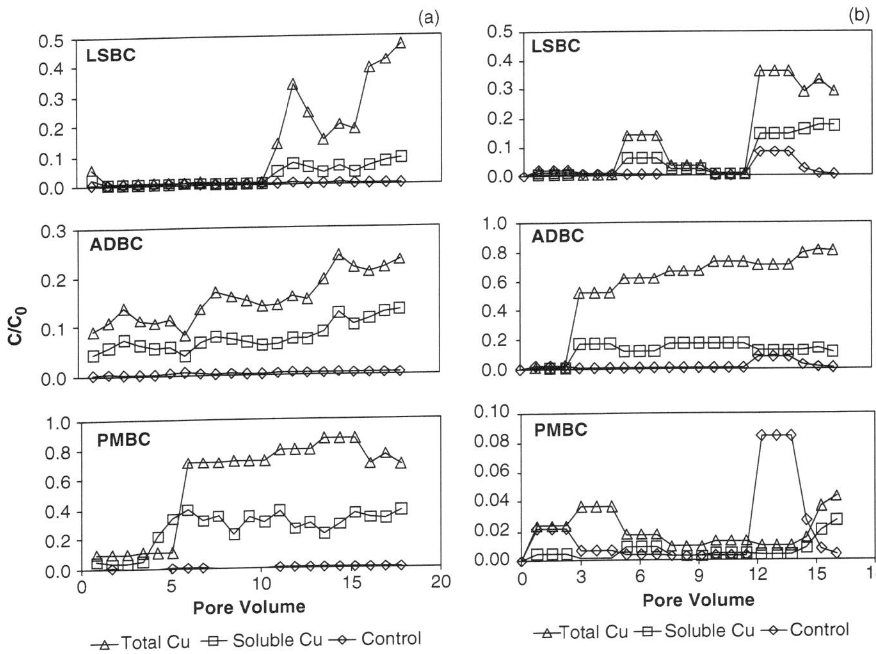

Fig. 4.8 Mean total and soluble Cu loads eluted with LSBC, ADBC, and PMBC nanocolloids from Argiudoll (**a**) and Udifluvent (**b**) soil monoliths

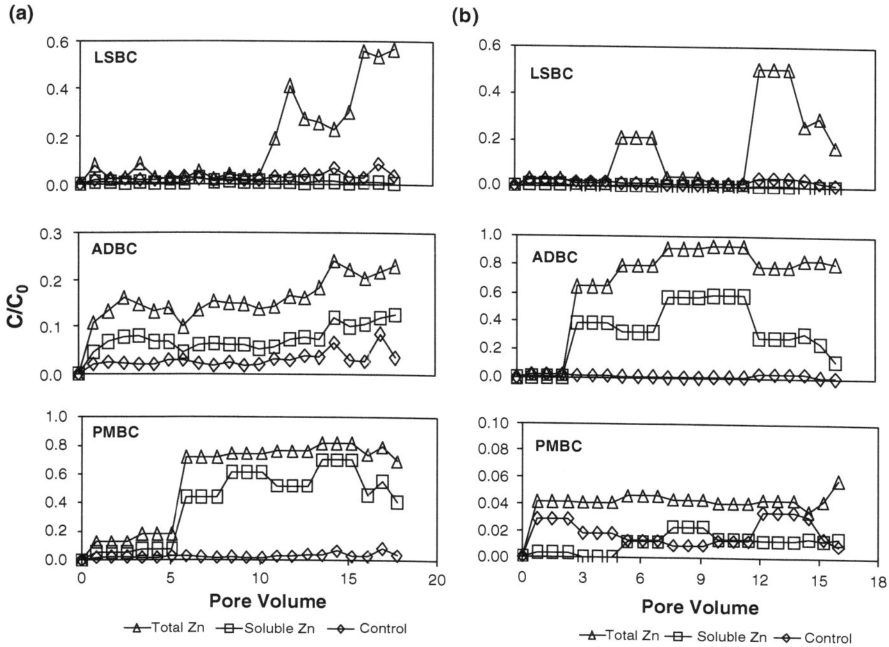

Fig. 4.9 Mean total and soluble Zn loads eluted with LSBC, ADBC, and PMBC nanocolloids from Argiudoll (**a**) and Udifluvent (**b**) soil monoliths

for all control treatments, suggesting total attenuation by the soil matrix. Total Cu elution was highest in association with the ADBC through the Udifluvent and the PMBC through the Argiudoll, with recoveries approaching 80% toward the end of the leaching cycle (Fig. 4.8). The high NOM, surface area, and electrophoretic mobility of these nanocolloids accounted for 56% of the variation in total Cu elution, as indicated by multiple regression analysis. In addition, the low ionic strength of their eluents would tend to promote double layer expansion and limit colloid attachment to the soil matrix. Soluble Cu accounted for 15–45% of the total Cu eluted through both soils and is attributed to the high NOM content of the nanocolloids that may induce formation of soluble organic-Cu complexes, particularly in the presence of high molecular weight hydrophilic organic constituents (Han and Thompson, 1999; Gove et al., 2001). Total Cu elution associated with LSBC through the Argiudoll and Udifluvent soil monoliths was irregular and nearly absent during the first half of the leaching cycle, reaching maxima of 0.4 and 0.5 C/C_0 toward the end of the leaching period as the EC of the eluents dropped below 30 μS/cm. Total Cu elution with ADBC through the Argiudoll monolith was moderate (0.25 C/C_0) but more symmetrical and that of PMBC through the Udifluvent irregular but lower than 0.1 C/C_0 throughout the leaching cycle. The latter trends were consistent with lower nanocolloid elution and greater attenuation by the soil matrices.

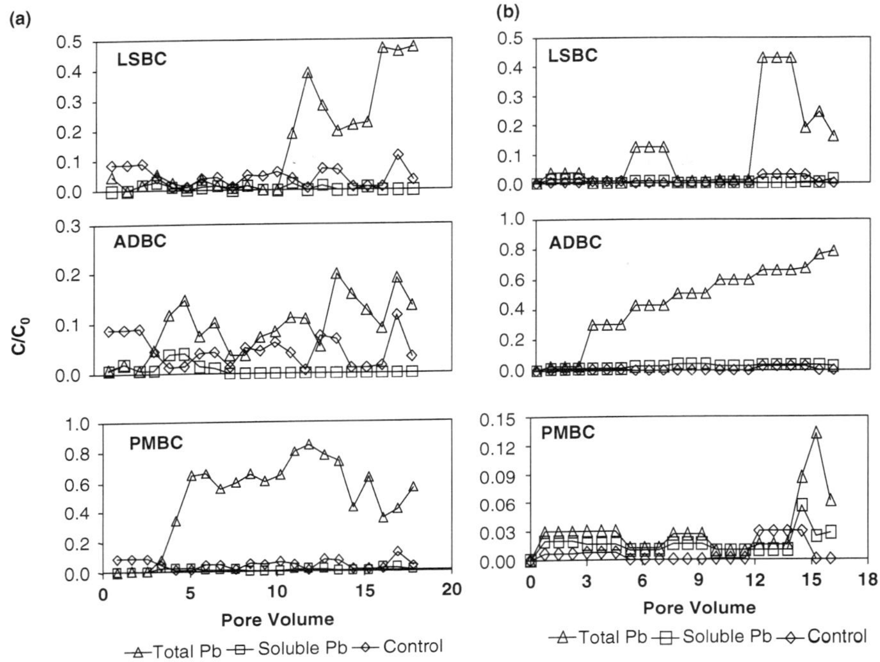

Fig. 4.10 Mean total and soluble Pb loads eluted with LSBC, ADBC, and PMBC nanocolloids from Argiudoll (**a**) and Udifluvent (**b**) soil monoliths

Breakthrough curves for total Zn eluted in the presence of bio-nanocolloids showed only subtle differences compared to Cu-BTCs, showing a strong relationship with colloid elution patterns. The ADBC through the Udifluvent and the PMBC through the Argiudoll appeared to be the most effective Zn carriers, with relatively symmetrical BTC and recoveries of 80–90% after 6 pore volumes of leaching. Surface charge, NOM, and surface area accounted for 70% of the variation in total Zn elution (Li and Shuman, 1997). Significant amounts of soluble Zn were also eluted in the presence of the above nanocolloids, amounting to nearly 50% of the eluted Zn load. Total Zn elution in association with LSBC through both soils was irregular, showing maxima of 0.55 C/C_0 toward the end of the leaching period as a result of decreasing EC levels. Total Zn elution in the presence of ADBC through the Argiudoll was moderate and more symmetrical, reaching maxima similar to that of Cu (0.25 C/C_0) at the end of the cycle. The decreased Zn elution with remaining nanocolloids was consistent with lower colloid elution trends.

Total Pb-BTCs were very similar to those of Cu and Zn. The highest recoveries (0.8 C/C_0) were observed in association with PMBC through the Argiudoll and with the ADBC through the Udifluvent. The LSBC-Pb elution patterns were almost identical to those of Zn, reaching maxima of 0.45 C/C_0 toward the end of the leaching period. The remaining nanocolloids showed irregular BTC with recoveries below

0.2 C/C_0. The soluble Pb fraction transported in the presence of bio-nanocolloids was considerably smaller than the soluble Cu and Zn fraction as a result of greater affinities of Pb for colloid surfaces (Denaix et al., 2001).

4.2.3.2 Herbicides

Breakthrough curves of atrazine eluted with and without (control) the mineral nanocolloids indicated very little attenuation (5–15%) within the soil matrix over the entire leaching cycle (Fig. 4.11) (Seta and Karathanasis, 1997b). In all cases, the presence of colloids enhanced atrazine elution above that of the control treatment. The nanocolloid-mediated increase in atrazine transport ranged from 2 to 18%, following the sequence smectitic > mixed > kaolinitic. Apparently, the larger surface area, NOM content, and electrophoretic mobility of the smectitic and the mixed nanocolloids, in conjunction with their smaller size, induced greater atrazine load transportability. In contrast, the lower surface reactivity of the larger kaolinitic particles could carry smaller atrazine loads and encounter more opportunities for sorption and filtration within the soil monolith matrix. However, the nanocolloid-bound eluted atrazine fraction was smaller than the eluted soluble fraction, suggesting that atrazine attachment to the nanocolloid surfaces was not the only mechanism for enhancing atrazine transportability. Exclusion of soluble atrazine molecules from certain reactive sites of the soil matrix due to temporary (flow-path collisions) or permanent (attachment) blockage by colloid particles may have promoted soluble herbicide transportability in the presence of nanocolloids.

Unlike the atrazine transport, metolachlor BTCs showed significant attenuation by the soil matrix, ranging from 45 to 70% over the entire leaching cycle (Fig. 4.12) (Seta and Karathanasis, 1996). This is the result of the higher adsorption coefficient (3- to 5-fold) of metolachlor over that of atrazine for soil and colloid surfaces. The presence of nanocolloids consistently enhanced metolachlor transport by 5–25%, following the sequence smectitic > mixed > kaolinitic. As was the case with the atrazine transport, the higher surface reactivity of the smectitic and mixed nanocolloids enhanced greater metolachlor load transportability, while the larger size and low reactivity of the kaolinitic nanocolloids inhibited metolachlor elution. Exclusion

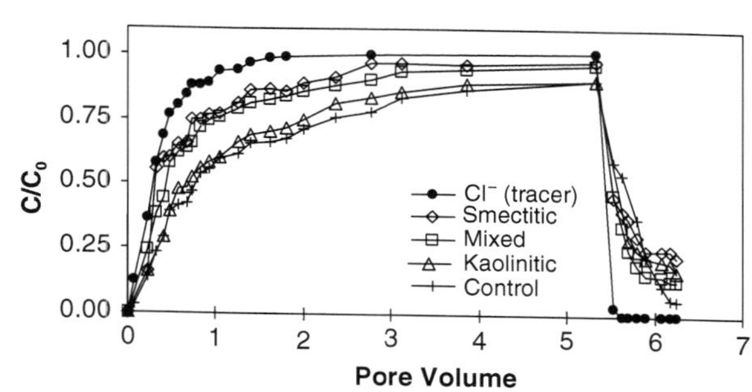

Fig. 4.11 Mean BTCs for atrazine eluted with smectitic, mixed, and kaolinitic nanocolloids from Argiudoll soil monoliths

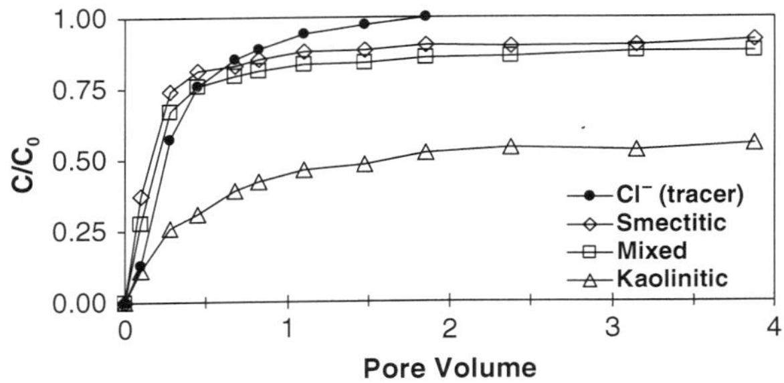

Fig. 4.12 Mean BTCs for metolachlor eluted with smectitic, mixed, and kaolinitic nanocolloids from Argiudoll soil monoliths

mechanisms may have also been responsible for significant increases in eluted soluble metolachlor fractions in the presence of nanocolloids.

4.2.4 Potential Mechanisms

The good agreement between metal/herbicide elution and nanocolloid elution patterns implies that a large portion of these potential pollutants eluted in the presence of migrating nanocolloids were sorbed or attached onto their surfaces. This was corroborated by strong correlations between metal/herbicide transport and nanocolloid surface properties and emphasizes the potential of mobile nanocolloids as contaminant carriers in subsurface soil environments (Karathanasis, 1999; Denaix et al., 2001). However, in all leaching experiments involving nanocolloids an increased load of the eluted metal/herbicide was in the soluble fraction, suggesting that the presence of nanocolloids enhanced the elution of both sorbed and soluble fractions. Correlation coefficients between the eluted soluble metal and eluted nanocolloid fractions were particularly high for Cu and Zn, explaining about 25% of the overall total metal elution. In contrast, essentially no elution of the metal soluble fraction occurred in the absence of nanocolloids (control treatments), suggesting complete attenuation by the soil matrix. This suggests that the mobile nanocolloids in the conducted experiments may have played a dual role in the metal transport process acting as carriers and facilitators. While the role of nanocolloids as metal carriers is envisioned through surface chemisorption and perhaps coprecipitation mechanisms, particularly for carbonate-containing colloids, their role as facilitators requires further documentation.

One possible mechanism contributing to the transport of elevated soluble metal loads in the presence of nanocolloids is certainly "complexation" and "ligand formation" between metals and dissolved organic carbon (DOC) constituents, particularly in the presence of bio-nanocolloids. Metal speciation simulations with MINTEQ suggested that between 20 and 60% of the eluted soluble metal fraction was in the form of organic ligands, while ion specific electrode analysis of selected metals indicated a range between 35 and 50%. The remaining soluble metal fraction eluted in the presence of bio- or mineral nanocolloids could have been transported via the action of physical or chemical exclusion mechanisms (Karathanasis et al., 2007).

While these mechanisms can only be speculated in this study, their functionality has been documented in a variety of physical and chemical processes, including transport phenomena. Examples of physical exclusion processes that may result in elution of extra soluble metal loads in the presence of nanocolloids include (a) pore clogging by migrating colloids that render some reactive soil matrix sites inaccessible for interaction with soluble or nanocolloid-sorbed metals; and (b) increased collisions between soluble metal species and charge-satisfied nanocolloid particles during the transport process that may block attachment or interaction opportunities with soil pore reactive sites (Bradford et al., 2007). Examples of chemical exclusion mechanisms contributing to higher soluble metal elution loads in the presence of nanocolloids include (a) inactivation of reactive soil pore wall sites by metal–colloid entities sharing metal ionic bonds, therefore excluding the sites from interactions with soluble metal species; and (b) soluble ionic metal species associated with the diffuse double layer of mobile nanocolloids may bypass attractive forces of soil matrix reactive sites due to their high coordination and kinetic energy.

4.3 Conclusions

Undoubtedly, the mesocosm-scale experiments employed in these studies cannot provide accurate assessments of the magnitude and variability in colloid generation and transport behavior to be expected under field-scale natural environmental settings. However, the findings provided unquestionable evidence for the ubiquitous presence of nanocolloids in soil environments, their increased stability and potential to migrate great distances through natural porous media, and their significant role as potential carriers and/or transport facilitators of organic and inorganic pollutants to groundwater aquifers. In all cases involving mineral or bio-nanocolloids, their composition, size, and surface properties were the primary parameters controlling their stability, mobility, and co-transportability of metals and herbicides. Decreases in mean nanocolloid diameter and increases in surface charge, surface area, and electrophoretic mobility drastically enhanced nanocolloid stability and transport through undisturbed soil monoliths, even at relatively high ionic strength levels, and significantly (up to 11-fold) increased metal and herbicide transport in subsurface soil environments. Substantial increases of both total metal and soluble metal loads in the presence of nanocolloids suggested the influence of strongly interactive chemical and physical processes between nanocolloid surfaces and soil matrices in controlling the co-transport behavior. Although the increased chemisorption affinity of the nanocolloid surfaces for the pollutants may be the dominant mechanism facilitating co-transport, organic complexation and physical exclusion processes may significantly increase the magnitude of the transported contaminant loads. These findings also emphasize the importance of investing more advanced technology resources toward in situ monitoring of environmental nanocolloid generation and transport processes as well as their improved characterization so that we can obtain a more realistic understanding of their role and behavior under natural conditions as potential pollutants or remediators. This is particularly important considering the inadequacy of batch and artificial simulation flow experiments alone to assess the dynamic nature of nanocolloid transport phenomena through natural environmental porous media.

References

Baalousha M, Lead JR (2007) Characterization of natural aquatic colloids (<5 nm) by flow-field flow fractionation and atomic force microscopy. Environ Sci Technol 41: 1111–1117.

Bradford SA, Torkzaban S, Walker SL (2007) Coupling of physical and chemical mechanisms of colloid straining in saturated porous media. Water Res 41: 3012–3024.

Camobreco VJ, Richards BK, Steenhuis TS, Peverly JH, McBride MB (1996) Movement of heavy metals through undisturbed and homogenized soil columns. Soil Sci 161: 740–750.

Christian P, Von der Kammer F, Baalousha M, Hofmann T (2008) Nanoparticles: structure, properties, preparation, and behaviour in environmental media. Ecotoxicology 17: 326–343.

Denaix L, Semlali RM, Douay F (2001) Dissolved and colloidal transport of Cd, Pb, and Zn in a silt loam soil affected by atmospheric industrial deposition. Environ Pollut 113: 29–38.

Gang C, Flury M (2005) Retention of mineral colloids in unsaturated porous media as related to their surface properties. Colloids Surf A Physicochem Eng Asp 256: 207–216.

Gove L, Cook CM, Nicholson EA, Beck AJ (2001) Movement of water and heavy metals (Zn, Cu, Pb, and Ni) through sand and sandy loam amended with biosolids under steady-state hydrological conditions. Biores Technol 78: 171–179.

Han N, Thompson ML (1999) Copper-binding ability of dissolved organic matter derived from anaerobically digested biosolids. J Environ Qual 28: 939–944.

Jacobsen OH, Moldrup P, Larsen C, Konnerup L, Petersen LW (1997) Particle transport in macropores of undisturbed soil columns. J Hydrol 196: 185–203.

Jekel MR (1986) The stabilization of dispersed mineral particles by adsorption of humic substances. Water Res 20: 1543–1554.

Kaplan DI, Bertsch PM, Adriano DC, Miller WP (1993) Soil-borne mobile colloids as influenced by water flow and organic carbon. Environ Sci Technol 27: 1192–1200.

Karathanasis AD (1999) Subsurface migration of Cu and Zn mediated by soil colloids. Soil Sci Soc Am J 63: 830–838.

Karathanasis AD (2000) Colloid-mediated transport of Pb through soil porous media. Intern J Environ Studies 57: 579–596.

Karathanasis AD (2003) Mineral controls in colloid-mediated transport of metals in soil environments. In: Kingery WLSelim HM(eds.) Geochemical and Hydrological Reactivity of Heavy Metals in Soils. CRC Press, pp. 25–49.

Karathanasis A, Johnson C, Matocha C (2007) Subsurface transport of heavy metals mediated by biosolid colloids in waste-amended soils. In Frimmel FH, Von der Kammer F, Fleming HC (eds.) Colloidal Transport in Porous Media, Springer, Berlin, pp. 175–201.

Karathanasis AD, Hajek BF (1982) Revised methods for rapid quantitative determination of minerals in soil clays. Soil Sci Soc Am J 46: 419–425.

Karathanasis AD, Johnson DMC (2006) Stability and transportability of biosolid colloids through undisturbed soil monoliths. Geoderma 130: 334–345.

Karathanasis AD, Johnson DMC, Matocha CJ (2005) Biosolid colloid-mediated transport of Cu, Zn, and Pb in waste-amended soils. J Environ Qual 34: 1153–1164.

Kretzschmar R, Robarge WP, Amoozegar A (1995) Influence of natural organic matter on transport of soil colloids through saprolite. Water Resour Res 31: 435–445.

Lead JR, Wilkinson KJ (2007) Environmental Colloids and Particles: Behaviour, Separation, and Characterization. J. Wiley and Sons, Chichester, UK, 687 pp.

Li Z, Shuman LM (1997) Mobility of Zn, Cd, and Pb in soils as affected by poultry litter extract-I. Leaching in soil columns. Environ Pollut 2: 219–226.

Liang P, Ding Q, Song F (2006) Application of multiwalled carbon nanotubes as solid phase extraction sorbent for preconcentration of trace Cu in water samples. J Sep Sci 28: 2339–2343.
Lyven B, Hassellov M, Turner DR, Haraldsson C, Andersson K (2003) Competition between iron- and carbon-based colloidal carriers for trace metals in a freshwater assessed using flow-field flow fractionation coupled to ICPMS. Geochim Cosmochim Acta 67: 3791–3802.
Miller JO (2008) Colloid-mediated transport of heavy metals in soils following reclamation with and without biosolid application. Ph.D. Dissertation, 129 pp., University of Kentucky.
NRCS (1996) Soil Survey Laboratory Methods Manual. Soil Survey Investigations Report No. 42. Version 3.0, USDA, National Soil Survey Center, Lincoln, NE.
Ouyang Y, Shinde D, Mansell RS, Harris W (1996) Colloid-enhanced transport of chemicals in subsurface environments: a review. Crit Rev Environ Sci Technol 26: 189–204.
Ranville JF, Chittleborough DJ, Beckett R (2005) Particle-size and element distribution of soil colloids: implications for colloid transport. Soil Sci Soc Am J 69: 1173–1184.
Roy SC, Dzombak DA (1997) Chemical factors influencing colloid-facilitated transport of contaminants in porous media. Environ Sci Technol 37: 656–664.
Ryan JN, Elimelech M (1996) Colloid mobilization and transport in groundwater. Colloids Surf A Physicochem Eng Asp 107: 1–56.
Seta AK, Karathanasis AD (1996) Colloid-facilitated transport of metolachlor through intact soil columns. J Environ Sci Health B 31(5): 949–968.
Seta AK, Karathanasis AD (1997a) Stability and transportability of water-dispersible soil colloids. Soil Sci Am J 61: 604–611.
Seta AK, Karathanasis AD (1997b) Atrazine adsorption by soil colloids and co-transport through subsurface environments. Soil Sci Soc Am J 61: 612–617.
USA-EPA (1994) Methods for the determination of metals in environmental samples. Method 200.2. EPA/600/R-94/111. Washington, DC, USA.
Wigginton NS, Haus KL, Hochella MF Jr. (2007) Aquatic environmental nanoparticles. J Environ Monitor 9: 1306–1316.
Zhou LX, Wong JW (2001) Effect of dissolved organic matter from sludge and sludge compost on soil copper sorption. J Environ Qual 30: 878–883.
Zhou Q, Ding Y, Xiao J (2006) Sensitive determination of thiamethoxam, imidacloprid, and acetamiprid in environmental water samples with solid-phase extraction packed with multi-walled carbon nanotubes prior to high-performance liquid chromatography. Anal Bioanal Chem 385: 1520–1525.

Chapter 5
Removal of Organic and Inorganic Pollutants and Pathogens from Wastewater and Drinking Water Using Nanoparticles – A Review

Carsten Prasse and Thomas Ternes

5.1 Introduction

Effective treatment processes for wastewater as well as for drinking water production are major prerequisites for a developing and growing economy. Therefore, it is crucial to develop and implement innovative water technologies treating water with high efficiency and low energy consumption. To date, major challenges for water treatment are pathogens and chemical pollutants. Municipal wastewater contains a multitude of bacteria, viruses, parasites as well as inorganic and organic chemicals. Many of the microorganisms and chemical pollutants are able to pass through municipal wastewater treatment plants (WWTP), at least to some extent. For instance, viruses were detected in 67% of 424 samples collected from unchlorinated secondary effluents (Crook, 1998).

On a global scale water-borne diseases are still a major cause of death in developing countries where access to safe drinking water is often limited. With the introduction of disinfection processes (mainly chlorine), water-borne infectious diseases have been significantly reduced. However, it is known that the application of disinfection agents such as chlorine, chlorine dioxide or ozone is associated with the formation of disinfection by-products, (e.g. trihalomethanes, halo-phenols, ketones, aldehydes) some with a high mutagenic and/or carcinogenic potential (Becher, 1999; Karnik et al., 2005a; Zhang et al., 2004; Zwiener et al., 2007). Chlorination also affects the taste and odour of drinking water. Therefore, there is still a need for new concepts to reduce the release of toxic by-products resulting from disinfection processes. The application of modern nanotechnology could be one approach to improve this situation.

There is an increasing number of chemicals which are emitted, intentionally or unintentionally, into rivers and streams. Today's waters contain a multitude of anthropogenic substances, and conventional treatment systems are often not capable

T. Ternes (✉)
Federal Institute of Hydrology (BfG), Am Mainzer Tor 1 D-56068 Koblenz, Germany
e-mail: Ternes@bafg.de

F.H. Frimmel, R. Niessner (eds.), *Nanoparticles in the Water Cycle*,
DOI 10.1007/978-3-642-10318-6_5, © Springer-Verlag Berlin Heidelberg 2010

of removing them properly, especially those which are resistant to microbial degradation. Many pharmaceuticals, ingredients of personal care products or additives of industrial products cannot be completely removed during wastewater treatment (Andersen et al., 2003; Carballa et al., 2004; Joss et al., 2006; Lissemore et al., 2006; Miao et al., 2004; Ternes, 1998). Thus, these organic compounds are permanently introduced into rivers and streams by WWTP discharges, while diffuse sources such as run-off from agricultural fields are possible but frequently contribute to a much smaller extent to the overall pollution (Glassmeyer et al., 2005; Halling-Sorensen et al., 1998; Lissemore et al., 2006).

As consequence, many of these emerging pollutants have been detected worldwide in aqueous matrices such as WWTP effluents, surface water, groundwater and drinking water.

Currently several processes can be distinguished for the removal of chemical pollutants: (a) physical barriers using membranes, (b) physical retention by adsorption on a solid phase with activated carbon (either as granular activated carbon (GAC) in filters, as powdered activated carbon (PAC) in conventional processes or in combination with membranes), (c) advanced oxidation processes (e.g. ozonation). Biotransformation is known to be relevant for organic contaminants. However, frequently mineralization of organic pollutants cannot be achieved. Nanotechnology offers opportunities to initiate or to enhance the phototransformation and the degradation/transformation of organic pollutants.

The current chapter aims to give an overview of the potential use of nanoparticles in processes and techniques for the treatment of wastewater and the production of drinking water.

5.2 Removal of Pollutants by Nanoparticles

Sorption, redox reactions, photocatalytic transformation and size exclusion are processes which can be improved or initiated by nanoparticles for the removal of organic and inorganic pollutants from wastewater and drinking water (Fig. 5.1).

Sorption onto nanoparticles depends on the sorption coefficient K_d, describing partitioning of a contaminant under equilibrium conditions (Schwarzenbach et al., 2003). Currently, there are no data available to estimate the rate of sorption and desorption. But since in most cases diffusion is fast compared to the hydraulic retention times in water treatment and fast compared to the biological removal of most compounds, equilibrium can be assumed for solid–liquid partitioning (Wang and Grady, 1995).

Redox reactions are favoured for persistent organic pollutants to initiate the transformation of the molecular structure. Frequently, the biodegradability of an organic molecule is increased by oxidation and the toxicity is diminished. In general, oxidation of a molecule introduces hydroxyl moieties in the molecular structure or causes the formation of keto groups which can react further. The reductive dehalogenation of organic compounds initiates the cleavage of a halogen atom (e.g. from a benzene

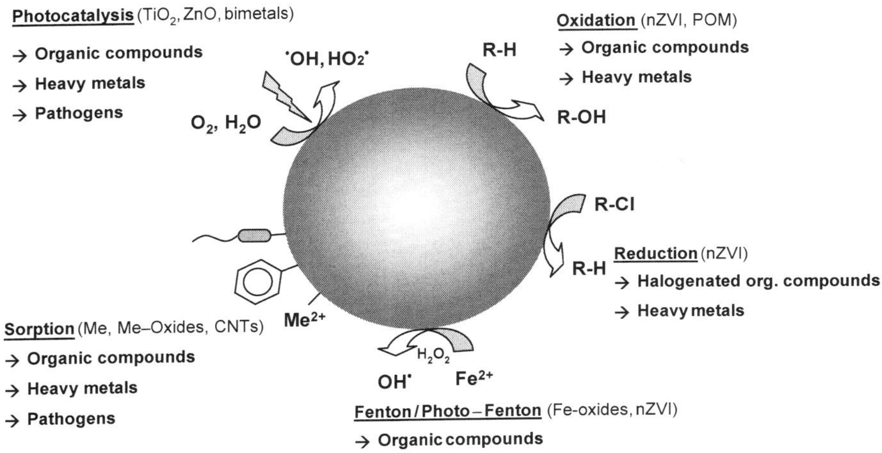

Fig. 5.1 Relevant processes for the removal of organic and inorganic pollutants from water

ring). Further biochemical or oxidative reactions are then able to attack the sites of the molecule where the halogen atom is cleaved. For metals it is well known that a change of the redox conditions has a major influence on its toxicity.

Size exclusion: The rejection of organic solutes by membrane filtration such as nanofiltration/reverse osmosis (NF/RO) is affected by different factors and parameters (Bellona et al., 2004). For dissolved organic pollutants key parameters are molecular weight (M_W), molecular size, acid dissociation constant (pK_a), hydrophobicity/hydrophilicity ($\log K_{OW}$) and diffusion coefficients, while for membranes the factors are the molecular weight cut-off, pore size, surface charge and morphology and hydrophobicity/hydrophilicity. Hydrophobic interactions are expected to play a major role for pollutants with higher K_{OW} values, whereas very polar and negatively charged compounds can add electrostatic rejection with the usually negatively charged NF membranes. In addition, nanoparticles with disinfection properties might play a major role in reducing the biofouling of membranes which is one of the main drawbacks of this technology.

Photocatalytic transformation (e.g. using TiO_2 solutions) is strongly effected by the natural water constituents, such as organic substances, nitrate and trace metals, which are a source of reactive oxygen species producing photosensitized reactions in solution. A photocatalyst is activated when the irradiation energy is equal to or greater than the band gap, i.e. the energy difference between the valence and the conduction bands. Briefly, the absorption of photons results in the elevation of electrons from the valence into the conduction band leading to the formation of excited state conduction band electrons and positive valence band holes. The formed charges can either recombine to produce heat or be used to reduce or oxidize species at the nanoparticle surface (Kwon et al., 2008). In general, OH-radicals are produced which are extremely strong oxidants, but are unselective leading to a variety of different oxidation products. Furthermore, formed electrons can react with O_2 to form O_2^- (Bhatkhande et al., 2002; Hoffmann et al., 1995).

5.2.1 Removal of Inorganic Pollutants

As(V) and Cr(VI) can be efficiently adsorbed on sorbents consisting of CeO_2, as well as either carbon nanotubes (CeO_2-CNT) or aligned carbon nanotubes (CeO_2-ACNT) (Di et al., 2006; Peng et al., 2005; Table 5.1). TiO_2 and ZnO nanoparticles as well as zero-valent iron nanoparticles (nZVI) have also been shown to remove arsenic from water by sorption, while nZVI leads additionally to the reduction of As(V) to As(III) and of Cr(VI) to Cr(III) (Jezequel and Chu, 2006; Kanel et al., 2006; Pena et al., 2005; Xu and Zhao, 2007; Zhang, 2003). However, the sorption of both As(III) and As(V) is heavily influenced by competing species. Giasuddin et al. (2007) observed a decrease in the surface area normalized adsorption rate constant of As(III) and As(V) to nZVI from 100 to 43 and 68%, respectively, when humic acids (20 mg L^{-1}) were added. Liu et al. (2008a) who investigated the effect of NOM on arsenic adsorption to TiO_2 obtained similar results. Lazaridis et al. (2005) described the removal of Cr(VI) by nanocrystalline akaganeite (β-FeO(OH)). Chromate (CrO_4^{2-}) and arsenate (AsO_4^{3-}) can be adsorbed by anion-SAMMS (self-assembled monolayers on mesoporous supports) or metal-capped ethylenediamin (EDA) SAMMS (Fryxell et al., 2002). These sorbents are synthesized by adsorbing functional molecules as a monolayer onto porous ceramic oxides (Mattigod et al., 1999). The photocatalytic oxidation of As(III) has been shown by several authors for TiO_2 nanoparticles (Dutta et al., 2005; Ferguson et al., 2005; Sharma et al., 2007) and Pt-TiO_2 nanoparticles (Ryu and Choi, 2004). In contrast to As(III), As(V) is less toxic and less mobile in the aquatic environment. It sorbs strongly to the surfaces of the photocatalysts and thereby being removed from the aqueous phase. Accordingly, the photocatalytic oxidation can be used for the removal and detoxification of these metals from water. Using TiO_2 nanoparticles the photocatalytic oxidation of organo arsenates (e.g. mono- and dimethylarsenic acid) has been shown by Xu et al. (2007). The authors also showed that organo arsenates have a high sorption affinity to TiO_2 nanoparticles.

Several heavy metals (e.g. Cu(II), Pb(II), Cd(II), Ni(II), Zn(II)) can be sorbed by multi-walled carbon nanotubes (MWCNTs) or single-walled carbon nanotubes (SWCNTs) pretreated by nitric acid or NaClO (Li et al., 2003; Lu and Liu, 2006; Lu et al., 2007; Rao et al., 2007). The pre-treatment activates the surfaces of the CNTs by introducing functional groups and increasing the number of negatively charged sites which are capable of sorbing positively charged metal ions.

Furthermore, zeolites have been shown to remove several heavy metals from metal electroplating wastewater (Alvarez-Ayuso et al., 2003). Even though appreciable work has focused on the preparation of zeolite crystals in the μm-range, these materials can also be synthesized in smaller sizes <100 nm (Madsen and Jacobsen,

Table 5.1 Removal of inorganic pollutants from water using nanoparticles

Process	Contaminants	Class of nanoparticles	Examples of nanoparticles	References
Sorption	Hg(II), Ag(I), Cd(II), Pb(II), Cu(II), Ni(II), Zn(II), As(V), Cr(VI), Cr(III), CrO_4^{2-}, AsO_4^{3-}	SAMMS	Thiol-SAMMS anion-SAMMS EDA-SAMMS	Fryxell et al. (2002) Mattigod et al. (1999)
		CNTs	MWCNTs SWCNTs CeO_2-CNTs	Li et al. (2003) Di et al. (2006)
		Metal oxides	TiO_2, ZnO, β-FeO(OH)	Liu, G. J. et al. (2008a) Lazaridis et al. (2005)
		Tunable Biopolymers	Elastin-based biopolymers	Kostal et al. (2001)
Photocatalytic oxidation	As(III), organo arsenates	Metal oxides Bimetals	TiO_2 Pt-TiO_2	Dutta et al. (2005) Xu et al. (2008) Ryu et al. (2004)
Photocatalytic reduction	Cu(II), Hg(II), Cd(II), Zn(II), Ni(II), Cr(VI)	Metal oxides	POM, TiO_2	Troupis et al. (2002) Chen et al. (1997) Kabra et al. (2008)
Reduction	As(V), Cr(VI), Ni(II), NO_3^-	Metals	nZVI	Jezequel and Chu (2006) Choe et al. (2000)
Ion exchange	Pb(II), Cd(II), Zn(II), Hardness	Metal oxides	$Zr(HPO_3S)_2$ HFO IX-fibers PPy-PPS	Zhang, Q. R. et al. (2008) Greenleaf et al. (2006) Weidlich et al. (2001)

1999). The use of tunable metal-binding biopolymers has been described for heavy metal remediation of aqueous matrices and appears to be a promising technique (Kostal et al., 2001, 2003). The polymers, consisting of elastin-like polypeptides, can be synthesized with various domains like histidine which have high affinities towards heavy metals and have been shown to efficiently adsorb Cd(II) and Hg(II). Recently, Zhang et al. (2008) described the fabrication of a polymeric cation exchanger containing nano-$Zr(HPO_3S)_2$ for the selective sorption of Pb(II), Cd(II) and Zn(II). Also SAMMS can be designed to adsorb Hg(II) with a very high efficiency (Mattigod et al., 2006). Actinides and lanthanides can be sequestered by a variety of phosphonate and hydroxypyridine functionalized SAMMS, while Thiol-SAMMS sorbed Ag(I), Cd(II), Pb(II), Cu(II), Hg(II) (Fryxell et al., 2004; Makhluf et al., 2005; Mattigod et al., 1999). Iron oxide-capped iron nanoparticles have been shown to be good sorbents for metal contaminants (Macdonald et al., 2007). Ni(II) can be removed using nZVI by sorption and reduction to Ni(0). At equilibrium the ratio of Ni(II)/Ni(0) was 1:1 (Li and Zhang, 2006). The photocatalytic reduction of Cu(II) on polyoxometalates (POMs) was shown by Troupis et al. (2002). The reduction is taking place in two one-electron reduction processes yielding Cu(I) as intermediate and Cu(0) as the final product. The precipitation of Cu(0) resulted in the efficient removal of copper. Similar findings were obtained by Chen and co-workers (1997) who investigated the photocatalytic reduction of Cu(II) and Hg(II) on TiO_2 nanoparticle surfaces using X-ray absorption fine structure (XAFS) spectroscopy. Other metal ions which have been successfully reduced via nanoparticles-mediated photocatalysis include Cd(II), Zn(II) and Ni(II) (Chenthamarakshan and Rajeshwar, 2000; Kabra et al., 2008). The simultaneous degradation of phenolic compounds and reduction of metal ions such as Cu(II) and Cr(VI) using nano-TiO_2 under UV-light has been investigated by Vinu and Madras (2008). The reduction of Cu(II) and Cr(VI) was accelerated in the presence of the phenolic compounds. However, the reduction of metals is suppressed when O_2 is present. Therefore, O_2 has to be removed before an effective reduction can take place. Nevertheless, the precipitation of metals to photocatalyst surfaces leads to a reduction of the catalytic activity (Prairie et al., 1993). To avoid the regular replacement of the catalyst, Byrne et al. (1999) used a spatial separation of both processes to circumvent the poisoning of TiO_2 with the photogenerated holes being consumed on a TiO_2 electrode while the reduction of metal ions by the excess electrons occurs in a separate place (cathode). Furthermore, it was shown that the surface precipitation of photodeposited metal particulates on POMs does not hinder the photocatalytic activity of these catalysts (Troupis et al., 2002).

Nitrate is converted to N_2 by nZVI particles (Choe et al., 2000) due to an acid-driven surface-mediated process. Carbon-supported palladium nanoparticles as well as titania-coated nano-hematite can be used for the electrocatalytic reduction of NO_3^- (Andrade et al., 2007; Penpolcharoen et al., 2001). Rengaraj and Li (2007) prepared a series of Bi^{3+}-doped TiO_2 catalysts with a doping concentration up to 2 wt% and evaluated their catalytic activity in the photochemical reduction of nitrate under UV-light. Maximum reduction of NO_3^- was observed with a doping concentration of 1.5 wt% Bi^{3+}. Under these conditions more than 83% of the initial nitrate was reduced after 150 min.

The reduction of water hardness, which is necessary to decrease membrane fouling in water treatment trains, can be achieved by hybrid ion exchange fibers that contain dispersed hydrated ferric oxide nanoparticles (Greenleaf et al., 2006). CO_2-sparged water can be used as regenerant to avoid the addition of aggressive chemicals, normally needed for regeneration of ion exchangers, as well as the formation of sludge during lime softening. Weidlich et al. (2001) investigated the use of an ion exchange electrode based on modified polypyrrole-nanoparticles (PPy) for softening of drinking water. The PPy-nanoparticles were modified with large counterions such as polystyrene sulphonate to increase the cation exchange capacity. The used polymers can be regenerated by electrochemical oxidation without the addition of chemicals.

5.2.2 Removal of Organic Contaminants

Sorption to nanoparticles has been found to be a crucial process for the removal of organic contaminants from aqueous phases. Metal and metal oxide nanoparticles possess several functional groups (e.g. hydroxyl moieties) at the surface enabling the sorption of organic pollutants. These functional groups can be protonated or deprotonated according to the ambient pH and hence the nanoparticle surfaces can be charged enabling an electrostatic interaction with oppositely charged molecules. For instance, amorphous titanium phosphate acts as a cation exchanger when removing cationic dyes such as rhodamine 6G (Maheria and Chudasama, 2007; Table 5.2). Meng et al. (2008) used SiO_2 and TiO_2 immobilized on hydrophobic clay for the removal of methyl orange from water. Due to their enlarged surface areas, nanoparticles have a higher sorption capacity than other sorbents such as granular activated carbon (GAC) (Standeker et al., 2007). The authors tested several silica monolith aerogels with different degrees of hydrophobicity as potential sorbents for the removal of various toxic organic pollutants from water. In comparison to GAC the aerogels exhibit sorption capacities which are up to 400 times higher. Also carbon nanomaterials have been shown to remove hydrophobic organic pollutants by sorption. For instance, it was observed that CNTs sorb strongly 1,2-dichlorobenzene in a wide pH range (Peng et al., 2003). Chen et al. (2007) observed an increase in the sorption affinity in the order of nonpolar aliphatic < nonpolar aromatics < nitroaromatics when investigating the sorption capacity of carbon nanotubes. Furthermore, Li et al. (2004) reported that multi-walled CNTs (MWCNTs) are better sorbents than carbon black for the sorption of volatile HOCs from water. However, natural waters constituents such as dissolved organic matter can significantly reduce the sorption affinity of PAHs due to the formation of a surface coating (Wang et al., 2008). Therefore, these substances should be removed prior to the addition of the carbon nanotubes.

Table 5.2 Removal of organic pollutants from water using nanoparticles

Process	Contaminants	Class of nanoparticle	Examples of nanoparticles	References
Sorption	Dyes, BTX, chloro-organics, volatile HOCs, aromatics	Metal oxides	SiO_2, TiO_2, Ti-phosphate	Meng et al. (2008) Maheria and Chudasama (2007)
		CNTs	SWNTs MWNTs	Peng et al. (2003)
			Graphitized-CNTs	Li et al. (2004)
Photocatalysis	Pesticides, drugs, dyes, aromatics, chloro-organics, humic acids	Metal oxides	TiO_2, α-Fe_2O_3, ZnO, WO_3, $SrTiO_3$, POM	Hoffmann et al. (1995) Mills et al. (1993) Doll and Frimmel (2005)
		Fullerenes	OH-fullerenes	Krishna et al. (2008)
		Doped metal oxides	N-, Ag-, carbon-, Fe-doped TiO_2, N-doped SiO_2, Mg-doped ZnO	Asahi et al. (2001) Tryba (2008) Hou et al. (2008) Qiu et al. (2008)
Fenton reaction	Pesticides, aromatics	Metal oxides	Fe-oxides	Nie et al. (2008) Zelmanov and Semiat (2008)
Oxidation	Pesticides, aromatics, alcohols	Metals	nZVI, nZVI+POM	Joo et al. (2005) Keenan and Sedlak (2008) Lee, C. et al. (2008)
		Bimetals	Fe-Pd	Joo and Zhao (2008)
Reduction	Chloro-organics	Metals	nZVI, Zn^0	Cheng and Wu (2000)
		Bimetals	Pt/Fe, Pd/Fe, Si/Fe, Ni/Fe, Au/Pd	Tee et al. (2005) Lee and Doong (2008) Nutt et al. (2005)

Engineered nanoparticles have also been applied for the degradation of organic contaminants. For various metals and metal oxides including TiO_2, α-Fe_2O_3, ZnO, WO_3 and $SrTiO_3$ it has been recognized that they are capable of catalyzing photochemical reactions (Bhatkhande et al., 2002; Hariharan, 2006; Hoffmann et al., 1995; Mills et al., 1993). These materials received special attention since they exhibit several beneficial properties such as high environmental stability and low toxicity. TiO_2 and ZnO have been predominantly used due to their high photoactivity and favourable band-gap energy. Recently developed photocatalysts include polyoxometalates (Carraro et al., 2008; Gkika et al., 2004) and polyhydroxy fullerenes (Krishna et al., 2008). The applications of metal sulphide semiconductors and iron oxides are very limited because they are unstable in water, undergo rapid corrosion and/or are toxic (Beydoun et al., 1999; Hoffmann et al., 1995). However, nanoparticles have been used for the photocatalytic degradation of a variety of environmental pollutants such as phenols, nitrophenols, organophosphorous pesticides, non-steroidal anti-inflammatory drugs and aromatic alcohols, dyes as well as aliphatic and chlorinated aromatic compounds (Augugliaro et al., 2008; Bromberg and Hatton, 2005; Hariharan, 2006; Lachheb et al., 2008; Liu, H. Q. et al., 2008b; Mendez-Arriaga et al., 2008; Nagaveni et al., 2004). In addition, humic acids can be successfully removed via photocatalysis (Espinoza et al., 2009; Liu, S. et al., 2008c).

Most nanoparticles show a high photocatalytic activity in the UV-range (e.g. $\lambda < 310$ nm for TiO_2) and require artificial light sources. However, there is a considerable interest in the use of visible light, i.e. sunlight (400–700 nm) mainly due to the potential reduction in energy consumption. The deposition or incorporation of metal ion dopants into nanoparticles can influence the performance of these photocatalysts due to the decrease of the charge carrier recombination rate and the shift of the absorption band to the visible region (Choi et al., 1994). Among these materials are N-, Co-, Ag-, In_2O_3-, carbon- and Fe-doped TiO_2 (Asahi et al., 2001; Barakat et al., 2005; Tryba, 2008; Yang et al., 2008), N-doped SiO_2 (Hou et al., 2008) or Mg-doped ZnO (Qiu et al., 2008). In addition, photo-Fenton reactions, requiring wavelengths of lower than 580 nm, would be a promising approach. However, the acidic pH 2–3 needed for the reaction is a limiting factor for most applications. During photo-Fenton reactions, Fe(II) is oxidized when H_2O_2 is added under acidic conditions yielding Fe(III), OH^- as well as OH-radicals. Malato et al. (2002) compared the degradation kinetics of different pesticides using TiO_2 as well as the photo-Fenton reaction. Complete mineralization of the investigated compounds was attained for both processes. In addition, for most of the substances investigated, rate constants of the photo-Fenton process were at least of one magnitude higher than for TiO_2 photocatalysis.

Nanoparticle-mediated oxidative degradation of organic substances can also be achieved without illumination. Zero-valent iron nanoparticles (nZVI) are capable of oxidizing organic compounds such as pesticides and aromatic compounds when used under aerobic conditions (Joo et al., 2005). Keenan and Sedlak (2008) investigated the influence of pH on the reactivity of nZVI under aerobic conditions. Under acidic conditions (pH < 5) primarily OH-radicals, generated through the

Fenton reaction, were responsible for degradation of the compounds investigated, whereas at higher pH values oxidant formation was primarily due to the oxidation of Fe(II). However, the low oxidation rates observed might be caused by the formation of hydroxide and oxide layers on the surfaces of the particles as well as by other compounds reacting with OH-radicals. To increase the oxidation rates using nZVi, Lee et al. (2008) applied both nZVI and polyoxometalates (POM). The addition of POMs greatly increased the yield of oxidants in the systems whereas a strong pH dependency was observed. This can be attributed to the mediation of the electron transfer from nZVI to oxygen by POM at low pH values which results in the increased formation of OH-radicals through the Fenton reaction. Furthermore, at neutral pH values POM serves as a complexing agent for iron thus preventing surface precipitation of iron hydroxides.

Chlorinated pollutants and several chloro pesticides (e.g. PCB, chlorophenols or chlorinated aliphatics) are not or only to a minor extent degraded under aerobic conditions. The use of zero-valent metals leading to a reduction of chlorinated organic compounds has been frequently studied (Casey et al., 2000; Cheng and Wu, 2000). Efficient reductants are metals such as iron, zinc, magnesium or tin (Boronina et al., 1995; Orth and Gillham, 1996; Roberts et al., 1996). Wang and Zhang (1997) observed that the reduction of trichloroethene (TCE) as well as polychlorinated biphenyls (PCBs) was significantly faster when using nanoparticulate zero-valent iron (nZVI) in comparison to commercial Fe powders with particle sizes in the lower μm-range. However, the degradation of organic pollutants is frequently incomplete when zero-valent metals are used, resulting in the formation of potentially hazardous by-products and the release of potentially toxic metal ions such as Zn(II) (Arnold and Roberts, 1998; O'Hannesin and Gillham, 1998; Orth and Gillham, 1996). As discussed above, surface activity decreases over time due to the oxidation of the zero-valent metals by molecular oxygen and the formation of a metal-hydroxide surface layer. Most of these problems can be avoided when bimetallic particles such as Pt/Fe, Pd/Fe, Si/Fe, Ni/Fe or Au/Pd are used (Joo and Zhao, 2008; Nutt et al., 2005; Tee et al., 2005; Wang and Zhang, 1997; Wei et al., 2006; Wu and Ritchie, 2008; Zheng et al., 2008). Lee et al. (2008) have been investigating the suitability of surface modified zero-valent silicon nanoparticles using Ni, Cu and Fe for dechlorination of tetrachloroethylene (PCE). Reactivity decreased in the order Ni/Si > Fe/Si > Cu/Si. Furthermore, an increase in the dechlorination efficiency was observed for Ni and Fe of up to 3.8 times compared to Si alone, whereas a decrease was observed for Cu. Also palladium has been shown to be a very efficient catalyst for the dechlorination of organic substances. Due to its high reactivity an efficient inhibition of the formation of by-products can be achieved (Cwiertny et al., 2007).

Arkas et al. (2006) used TiO_2 filters impregnated with various compounds such as alkylated dendritic polymers or cyclodextrins. These substances, called 'nanosponges', are able to encapsulate organic molecules and therefore increase the filtration capacity for substances (e.g. PAHs) significantly. DeFriend et al. (2003) reported the application of alumina membranes consisting of acetic acid surface stabilized alumina nanoparticles (A-alumoxanes) with pore sizes between 7 and 25 nm. The molecular weight cut-off of these membranes was shown to be <1,000 Da and they were successfully applied for the separation of various synthetic dyes. Holt et al. (2006) investigated the water flow through microfabricated membranes in which aligned carbon nanotubes with diameters of less than 2 nm served as pores. The water permeabilities of these membranes were several orders of magnitude higher than those of commercial polycarbonate membranes even though the nominal pore sizes were much smaller. Finally, it has been shown that membrane fouling can be significantly decreased when membranes are coated with nanoparticles. This was illustrated by Bae and Tak (2005) who investigated the filtration of sewage sludge by ultrafiltration membranes which were coated with TiO_2 nanoparticles. A significant reduction in membrane fouling was observed which could even be enhanced further when membranes were illuminated with UV-light (Rahimpour et al., 2008).

The use of nanomaterial technology in combination with conventional treatment techniques has also been investigated. For instance, Essam et al. (2007) investigated the application of TiO_2/UV catalysis as pre-treatment prior to biological treatment for the removal of chlorophenols. The combined UV/TiO_2/H_2O_2 - biological treatment allowed for the complete pollutant removal, complete detoxification, >98% dechlorination rate and nearly complete COD removal. Similar results were obtained by Suryaman et al. (2006) for the removal of phenol from wastewater. Karnik et al. (2005b) used a commercial ultrafiltration membrane which was afterwards coated with iron nanoparticles for the treatment of wastewater after ozonation. A reduction of dissolved organic carbon of more than 85% was achieved and also concentrations of disinfection by-products were significantly reduced (up to 90 and 85% for trihalomethanes and halo acetic acid, respectively, and more than 50% for aldehydes, ketones and ketoacids). Bosc et al. (2005) used a bilayer membrane in which the separative function is provided by the non-photoactive top layer, whereas the opposite side is irradiated with UV-light. This setup allows for both the filtration of larger organic molecules and organic colloids and the degradation of smaller molecules which are able to penetrate through the membrane.

5.3 Removal of Pathogens

The removal of pathogens using nanotechnology is an emerging area of research and it is a promising alternative to existing processes such as chlorination. For instance, the use of solar energy for disinfection of water can be economically highly attractive for developing countries. Nanotechnology can be used to produce nanofiltration membranes removing viruses and water-borne pathogens which are generally larger than 50–60 nm. Examples were shown by several studies using carbon nanotube filters to eliminate bacterial and viral pollutants such as *Escherichia coli* or Poliovirus sabin 1 (Brady-Estevez et al., 2008; Srivastava et al., 2004). CNT filters can be repeatedly cleaned (e.g. by ultrasonication) after each filtration process to regain their filtering efficiency. The combined use of ceramic membranes and nanoparticles by applying a nanoparticle surface coating to achieve smaller pore sizes is a possibility to enlarge the flux through the membranes. Ke et al. (2007) applied ceramic membranes with a separation layer of TiO_2 and AlOOH nanofibers and compared their performance with conventional membranes. At comparable removal efficiencies flow rates up to 100 times higher were obtained with the new membranes which can be explained by their increased porosity. Link et al. (2007) reported the suitability of nanoparticles to remove viruses by sorption. For this they used low-cost nanoparticles such as tricalcium phosphate which were mixed with DNA in solution prior to being added to the target cell population.

Silver has long been recognized as potential biocide with Ag^+ ions being highly toxic for bacteria (Choi et al., 2008). Hence, the antibacterial effects of Ag nanoparticles have been investigated in several studies (Hwang et al., 2008; Sondi and Salopek-Sondi, 2004; Yoon et al., 2008). A reduction of three orders of magnitude was observed for the number of *E. coli* investigated with silver-embedded granular activated carbon whereas no inhibition of microbial growth occurred when GAC was used alone (Bandyopadhyaya et al., 2008). In addition to silver, antibacterial activities of nanoparticles have also been found for Cu, ZnO, CeO_2 or CNTs (Jones et al., 2008; Kang et al., 2008; Thill et al., 2006; Yoon et al., 2008; Zhang et al., 2007). Furthermore, non-biocidal nanoparticles can be activated by attaching various molecules to their surfaces. For instance, Lin et al. (2002) attached poly(4-vinyl-*N*-alkylpyridinium) molecules to Fe_3O_4 (magnetit) nanoparticles and attained bactericidal efficiencies against *Staphylococcus aureus* of up to 93%. The bactericidal activity of nanoparticles seems to be primarily influenced by their concentrations and shapes. Pal et al. (2007) observed the highest activity with regard to *E.coli* with planar silver nanoparticles, whereas spherical or rod-shaped nanoparticles had significantly lower antibactericidal activities.

Nanoparticle-mediated photocatalysis can also be used for disinfection of water. All reactive oxygen species (ROS) such as $^{\bullet}OH$, H_2O_2 and $^1O_2$ are strong oxidizing agents and therefore are excellent disinfectants (Tsuang et al., 2008). This is especially true for OH-radicals which are capable to oxidize all kinds of biological substances such as proteins and carbohydrates, and hence might induce peroxidation of cell membranes (Maness et al., 1999; Srinivasan and Somasundaram, 2003). For this reason TiO_2 has been utilized for the disinfection of all kinds of pathogens such as bacteria, fungi or viruses (Bekbolet and Araz, 1996; Paleologou et al., 2007). Furthermore, it was shown that TiO_2 inhibits the re-growth of pathogens (Rincon and Pulgarin, 2007). Mitoraj and co-authors (2007) investigated the inactivation of bacteria and fungi using carbon modified TiO_2 under visible light. They observed photoinactivation in the order Gram-negative bacteria > Gram-positive bacteria >> fungi. An increased antimicrobial activity of sulphur- and nitrogen-doped TiO_2-nanoparticles was reported by Yu et al. (2005) and Asahi et al. (2001). Surface modifications of nanoparticles are especially interesting since they showed an elevated antibacterial activity under visible light, making them suitable for the usage under solar irradiation. Also photo-Fenton reactions have been reported as a promising treatment technique (Rincon and Pulgarin, 2007). More recently, carbon-based nanoparticles have been investigated for their efficiency in pathogen removal. Badireddy et al. (2007) observed the formation of ROS in fullerol (polyhydroxylated fullerene) in the presence of UV light which resulted in the inactivation of viruses.

5.4 Nano-sensors for Water Quality Monitoring

There is a need for quick measurements of chemical pollutants during water treatment in order to optimize operational parameters. Nanoscale sensors have been designed combining desired attributes such as small sizes, in situ applicability, high efficiency or sufficient accuracies. Nano-sensors currently contain carbon nanotubes (CNT), specific metal nanoparticles, quantum dots (QD) and nanoshells. Enzymatic CNT-based biosensors have been used for the detection of several phenolic compounds and organophosphate pesticides (Balasubramanian and Burghard, 2006). Wu et al. (2008) described a highly sensitive multi-responsive chemosensor for the selective detection of Hg^{2+} in natural water. Due to the high proportionality between fluorescence intensities and the Hg^{2+} levels a limit of detection of 1 $\mu g\ L^{-1}$ was achieved. Silver nanoparticles have been used to monitor ammonia which is one of the key toxicants for WWTPs because of its fish toxicity (Dubas and Pimpan, 2008). The silver nanoparticle solution changes its colour from purple to yellow when ammonia is present and a linear response was identified for the range from 5 to 100 $mg\ L^{-1}$. Cui et al. (2001) reported the use of boron-doped silicon nanowires (SiNWs) to detect streptavidin, a bacterial protein, down to picomolar concentrations. Furthermore, these nanowires are also capable of detecting pH changes. Recently, considerable interest has been placed on the use of quantum dots (QD). It was shown that the luminescence of QD is extremely sensitive to their surface states (Banerjee et al., 2008). Therefore, QD have been used for the optical sensing of organic molecules as well as ions (Chen and Rosenzweig, 2002; Isarov and Chrysochoos, 1997). Furthermore, they can be used for the biolabelling of bacteria and the detection of water-borne pathogens (Decho et al., 2008; Yang and Li, 2006). Nanoshells, normally consisting of a non-conducting core and a metal surface coating, absorb light in different wavelengths which can be utilized to detect chemical and biological analytes such as immunoglobulin (Hirsch et al., 2003). The use of silica nanoparticles coated with a shell of silver in nanosensing applications of hemoglobin was also shown by Gordon and Ziolkowski (2007).

5.5 Removal of Nanoparticles from Treated Waters

The growing use and applications of engineered nanoparticles in a variety of industrial products and their potential for wastewater purification and drinking water treatment raise the question how these nanoparticles can be removed in the urban water cycle. Conventional methods for the removal of particulate matter during wastewater treatment include sedimentation and filtration. However, due to the small sizes of nanoparticles the sedimentation velocities are relatively low and a significant sedimentation will not occur as long as there is no formation of larger aggregates or flocculants are not added. The stability of nanoparticle dispersions depends on the properties of the surrounding solution, namely pH and ionic strength which influence the surface charge of the particles and thus the repulsive forces between them. Furthermore, the stability can be increased by surface modifications of the particles (e.g. due to sorption of molecules). For example, an increasing ionic strength of a nanoparticle solution leads to a reduction of the zeta potential and a decreased thickness of the diffuse part of the electrical double layer. The solution becomes unstable and the particles agglomerate.

The influence of sorption on the dispersive behaviour of nanoparticles is also utilized in many industrial applications. In many commercial nanoparticles suspensions, surfactants are used to obtain stable dispersions (Mefford et al., 2008; Scheer and Schweizer, 2008). This might have strong environmental impacts. If the surfactant–nanoparticle bonds are strong enough to persist in wastewater and the aqueous environment, an inhibition of the agglomeration should be the consequence leading to a very limited sedimentation during wastewater treatment and an enhanced groundwater mobility. Limbach et al. (2008) investigated the removal of oxide nanoparticles in a laboratory-scaled WWTP with and without the addition of industrial dispersion stabilizing agents such as acryl polymers and benzyl sulphnic acids. The addition of surfactants resulted in a high dispersion stability over a wide pH range (3–12), even close to the pH_{PZC}. However, high levels of unagglomerated nanoparticles were also observed without the addition of a surfactant. This can be explained by the distinct influence of the wastewater composition on the surface charge of the selected nanoparticles and the adsorption of various wastewater constituents to the nanoparticle surfaces resulting in the formation of stable dispersions.

Only few studies have investigated the influence of coagulants on the aggregation behaviour of nanoparticles. In conventional WWTPs various coagulants such as aluminium sulphate, aluminium hydroxide chloride, polysterene sulphonate, iron chloride or negatively and positvely charged polymers are used for the removal of suspended matter (Duan and Gregory, 2003; Flood et al., 2008; Landim et al., 2007; Menezes et al., 1996; Rossini et al., 1999; Wang and Tang, 2006). Chang et al. (2006) investigated the influence of polyaluminium chloride (PACl) addition as a coagulant on the removal of nanoparticles from an aqueous solution. Even high concentrations of the coagulants (10.4 mg L^{-1} as Al) yielded only a reduction of the silt density index (SDI) from 65 to 27, indicating that a significant proportion of particles remained in solution. Zhang, Y. et al. (2008) tested aluminium sulphate (alum) as potential coagulant and investigated the influence of high electrolyte concentrations. At an alum dosage of 20 mg L^{-1}, 20–80% of commercial nanoparticles (TiO_2, Fe_2O_3, ZnO, NiO and SiO_2) were removed in nanopure water and tap water by sedimentation (following coagulation–flocculation). A further increase of the coagulant concentrations up to 60 mg L^{-1} did not enhance the removal of nanoparticles. A reduction of the zeta potential below 10 mV was observed for all investigated nanoparticles when the ionic strength was increased up to 10 mg L^{-1} (using $MgCl_2$). Nevertheless, after 30 min and 8 h, only 20–30% and 40–80% of the nanoparticles were removed, respectively. Common technologies such as flocculation might be inappropriate to remove engineered nanoparticles from water, which indicates the need to find new technical solutions (Reijnders, 2006).

Several organic nanoparticles (e.g. fullerenes) exhibit strong hydrophobic properties ($K_{ow} > 6$) (Jafvert and Kulkarni, 2008). Therefore, it is likely that these particles are removed from the water phase by sorbing to suspended solids as long as they are not biodegraded. However, carbon nanomaterials such as CNTs can undergo environmental degradation, e.g. when exposed to light yielding to a modification of the surface and the introduction of hydroxyl groups in the molecule (Cho et al., 2008; Savage et al., 2003). Due to the enhanced polarity caused by the functional groups, the sorption affinity to suspended solids is reduced and the mobility is increased.

Electrocoagulation has been applied for the removal of fine dispersed particles from various wastewaters (Ge et al., 2004; Holt et al., 2002). This technique is based on the continuous release of metal ions into the solution by anodic electrodes, typically made of iron and aluminum. These metal ions form hydroxides which are capable of destabilizing the particles in dispersion. However, electrocoagulation strongly depends on the type of the nanoparticles (i.e. predominantly silica or alumina particles) and the electric conductivity of the suspension (Den and Huang, 2006). The magnetic properties of nanoparticles such as hematite can be utilized for their removal when applying a magnetic field. However, this technique can also be employed for non-magnetic particles such as SiO_2 via magnetic seeding aggregation (Chin et al., 2006). Hematite and silica particles are oppositely charged and thus, aggregation can occur enabling both to be removed together from solution with magnetic fields.

Up to now, membrane filtration (e.g. nanofiltration and reverse osmosis) has been already applied for the removal of small bacteria and viruses from water. Hence, this technique can also be used for the removal of nanoparticles.

5.6 Feasibility and Economic Considerations

It becomes obvious that nanoparticles can be used in numerous processes and techniques for the treatment of wastewater or the production of drinking water. So far most of them have only been investigated in laboratory scale and not all of them are likely to be realistic alternatives or supplements for existing treatment technologies, e.g. due to economic reasons. This makes it very difficult if not impossible to predict future developments at this stage. Criteria for the feasibility and applicability of nanoparticles in water treatment are the following:

(a) Transferability of the lab-scale experiments to pilot and full-scale: at this stage this is in most cases totally missing.
(b) Prediction of investment and operational costs: very difficult without appropriate pilot and full-scale results.
(c) A clear benefit from toxicological assessments: this is of special importance due to the potential toxic effects of nanoparticles themselves.
(d) Competitive to current exisiting techniques and processes.

Among the techniques which have already been investigated during pilot plant studies are the photocatalytic treatment of wastewater with nanoparticles and the application of (n)ZVI for the treatment of groundwater contaminated with chlorinated solvents. Especially photocatalysis has the great advantage that it is very cost effective because the solar photocatalysts can be recovered and reused and the energy is supplied by sunlight.

The fixation of nanoparticles, e.g. in membranes, is another promising approach. This allows for the combination of several removal processes, e.g. via filtration and photocatalytic treatment when photocatalytical active nanoparticles are used. Furthermore, the modification of the membrane surfaces with nanoparticles increases their permeability and prevents biofouling. Magnetic nanoparticles are a promising alternative because they can be easily regained from treated waters by magnets. The sorption capacities of existing sorbents can also be improved when nanoparticles are embedded.

However, based on the current knowledge about the elevated production costs of nanoparticles and the low costs when precipitates or flocculants are applied for removing inorganic pollutants, it is questionable whether nanoparticles will be used in water treatment for sorption purposes in the near future. The sorption and/or catalytic-enhanced (photo)degradation of organic pollutants is the most promising application for nanoparticles in water treatment. Although pilot and full-scale experiments are still widely missing, the initiative transformation, e.g. by dehalogenation or subsequent oxidation, might lead in combination with biological processes to a significant degradation of organic pollutants.

5.7 Conclusions

Nanotechnology has a reasonable potential for the treatment of wastewater and production of drinking water regarding the removal of biological and chemical pollutants. Nanoparticles, already existing or newly designed ones, might possess several times higher sorption affinities to polar organic molecules and metals than activated carbon which makes them applicable as sorbents for the removal of both polar and hydrophobic contaminants. In any case, the nanoparticles by themselves need to be separated from the treated water if they are not immobilized to a supporting material. The immobilization of nanoparticles, e.g. on activated carbon, might be a possible approach to solve these problems. Additionally, nanoparticles can enhance reductive dehalogenation (e.g. by bimetalic Au/Pd) or photocatalytic oxidation (e.g. with TiO_2 suspensions) of very persistent organic compounds such as halogenated alkanes, benzenes and PAHs. Hence, nanoparticles can be used for the transformation of all kinds of persistent molecules into more biodegradable compounds. Another major advantage of nanotechnology is the disinfection properties of several nanoparticles enabling the inhibition of pathogens and lowering the formation of toxic disinfection by-products. Due to their extremely high potency in combination with the high specificity, nanoparticles are ideal materials for water treatment and may contribute to solving future challenges in the area of water treatment technologies. However, it can be assumed that only the combined benefits such as disinfective, sorptive to organic and inorganic pollutants and initializing biodegradation/photodegradation will be competitive to already existing technologies.

Acknowledgement We thank Corinna Brinkmann and Jennifer Kormos for reviewing the manuscript.

References

Alvarez-Ayuso E, Garcia-Sanchez A, Querol X (2003) Purification of metal electroplating waste waters using zeolites. Water Res 37(20): 4855–4862.

Andersen H, Siegrist H, Halling-Sorensen B, Ternes TA (2003) Fate of estrogens in a municipal sewage treatment plant. Environ Sci Technol 37(18): 4021–4026.

Andrade FV, Deiner LJ, Varela H, de Castro JFR, Rodrigues IA, Narta FC (2007) Electrocatalytic reduction of nitrate over palladium nanoparticle catalysts. J Electrochem Soc 154(9): F159–F164.

Arkas M, Allabashi R, Tsiourvas D, Mattausch EM, Perfler R (2006) Organic/inorganic hybrid filters based on dendritic and cyclodextrin "nanosponges" for the removal of organic pollutants from water. Environ Sci Technol 40(8): 2771–2777.

Arnold WA, Roberts AL (1998) Pathways of chlorinated ethylene and chlorinated acetylene reaction with Zn(0). Environ. Sci. Technol 32(19): 3017–3025.

Asahi R, Morikawa T, Ohwaki T, Aoki K, Taga Y (2001) Visible-light photocatalysis in nitrogen-doped titanium oxides. Science 293(5528): 269–271.

Augugliaro V, Caronna T, Loddo V, Marci G, Palmisano G, Palmisano L, Yurdakal S (2008) Oxidation of aromatic alcohols in irradiated aqueous suspensions of commercial and home-prepared ruffle TiO2: a selectivity study. Chem-Eur J 14(15): 4640–4646.

Badireddy AR, Hotze EM, Chellam S, Alvarez P, Wiesner MR (2007) Inactivation of bacteriophages via photosensitization of fullerol nanoparticles. Environ Sci Technol 41(18): 6627–6632.

Bae TH, Tak TM (2005) Effect of TiO_2 nanoparticles on fouling mitigation of ultrafiltration membranes for activated sludge filtration. J Membr Sci 249(1-2): 1–8.

Balasubramanian K, Burghard M (2006) Biosensors based on carbon nanotubes. Anal Bioanal Chem 385(3): 452–468.

Bandyopadhyaya R, Sivaiah MV, Shankar PA (2008) Silver-embedded granular activated carbon as an antibacterial medium for water purification. J Chem Technol Biotechnol 83(8): 1177–1180.

Banerjee S, Kara S, Santra S (2008) A simple strategy for quantum dot assisted selective detection of cadmium ions. Chem Commun 26: 3037–3039.

Barakat MA, Schaeffer H, Hayes G, Ismat-Shah S (2005) Photocatalytic degradation of 2-chlorophenol by co-doped TiO_2 nanoparticles. Appl Catal B-Environ 57(1): 23–30.

Becher G (1999) Drinking water chlorination and health. Acta Hydrochim Hydrobiol 27(2): 100–102.

Bekbolet M, Araz CV (1996) Inactivation of *Escherichia coli* by photocatalytic oxidation. Chemosphere 32(5): 959–965.

Bellona C, Drewes JE, Xu P, Amy G (2004) Factors affecting the rejection of organic solutes during NF/RO treatment – a literature review. Water Res 38(12): 2795–2809.

Beydoun D, Amal R, Low G, McEvoy S (1999) Role of nanoparticles in photocatalysis. J Nanopart Res 1: 439–458.

Bhatkhande DS, Pangarkar VG, Beenackers A (2002) Photocatalytic degradation for environmental applications – a review. J Chem Technol Biotechnol 77(1): 102–116.

Boronina T, Klabunde KJ, Sergeev G (1995) Destruction of organohalides in water using metal particles – Carbon tetrachloride/water reactions with magnesium, tin, and zinc. Environ Sci Technol 29(6): 1511–1517.
Bosc F, Ayral A, Guizard C (2005) Mesoporous anatase coatings for coupling membrane separation and photocatalyzed reactions. J Membr Sci 265(1–2): 13–19.
Brady-Estevez AS, Kang S, Elimelech M (2008) A single-walled-carbon-nanotube filter for removal of viral and bacterial pathogens. Small 4(4): 481–484.
Bromberg L, Hatton TA (2005) Nerve agent destruction by recyclable catalytic magnetic nanoparticles. Ind Eng Chem Res 44(21): 7991–7998.
Byrne JA, Eggins BR, Byers W, Brown NMD (1999) Photoelectrochemical cell for the combined photocatalytic oxidation of organic pollutants and the recovery of metals from waste waters. Appl Catal B-Environ 20(2): L85–L89.
Carballa M, Omil F, Lema JM, Llompart M, Garcia-Jares C, Rodriguez I, Gomez M, Ternes T (2004) Behavior of pharmaceuticals, cosmetics and hormones in a sewage treatment plant. Water Res 38(12): 2918–2926.
Carraro M, Sartorel A, Scorrano G, Carofiglio T, Bonchio M (2008) Catalytic strategies for sustainable oxidations in water. Synthesis 12: 1971–1978.
Casey FXM, Ong SK, Horton R (2000) Degradation and transformation of trichloroethylene in miscible displacement experiments through zerovalent metals. Environ Sci Technol 34(23): 5023–5029.
Chang MR, Lee DJ, Lai JY (2006) Coagulation and filtration of nanoparticles in wastewater from Hsinchu Science-based Industrial Park (HSIP). Sep Sci Technol 41(7): 1303–1311.
Chen LX, Rajh T, Wang ZY, Thurnauer MC (1997) XAFS studies of surface structures of TiO_2 nanoparticles and photocatalytic reduction of metal ions. J Phys Chem B 101(50): 10688–10697.
Chen W, Duan L, Zhu DQ (2007) Adsorption of polar and nonpolar organic chemicals to carbon nanotubes. Environ Sci Technol 41(24): 8295–8300.
Chen YF, Rosenzweig Z (2002) Luminescent CdS quantum dots as selective ion probes. Anal Chem 74(19): 5132–5138.
Cheng SF, Wu SC (2000) The enhancement methods for the degradation of TCE by zero-valent metals. Chemosphere 41(8): 1263–1270.
Chenthamarakshan CR, Rajeshwar K (2000) Photocatalytic reduction of divalent zinc and cadmium ions in aqueous TiO2 suspensions: an interfacial induced adsorption-reduction pathway mediated by formate ions. Electrochem Commun 2(7): 527–530.
Chin CJM, Chen PW, Wang LJ (2006) Removal of nanoparticles from CMP wastewater by magnetic seeding aggregation. Chemosphere 63(10): 1809–1813.
Cho HH, Smith BA, Wnuk JD, Fairbrother DH, Ball WP (2008) Influence of surface oxides on the adsorption of naphthalene onto multiwalled carbon nanotubes. Environ Sci Technol 42(8): 2899–2905.
Choe S, Chang YY, Hwang KY, Khim J (2000) Kinetics of reductive denitrification by nanoscale zero-valent iron. Chemosphere 41(8): 1307–1311.
Choi O, Deng KK, Kim NJ, Ross L, Surampalli RY, Hu ZQ (2008) The inhibitory effects of silver nanoparticles, silver ions, and silver chloride colloids on microbial growth. Water Res 42(12): 3066–3074.
Choi WY, Termin A, Hoffmann MR (1994) The role of metal-ion dopants in quantum-sized TiO_2 - Correlation between photoreactivity and charge-carrier recombination dynamics. J Phys Chem 98(51): 13669–13679.
Crook J (1998) Water reclamation and reuse criteria. In: Asano T, (ed.) Wastewater Reclamation and Reuse, Water Quality Management Library, vol. 10. Technomic Publishing, Inc., Lancaster, PA, pp. 627–696.
Cui Y, Wei QQ, Park HK, Lieber CM (2001) Nanowire nanosensors for highly sensitive and selective detection of biological and chemical species. Science 293(5533): 1289–1292.

Cwiertny DM, Bransfield SJ, Roberts AL (2007) Influence of the oxidizing species on the reactivity of iron-based bimetallic reductants. Environ Sci Technol 41(10): 3734–3740.
Decho AW, Beckman EM, Chandler GT, Kawaguchi T (2008) Application of photostable quantum dots for indirect immunofluorescent detection of specific bacterial serotypes on small marine animals. Nanotechnology 19: 23.
DeFriend KA, Wiesner MR, Barron AR (2003) Alumina and aluminate ultrafiltration membranes derived from alumina nanoparticles. J Membr Sci 224(1–2): 11–28.
Den W, Huang CP (2006) Electrocoagulation of silica nanoparticles in wafer polishing wastewater by a multichannel flow reactor: a kinetic study. J Environ Eng-ASCE 132(12): 1651–1658.
Di ZC, Ding J, Peng XJ, Li YH, Luan ZK, Liang J (2006) Chromium adsorption by aligned carbon nanotubes supported ceria nanoparticles. Chemosphere 62(5): 861–865.
Doll TE, Frimmel FH (2005) Removal of selected persistent organic pollutants by heterogeneous photocatalysis in water. Catal Today 101(3–4): 195–202.
Duan JM, Gregory J (2003) Coagulation by hydrolysing metal salts. Adv Colloid Interface Sci 100: 475–502.
Dubas ST, Pimpan V (2008) Green synthesis of silver nanoparticles for ammonia sensing. Talanta 76(1): 29–33.
Dutta PK, Pehkonen SO, Sharma VK, Ray AK (2005) Photocatalytic oxidation of arsenic(III): evidence of hydroxyl radicals. Environ Sci Technol 39(6): 1827–1834.
Espinoza LAT, ter Haseborg E, Weber M, Frimmel FH (2009) Investigation of the photocatalytic degradation of brown water natural organic matter by size exclusion chromatography. Appl Catal B-Environ 87(1–2): 56–62.
Essam T, Amin MA, El Tayeb O, Mattiasson B, Guieysse B (2007) Sequential photochemical-biological degradation of chlorophenols. Chemosphere 66(11): 2201–2209.
Ferguson MA, Hoffmann MR, Hering JG (2005) TiO_2-photocatalyzed As(III) oxidation in aqueous suspensions: reaction kinetics and effects of adsorption. Environ Sci Technol 39(6): 1880–1886.
Flood C, Cosgrove T, Espidel Y, Howell I, Revell P (2008) Effects of surfactants and electrolytes on adsorbed layers and particle stability. Langmuir 24(14): 7323–7328.
Fryxell GE, Lin YH, Wu H, Kemner KM (2002) Environmental applications of self-assembled monolayers on mesoporous supports (SAMMS). In: Sayari, A, Jaroniec, M (eds.) Nanoporous Materials III, Studies in Surface Science and Catalysis, vol. 141. Elsevier Science Bv., Amsterdam, pp. 583–590.
Fryxell GE, Wu H, Lin YH, Shaw WJ, Birnbaum JC, Linehan JC, Nie ZM, Kemner K, Kelly S (2004) Lanthanide selective sorbents: self-assembled monolayers on mesoporous supports (SAMMS). J Mater Chem 14(22): 3356–3363.
Ge JT, Qu JH, Lei PJ, Liu HJ (2004) New bipolar electrocoagulation-electroflotation process for the treatment of laundry wastewater. Sep Purif Technol 36(1): 33–39.
Giasuddin ABM, Kanel SR, Choi H (2007) Adsorption of humic acid onto nanoscale zerovalent iron and its effect on arsenic removal. Environ Sci Technol 41(6): 2022–2027.
Gkika E, Kormali P, Antonaraki S, Dimoticali D, Papaconstantinou E, Hiskia A (2004) Polyoxometalates as effective photocatalysis in water purification from pesticides. Int J Photoenergy 6(4): 227–231.
Glassmeyer ST, Furlong ET, Kolpin DW, Cahill JD, Zaugg SD, Werner SL, Meyer MT, Kryak DD (2005) Transport of chemical and microbial compounds from known wastewater discharges: potential for use as indicators of human fecal contamination. Environ Sci Technol 39(14): 5157–5169.
Gordon JA, Ziolkowski RW (2007) Investigating functionalized active coated nanoparticles for use in nano-sensing applications. Opt Express 15(20): 12562–12582.
Greenleaf JE, Lin JC, Sengupta AK (2006) Two novel applications of ion exchange fibers: arsenic removal and chemical-free softening of hard water. Environ Prog 25(4): 300–311.
Halling-Sorensen B, Nielsen SN, Lanzky PF, Ingerslev F, Lutzhoft HCH, Jorgensen SE (1998) Occurrence, fate and effects of pharmaceutical substances in the environment - A review. Chemosphere 36(2): 357–394.

Hariharan C (2006) Photocatalytic degradation of organic contaminants in water by ZnO nanoparticles: revisited. Appl Catal A-Gen 304(1): 55–61.
Hirsch LR, Jackson JB, Lee A, Halas NJ, West J (2003) A whole blood immunoassay using gold nanoshells. Anal Chem 75(10): 2377–2381.
Hoffmann MR, Martin ST, Choi WY, Bahnemann DW (1995) Environmental applications of semiconductor photocatalysis. Chem Rev 95(1): 69–96.
Holt JK, Park HG, Wang YM, Stadermann M, Artyukhin AB, Grigoropoulos CP, Noy A, Bakajin O (2006) Fast mass transport through sub-2-nanometer carbon nanotubes. Science 312(5776): 1034–1037.
Holt PK, Barton GW, Wark M, Mitchell CA (2002) A quantitative comparison between chemical dosing and electrocoagulation. Colloid Surf A-Physicochem Eng Asp 211(2–3): 233–248.
Hou YD, Wang XC, Wu L, Chen XF, Ding ZX, Wang XX, Fu XZ (2008) N-doped SiO_2/TiO_2 mesoporous nanoparticles with enhanced photocatalytic activity under visible-light irradiation. Chemosphere 72(3): 414–421.
Hwang ET, Lee JH, Chae YJ, Kim YS, Kim BC, Sang BI, Gu MB (2008) Analysis of the toxic mode of action of silver nanoparticles using stress-specific bioluminescent bacteria. Small 4(6): 746–750.
Isarov AV, Chrysochoos J (1997) Optical and photochemical properties of nonstoichiometric cadmium sulfide nanoparticles: surface modification with copper(II) ions. Langmuir 13(12): 3142–3149.
Jafvert CT, Kulkarni PP (2008) Buckminsterfullerene's (C-60) octanol-water partition coefficient (K_{ow}) and aqueous solubility. Environ Sci Technol 42(16): 5945–5950.
Jezequel H, Chu KH (2006) Removal of arsenate from aqueous solution by adsorption onto titanium dioxide nanoparticles. J Environ Sci Health A Tox Hazard Subst Environ Eng 41(8): 1519–1528.
Jones N, Ray B, Ranjit KT, Manna AC (2008) Antibacterial activity of ZnO nanoparticle suspensions on a broad spectrum of microorganisms. FEMS Microbiol Lett 279(1): 71–76.
Joo SH, Feitz AJ, Sedlak DL, Waite TD (2005) Quantification of the oxidizing capacity of nanoparticulate zero-valent iron. Environ Sci Technol 39(5): 1263–1268.
Joo SH, Zhao D (2008) Destruction of lindane and atrazine using stabilized iron nanoparticles under aerobic and anaerobic conditions: effects of catalyst and stabilizer. Chemosphere 70(3): 418–425.
Joss A, Zabczynski S, Gobel A, Hoffmann B, Loffler D, McArdell CS, Ternes TA, Thomsen A, Siegrist H (2006) Biological degradation of pharmaceuticals in municipal wastewater treatment: proposing a classification scheme. Water Res 40(8): 1686–1696.
Kabra K, Chaudhary R, Sawhney RL (2008) Solar photocatalytic removal of Cu(II), Ni(II), Zn(II) and Pb(II): speciation modeling of metal-citric acid complexes. J Hazard Mater 155(3): 424–432.
Kanel SR, Greneche JM, Choi H (2006) Arsenic(V) removal from groundwater using nano scale zero-valent iron as a colloidal reactive barrier material. Environ Sci Technol 40(6): 2045–2050.
Kang S, Herzberg M, Rodrigues DF, Elimelech M (2008) Antibacterial effects of carbon nanotubes: size does matter. Langmuir 24(13): 6409–6413.
Karnik BS, Davies SH, Baumann MJ, Masten SJ (2005a) The effects of combined ozonation and filtration on disinfection by-product formation. Water Res 39(13): 2839–2850.
Karnik BS, Davies SH, Baumann MJ, Masten SJ (2005b) Fabrication of catalytic membranes for the treatment of drinking water using combined ozonation and ultrafiltration. Environ Sci Technol 39(19): 7656–7661.
Ke XB, Zhu HY, Gao XP, Liu JW, Zheng ZF (2007) High-performance ceramic membranes with a separation layer of metal oxide nanofibers. Adv Mater 19(6): 785–790.
Keenan CR, Sedlak DL (2008) Factors affecting the yield of oxidants from the reaction of nanoparticulate zero-valent iron and oxygen. Environ Sci Technol 42(4): 1262–1267.
Kostal J, Mulchandani A, Chen W (2001) Tunable biopolymers for heavy metal removal. Macromolecules 34(7): 2257–2261.

Kostal J, Mulchandani A, Gropp KE, Chen W (2003) A temperature responsive biopolymer for mercury remediation. Environ Sci Technol 37(19): 4457–4462.

Krishna V, Yanes D, Imaram W, Angerhofer A, Koopman B, Moudgil B (2008) Mechanism of enhanced photocatalysis with polyhydroxy fullerenes. Appl Catal B-Environ 79(4): 376–381.

Kwon S, Fan M, Cooper AT, Yang HQ (2008) Photocatalytic applications of micro- and nano-TiO_2 in environmental engineering. Crit Rev Environ Sci Technol 38(3): 197–226.

Lachheb H, Houas A, Herrmann JM (2008) Photocatalytic degradation of polynitrophenols on various commercial suspended or deposited titania catalysts using artificial and solar light. Int J Photoenergy Article ID 497895, 9 pages. doi: 10.1155/2008/497895.

Landim AS, Rodrigues G, de Assuncao RMN (2007) Use of polystyrene sulfonate produced from waste plastic cups as an auxiliary agent of coagulation, flocculation and flotation for water and wastewater treatment in Municipal Department of Water and Wastewater in Uberlandia-MG, Brazil. Polym Bull 58(2): 457–463.

Lazaridis NK, Bakoyannakis DN, Deliyanni EA (2005) Chromium(VI) sorptive removal from aqueous solutions by nanocrystalline akaganeite. Chemosphere 58(1): 65–73.

Lee C, Keenan CR, Sedlak DL (2008) Polyoxometalate-enhanced oxidation of organic compounds by nanoparticulate zero-valent iron and ferrous ion in the presence of oxygen. Environ Sci Technol 42(13): 4921–4926.

Lee CC, Doong RA (2008) Dechlorination of tetrachloroethylene in aqueous solutions using metal-modified zerovalent silicon. Environ Sci Technol 42(13): 4752–4757.

Li QL, Yuan DX, Lin QM (2004) Evaluation of multi-walled carbon nanotubes as an adsorbent for trapping volatile organic compounds from environmental samples. J Chromatogr A 1026(1–2): 283–288.

Li XQ, Zhang WX (2006) Iron nanoparticles: the core-shell structure and unique properties for Ni(II) sequestration. Langmuir 22(10): 4638–4642.

Li YH, Luan ZK, Xiao X, Zhou XW, Xu CL, Wu DH, Wei BQ (2003) Removal of Cu^{2+} ions from aqueous solutions by carbon nanotubes. Adsorpt Sci Technol 21(5): 475–485.

Limbach LK, Bereiter R, Mueller E, Krebs R, Gaelli R, Stark WJ (2008) Removal of oxide nanoparticles in a model wastewater treatment plant: influence of agglomeration and surfactants on clearing efficiency. Environ Sci Technol 42(15): 5828–5833.

Lin J, Qiu SY, Lewis K, Klibanov AM (2002) Bactericidal properties of flat surfaces and nanoparticles derivatized with alkylated polyethylenimines. Biotechnol Prog 18(5): 1082–1086.

Link N, Brunner TJ, Dreesen IAJ, Stark WJ, Fussenegger M (2007) Inorganic nanoparticles for transfection of mammalian cells and removal of viruses from aqueous solutions. Biotechnol Bioeng 98(5): 1083–1093.

Lissemore L, Hao CY, Yang P, Sibley PK, Mabury S, Solomon KR (2006) An exposure assessment for selected pharmaceuticals within a watershed in southern Ontario. Chemosphere 64(5): 717–729.

Liu GJ, Zhang XR, Talley JW, Neal CR, Wang HY (2008a) Effect of NOM on arsenic adsorption by TiO_2 in simulated As(III)-contaminated raw waters. Water Res 42(8-9): 2309–2319.

Liu HQ, Yang JX, Liang JH, Huang YX, Tangz CY (2008b) ZnO nanofiber and nanoparticle synthesized through electrospinning and their photocatalytic activity under visible light. J Am Ceram Soc 91(4): 1287–1291.

Liu S, Lim M, Fabris R, Chow C, Chiang K, Drikas M, Amal R (2008c) Removal of humic acid using TiO_2 photocatalytic process - Fractionation and molecular weight characterisation studies. Chemosphere 72(2): 263–271.

Lu CY, Liu CT (2006) Removal of nickel(II) from aqueous solution by carbon nanotubes. J Chem Technol Biotechnol 81(12): 1932–1940.

Lu ZH, Sun KN, Ren S, Jiao MC (2007) Surface modification and dispersion of multi-walled carbon nanotubes. Rare Metal Mat Eng 36: 100–103.

Macdonald JE, Kelly JA, Veinot JGC (2007) Iron/Iron oxide nanoparticle sequestration of catalytic metal impurities from aqueous media and organic reaction products. Langmuir 23(19): 9543–9545.

Madsen C, Jacobsen CJH (1999) Nanosized zeolite crystals – convenient control of crystal size distribution by confined space synthesis. Chem Commun 8: 673–674.

Maheria KC, Chudasama UV (2007) Sorptive removal of dyes using titanium phosphate. Ind Eng Chem Res 46(21): 6852–6857.

Makhluf S, Dror R, Nitzan Y, Abramovich Y, Jelinek R, Gedanken A (2005) Microwave-assisted synthesis of nanocrystalline MgO and its use as a bacteriocide. Adv Funct Mater 15(10): 1708–1715.

Malato S, Blanco J, Caceres J, Fernandez-Alba AR, Aguera A, Rodriguez A (2002) Photocatalytic treatment of water-soluble pesticides by photo-Fenton and TiO_2 using solar energy. Catal Today 76(2-4): 209–220.

Maness PC, Smolinski S, Blake DM, Huang Z, Wolfrum EJ, Jacoby WA (1999) Bactericidal activity of photocatalytic TiO_2 reaction: toward an understanding of its killing mechanism. Appl Environ Microbiol 65(9): 4094–4098.

Mattigod SV, Feng XD, Fryxell GE, Liu J, Gong ML (1999) Separation of complexed mercury from aqueous wastes using self-assembled mercaptan on mesoporous silica. Sep Sci Technol 34(12): 2329–2345.

Mattigod SV, Parker K, Fryxell GE (2006) Correlation of heavy metal binding capacity of thiol-SAMMS using the Misono softness parameter. Inorg Chem Commun 9(1): 96–98.

Mefford OT, Vadala ML, Goff JD, Carroll MRJ, Mejia-Ariza R, Caba BL, St Pierre TG, Woodward RC, Davis RM, Riffle JS (2008) Stability of polydimethylsiloxane-magnetite nanoparticle dispersions against flocculation: interparticle interactions of polydisperse materials. Langmuir 24(9): 5060–5069.

Mendez-Arriaga F, Esplugas S, Gimenez J (2008) Photocatalytic degradation of non-steroidal anti-inflammatory drugs with TiO_2 and simulated solar irradiation. Water Res 42(3): 585–594.

Menezes FM, Amal R, Luketina D (1996) Removal of particles using coagulation and flocculation in a dynamic separator. Powder Technol 88(1): 27–31.

Meng XF, Qian ZZ, Wang HT, Gao XW, Zhang SM, Yang MS (2008) Sol-gel immobilization of SiO_2/TiO_2 on hydrophobic clay and its removal of methyl orange from water. J Sol-Gel Sci Technol 46(2): 195–200.

Miao XS, Bishay F, Chen M, Metcalfe CD (2004) Occurrence of antimicrobials in the final effluents of wastewater treatment plants in Canada. Environ Sci Technol 38(13): 3533–3541.

Mills A, Davies RH, Worsley D (1993) Water-purification by semiconductor photocatalysis. Chem Soc Rev 22(6): 417–425.

Mitoraj D, Janczyk A, Strus M, Kisch H, Stochel G, Heczko PB, Macyk W (2007) Visible light inactivation of bacteria and fungi by modified titanium dioxide. Photochem Photobiol Sci 6(6): 642–648.

Nagaveni K, Sivalingam G, Hegde MS, Madras G (2004) Photocatalytic degradation of organic compounds over combustion-synthesized nano-TiO_2. Environ Sci Technol 38(5): 1600–1604.

Nie YL, Hu C, Zhou L, Qu JH (2008) An efficient electron transfer at the Fe^0/iron oxide interface for the photoassisted degradation of pollutants with H_2O_2. Appl Catal B-Environ 82(3–4): 151–156.

Nutt MO, Hughes JB, Wong MS (2005) Designing Pd-on-Au bimetallic nanoparticle catalysts for trichloroethene hydrodechlorination. Environ Sci Technol 39(5): 1346–1353.

O'Hannesin SF, Gillham RW (1998) Long-term performance of an in situ "iron wall" for remediation of VOCs. Ground Water 36(1): 164–170.

Orth WS, Gillham RW (1996) Dechlorination of trichloroethene in aqueous solution using FeO. Environ Sci Technol 30(1): 66–71.

Pal S, Tak YK, Song JM (2007) Does the antibacterial activity of silver nanoparticles depend on the shape of the nanoparticle? A study of the gram-negative bacterium *Escherichia coli*. Appl Environ Microbiol 73(6): 1712–1720.

Paleologou A, Marakas H, Xekoukoulotakis NP, Moya A, Vergara Y, Kalogerakis N, Gikas P, Mantzavinos D (2007) Disinfection of water and wastewater by TiO_2 photocatalysis, sonolysis and UV-C irradiation. Catal Today 129(1-2): 136–142.

Pena ME, Korfiatis GP, Patel M, Lippincott L, Meng XG (2005) Adsorption of As(V) and As(III) by nanocrystalline titanium dioxide. Water Res 39(11): 2327–2337.
Peng XJ, Li YH, Luan ZK, Di ZC, Wang HY, Tian BH, Jia ZP (2003) Adsorption of 1,2-dichlorobenzene from water to carbon nanotubes. Chem Phys Lett 376(1–2): 154–158.
Peng XJ, Luan ZK, Ding J, Di ZH, Li YH, Tian BH (2005) Ceria nanoparticles supported on carbon nanotubes for the removal of arsenate from water. Mater Lett 59(4): 399–403.
Penpolcharoen M, Amal R, Brungs M (2001) Degradation of sucrose and nitrate over titania coated nano-hematite photocatalysts. Journal of Nanoparticle Research 3(4): 289–302.
Prairie MR, Evans LR, Stange BM, Martinez SL (1993) An investigation of TiO_2 photocatalysis for the treatment of water contaminated with metals and organic-chemicals. Environ Sci Technol 27(9): 1776–1782.
Qiu XQ, Li LP, Zheng J, Liu JJ, Sun XF, Li GS (2008) Origin of the enhanced photocatalytic activities of semiconductors: a case study of ZnO doped with Mg^{2+}. J Phys Chem C 112(32): 12242–12248.
Rahimpour A, Madaeni SS, Taheri AH, Mansourpanah Y (2008) Coupling TiO_2 nanoparticles with UV irradiation for modification of polyethersulfone ultrafiltration membranes. J Membr Sci 313(1–2): 158–169.
Rao GP, Lu C, Su F (2007) Sorption of divalent metal ions from aqueous solution by carbon nanotubes: a review. Sep Purif Technol 58(1): 224–231.
Reijnders L (2006) Cleaner nanotechnology and hazard reduction of manufactured nanoparticles. J Clean Prod 14(2): 124–133.
Rengaraj S, Li XZ (2007) Enhanced photocatalytic reduction reaction over Bi^{3+}-TiO_2 nanoparticles in presence of formic acid as a hole scavenger. Chemosphere 66(5): 930–938.
Rincon AG, Pulgarin C (2007) Absence of *E. coli* regrowth after Fe^{3+} and TiO_2 solar photoassisted disinfection of water in CPC solar photoreactor. Catal Today 124(3–4): 204–214.
Roberts AL, Totten LA, Arnold WA, Burris DR, Campbell TJ (1996) Reductive elimination of chlorinated ethylenes by zero valent metals. Environ Sci Technol 30(8): 2654–2659.
Rossini M, Garrido JG, Galluzzo M (1999) Optimization of the coagulation-flocculation treatment: influence of rapid mix parameters. Water Res 33(8): 1817–1826.
Ryu J, Choi W (2004) Effects of TiO_2 surface modifications on photocatalytic oxidation of arsenite: the role of superoxides. Environ Sci Technol 38(10): 2928–2933.
Savage T, Bhattacharya S, Sadanadan B, Gaillard J, Tritt TM, Sun YP, Wu Y, Nayak S, Car R, Marzari N, Ajayan PM, Rao AM (2003) Photoinduced oxidation of carbon nanotubes. J Phys-Condes Matter 15(35): 5915–5921.
Scheer EN, Schweizer KS (2008) Haloing, flocculation, and bridging in colloid-nanoparticle suspensions. J Chem Phys 128: 16.
Schwarzenbach RP, Gschwend PM, Imboden DM (2003) Environmental organic chemistry. New Jersey, Wiley-Interscience.
Sharma VK, Dutta PK, Ray AK (2007) Review of kinetics of chemical and photocatalytical oxidation of Arsenic(III) as influenced by pH. J Environ Sci Health A Tox Hazard Subst Environ Eng 42(7): 997–1004.
Sondi I, Salopek-Sondi B (2004) Silver nanoparticles as antimicrobial agent: a case study on *E. coli* as a model for Gram-negative bacteria. J Colloid Interface Sci 275(1): 177–182.
Srinivasan C, Somasundaram N (2003) Bactericidal and detoxification effects of irradiated semiconductor catalyst, TiO_2. Curr Sci 85(10): 1431–1438.
Srivastava A, Srivastava ON, Talapatra S, Vajtai R, Ajayan PM (2004) Carbon nanotube filters. Nat Mater 3(9): 610–614.
Standeker S, Novak Z, Knez Z (2007) Adsorption of toxic organic compounds from water with hydrophobic silica aerogels. J Colloid Interface Sci 310(2): 362–368.
Suryaman D, Hasegawa K, Kagaya S (2006) Combined biological and photocatalytic treatment for the mineralization of phenol in water. Chemosphere 65(11): 2502–2506.
Tee YH, Grulke E, Bhattacharyya D (2005) Role of Ni/Fe nanoparticle composition on the degradation of trichloroethylene from water. Ind Eng Chem Res 44(18): 7062–7070.

Ternes TA (1998) Occurrence of drugs in German sewage treatment plants and rivers. Water Res 32(11): 3245–3260.
Thill A, Zeyons O, Spalla O, Chauvat F, Rose J, Auffan M, Flank AM (2006) Cytotoxicity of CeO_2 nanoparticles for *Escherichia coli*. Physico-chemical insight of the cytotoxicity mechanism. Environ Sci Technol 40(19): 6151–6156.
Troupis A, Hiskia A, Papaconstantinou E (2002) Photocatalytic reduction and recovery of copper by polyoxometalates. Environ Sci Technol 36(24): 5355–5362.
Tryba B (2008) Increase of the photocatalytic activity of TiO_2 by carbon and iron modifications. Int J Photoenergy 2008: 15.
Tsuang YH, Sun JS, Huang YC, Lu CH, Chang WHS, Wang CC (2008) Studies of photokilling of bacteria using titanium dioxide nanoparticles. Artif Organs 32(2): 167–174.
Vinu R, Madras G (2008) Kinetics of simultaneous photocatalytic degradation of phenolic compounds and reduction of metal ions with nano-TiO_2. Environ Sci Technol 42(3): 913–919.
Wang CB, Zhang WX (1997) Synthesizing nanoscale iron particles for rapid and complete dechlorination of TCE and PCBs. Environ Sci Technol 31(7): 2154–2156.
Wang DS, Tang HX (2006) Quantitative model of coagulation with inorganic polymer flocculant PACI: application of the PCNM. J Environ Eng-ASCE 132(5): 434–441.
Wang XL, Grady CPL (1995) Effects of biosorption and dissolution on the biodegradation of Di-N-Butyl Phthalate. Water Environ Res 67(5): 863–871.
Wang XL, Lu JL, Xing BS (2008) Sorption of organic contaminants by carbon nanotubes: influence of adsorbed organic matter. Environ Sci Technol 42(9): 3207–3212.
Wei JJ, Xu XH, Liu Y, Wang DH (2006) Catalytic hydrodechlorination of 2,4-dichlorophenol over nanoscale Pd/Fe: reaction pathway and some experimental parameters. Water Res 40(2): 348–354.
Weidlich C, Mangold KM, Juttner K (2001) Conducting polymers as ion-exchangers for water purification. Electrochim Acta 47(5): 741–745.
Wu DY, Huang W, Lin ZH, Duan CY, He C, Wu S, Wang DH (2008) Highly sensitive multiresponsive chemosensor for selective detection of Hg^{2+} in natural water and different monitoring environments. Inorg Chem 47(16): 7190–7201.
Wu LF, Ritchie SMC (2008) Enhanced dechlorination of trichloroethylene by membrane-supported Pd-coated iron nanoparticles. Environ Prog 27(2): 218–224.
Xu TL, Cai Y, O'Shea KE (2007) Adsorption and photocatalyzed oxidation of methylated arsenic species in TiO_2 suspensions. Environ Sci Technol 41(15): 5471–5477.
Xu YH, Zhao DY (2007) Reductive immobilization of chromate in water and soil using stabilized iron nanoparticles. Water Res 41(10): 2101–2108.
Xu ZH, Jing CY, Li FS, Meng XG (2008) Mechanisms of photocatalytical degradation of monomethylarsonic and dimethylarsinic acids using nanocrystalline titanium dioxide. Environ Sci Technol 42(7): 2349–2354.
Yang LJ, Li YB (2006) Simultaneous detection of *Escherichia coli* O157:H7 and *Salmonella typhimurium* using quantum dots as fluorescence labels. Analyst 131(3): 394–401.
Yang X, Wang YH, Xu LL, Yu XD, Guo YH (2008) Silver and indium oxide codoped TiO_2 nanocomposites with enhanced photocatalytic activity. J Phys Chem C 112(30): 11481–11489.
Yoon KY, Byeon JH, Park CW, Hwang J (2008) Antimicrobial effect of silver particles on bacterial contamination of activated carbon fibers. Environ Sci Technol 42(4): 1251–1255.
Yu JC, Ho WK, Yu JG, Yip H, Wong PK, Zhao JC (2005) Efficient visible-light-induced photocatalytic disinfection on sulfur-doped nanocrystalline titania. Environ Sci Technol 39(4): 1175–1179.
Zelmanov G, Semiat R (2008) Iron(3) oxide-based nanoparticles as catalysts in advanced organic aqueous oxidation. Water Res 42(1–2): 492–498.
Zhang LL, Jiang YH, Ding YL, Povey M, York D (2007) Investigation into the antibacterial behaviour of suspensions of ZnO nanoparticles (ZnO nanofluids). J Nanopart Res 9(3): 479–489.

Zhang QR, Pan BC, Pan BJ, Zhang WM, Jia K, Zhang QX (2008) Selective sorption of lead, cadmium and zinc ions by a polymeric cation exchanger containing nano-$Zr(HPO_3S)_2$. Environ Sci Technol 42(11): 4140–4145.

Zhang WX (2003) Nanoscale iron particles for environmental remediation: an overview. J Nanopart Res 5(3–4): 323–332.

Zhang XR, Minear RA, Guo YB, Hwang CJ, Barrett SE, Ikeda K, Shimizu Y, Matsui S (2004) An electrospray ionization-tandem mass spectrometry method for identifying chlorinated drinking water disinfection byproducts. Water Res 38(18): 3920–3930.

Zhang Y, Chen YS, Westerhoff P, Hristovski K, Crittenden JC (2008) Stability of commercial metal oxide nanoparticles in water. Water Res 42(8–9): 2204–2212.

Zheng TH, Zhan JJ, He JB, Day C, Lu YF, McPherson GL, Piringer G, John VT (2008) Reactivity characteristics of nanoscale zerovalent iron-silica composites for trichloroethylene remediation. Environ Sci Technol 42(12): 4494–4499.

Zwiener C, Richardson SD, De Marini DM, Grummt T, Glauner T, Frimmel FH (2007) Drowning in disinfection byproducts? Assessing swimming pool water. Environ Sci Technol 41(2): 363–372.

Chapter 6
Adsorption/Desorption Behavior of Charged Polymer Nanoparticles on a Mineral Surface in an Aqueous Environment

Hartmut Gliemann, Matthias Ballauff, and Thomas Schimmel

6.1 Introduction

Colloidal particles are often stabilized by long polymeric chains grafted to their surface (Israelachvili, 1992). If two such particles are dispersed in a good solvent for the chains and these particles approach each other, a repulsive interaction results. The steric interaction thus effected has been studied for decades and is well understood by now (Napper, 1983; Fleer et al., 1993). It can be enhanced even more if the polymers attached to the surface carry charges. The resulting electrosteric interaction can be understood in terms of the increased osmotic pressure of the counterions if the polyelectrolyte chains attached to the surfaces of the particles are to share the same volume (Jusufi et al., 2002a, b). The great practical importance of electrosteric interaction is related to the fact that most industrial polymer particles are stabilized in this way (Distler, 1999).

The application of polymer particles, however, does not only require a fundamental understanding of mutual interaction. Controlling the interaction of the particles with solid surfaces is of comparable importance when looking (i) at polymer particles as the base of paints and coatings or – considering environmental aspects – (ii) at the adsorption to or the release from soil material (Distler, 1999). In the latter case the attractive or repulsive interactions between particles and soil material depend on the hydrogeological and hydrochemical conditions of the soil. These interactions are of crucial relevance, as they influence the transport properties of nanoparticles in soil and, thus, affect, e.g., the contamination of soil and ground water with (nano)particulate pollutant systems or with pollutants (e.g., heavy metal ions) immobilized on the particle surface (Baumann et al., 2006; Keller and Auset, 2007). A comprehensive investigation of these problems requires (i) model particles with defined properties and (ii) model surfaces which can interact with the particles.

H. Gliemann (✉)
Institute of Functional Interfaces, Karlsruhe Institute of Technology (KIT), D-76131 Karlsruhe, Germany
e-mail: hartmut.gliemann@kit.edu

F.H. Frimmel, R. Niessner (eds.), *Nanoparticles in the Water Cycle*,
DOI 10.1007/978-3-642-10318-6_6, © Springer-Verlag Berlin Heidelberg 2010

In our experiments, polymer core particles onto which the polyelectrolyte chains are firmly attached by a chemical bond are used as model particles. The chemical grafting of the polymeric chains is necessary to avoid the disintegration upon strong interaction with a solid substrate. Moreover, the polyelectrolyte chains must be densely grafted to the particles. Colloidal objects carrying only a small number of chains can approach solid substrates so closely so that their van der Waals interaction with the substrate becomes the dominating effect (Israelachvili, 1992; Russel et al., 1989).

Colloidal particles with attached polyelectrolyte chains can conveniently be prepared by photo-emulsion polymerization (Guo et al., 1999). By this method the polyelectrolyte chains can be affixed rather densely to the surface of polystyrene particles so that the overall dimensions are much larger than their average distance on the surface of the particles. Hence, a spherical polyelectrolyte brush results which has overall dimensions in the colloidal domain. More detailed information concerning preparation, characterization, and stability of the spherical polyelectrolyte brushes (SPB) are found in Ballauff (2007), Borisov et al. (1991), Das et al. (2002), Guo and Ballauff (2000, 2001), Marra et al. (2003), Pincus (1991).

Evidently, these SPB provide a good model system for a systematic study of the interaction of sterically stabilized particles with solid substrates, particularly with mineral substrates. Here we present a basic study of anionic and cationic SPB contacting a negatively charged mica substrate. Ex situ and in situ atomic force microscopy (AFM) in the intermittent contact mode (Zhong et al., 1993) has been used as tool to investigate the topography and the phase contrast (Pang et al., 2000) of the samples. The intermittent contact mode is usually applied for soft, scratch sensitive materials as polymers (Almeida et al., 2005; Berlinger et al., 2001; Gliemann et al., 2007) or for material which is loosely bound to the substrate. As the phase contrast as well as the friction contrast depends on the interaction forces between the AFM tip and the surface, chemical material contrasts on the substrate can be investigated (Pfrang et al., 2003; Müller et al., 2006). With the application of in situ AFM techniques it is possible to carry out investigations under realistic conditions (Almeida et al., 2003; Montero-Pancera et al., 2005, 2006a, b). AFM turned out to be a powerful method to study the local organization of particles on solid substrates (Evers et al., 2000, 2002; Kaufhold et al., 2009; Linke-Schaetzel et al., 2004; Rakers et al., 1997; Wen et al., 2002).

Two different SPB have been studied: (1) an anionic SPB system consisting of chains of sodium polystyrene–sulfonate and (2) a cationic SPB system with chains of poly((2-acryloxy)ethyl)-trimethylammonium-chloride. The synthesis is described in more detail in Ballauff (2007). Figure 6.1 shows schematically the structures of both types of SPB.

The choice of these particular polyelectrolytes derives from their excellent solubility in water and their use as technical flocculation agents. The polyelectrolyte chains are attached to a polystyrene core of 90 nm (for cationic SPB) and 136 nm (for anionic SPB) diameter, respectively, with low polydispersity (Guo et al., 1999). Since the chains can be cleaved off after synthesis, their molecular weight and

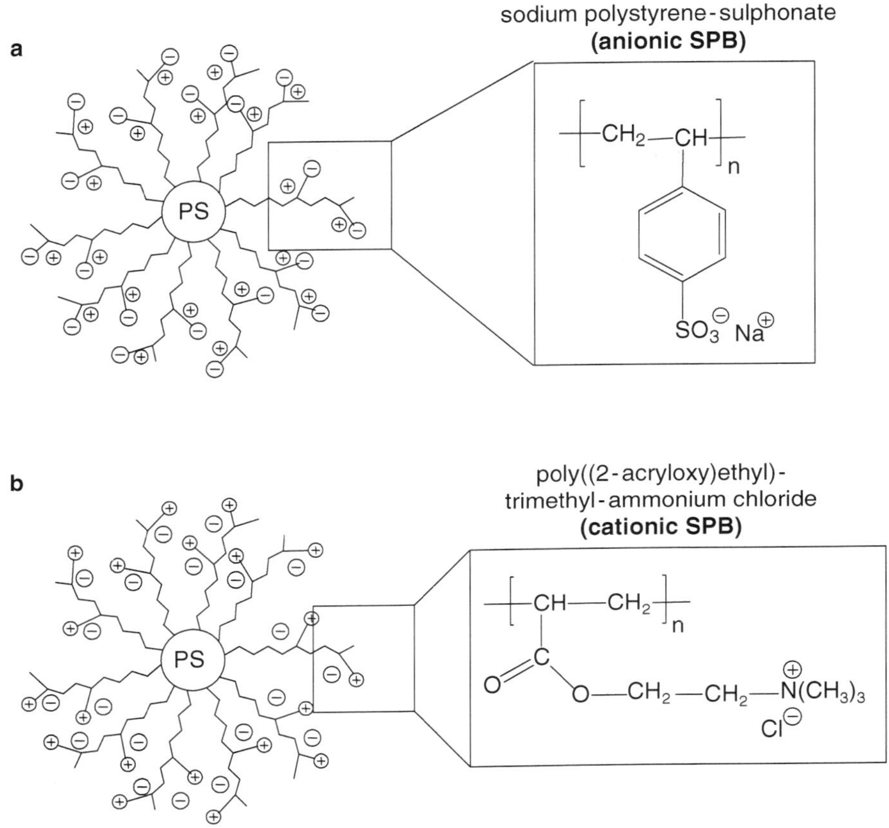

Fig. 6.1 Scheme of the anionic and cationic spherical polyelectrolyte brushes (SPB) used in this study. Long polyelectrolyte chains are densely attached to solid polystyrene particles so that a dense shell of charged chains results on the surface of the cores

molecular weight distribution can be determined easily (Guo et al., 1999). Moreover, the total charge of the SPB brush can be determined by titration. Hence, all pertinent parameters of the particles are known. Using these particles it will be shown that firmly attached polyelectrolyte chains are well suited to control the interaction of polymer particles with solid substrates, particularly with mineral surfaces. Two approaches are chosen to investigate the adsorption of the two different types of SPB: In a first approach the negatively and positively charged particles adsorbed on mica were investigated by ex situ AFM techniques. In the second, in situ AFM experiments were carried out to investigate the adsorption and desorption of cationic particles on mica substrates in an aqueous environment (Gliemann et al., 2006; Mei et al., 2003). Table 6.1 summarizes the characteristic data of both SPB.

Table 6.1 Characterization of the cationic and anionic core-shell nanoparticles

SPB	R (nm)	M_w (g/mcl)	L_c (nm)	σ (nm^{-2})	L_c/R	D (nm)
Cationic	45	89,700	116	0.049	2.57	5.1
Anionic	68	88,000	108	0.039	1.59	5.7

R: core radius; M_w: molecular weight of grafted chains; L_c: contour length of grafted chains; σ: graft density on core surface; D: the average distance between two neighboring graft points.

6.2 Atomic Force Microscopy: A Key to Investigate Topography and Material Contrast on the Nanometer Scale

The development and investigation of nanostructured surfaces and interfaces of materials including nanoscaled particular systems such as colloids is of increasing scientific and technological interest. To characterize the properties of nanostructured materials, surfaces, and interfaces on the nanometer scale, dedicated methods are required. Apart from the topographical properties such as the three-dimensional surface structure or the surface roughness, the chemical contrast and local material inhomogeneities of the surfaces have to be investigated with high lateral resolution on the nanometer scale. However, conventional characterization methods are often limited with respect to their lateral resolution.

The scanning probe methods, which go back to the development of the scanning tunneling microscope (STM) by Binnig and Rohrer in the 1980s (Binnig and Rohrer, 1982), offer the opportunity to investigate materials with a spatial resolution down to the atomic scale. Scanning tunneling microscopy, however, is limited to the investigation of conductive samples. Further development lead to the invention of the atomic force microscope (AFM) by Binnig et al. (1986), which also allows the investigation of non-conducting surfaces. In the AFM, an ultrafine tip is used to scan the surfaces of samples line by line, similar to an ultrafine profilometer. A basic AFM setup is schematically shown in Fig. 6.2.

If the AFM is operated in the *contact mode*, the tip, which is located at the end of an elastic cantilever, is in permanent contact with the substrate. The sample is mounted on a piezo stage (piezo I in Fig. 6.2) which allows to move the sample relative to the AFM tip in all three spatial dimensions by applying voltages to the corresponding piezo electrodes. The feedback controlling the height of the sample relative to the tip is achieved by setting a defined value of the vertical deflection of the cantilever when the tip is elastically pressed to the substrate surface. The deflection of the cantilever is most commonly detected by a laser beam deflection technique: A laser beam is focused to the reflecting back of the cantilever. The laser is aligned in a way that the spot of the reflected laser beam hits the four-segment photodiode exactly on the horizontal line between the segments A/B and C/D, respectively. The use of a four-segment photodiode thus allows to detect bending and distorsion of the cantilever independently.

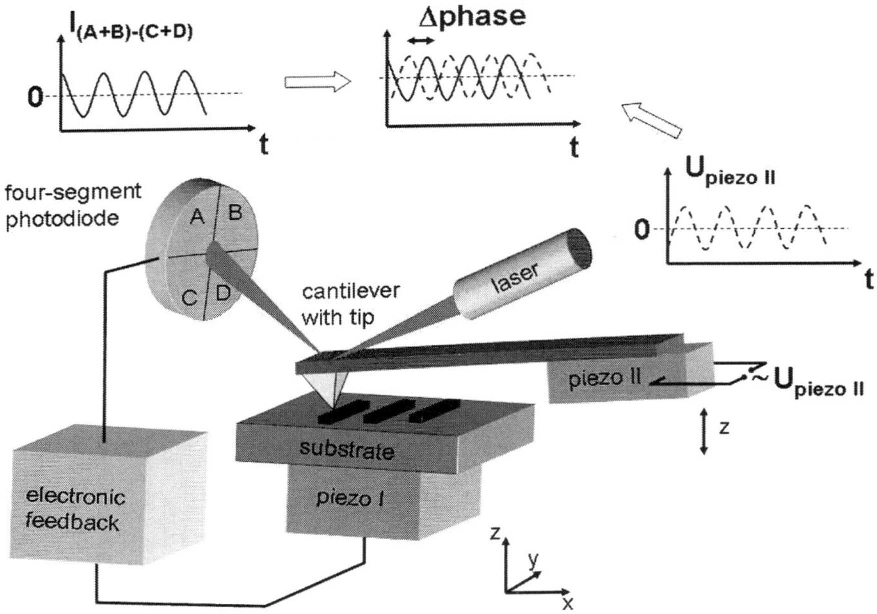

Fig. 6.2 Scheme of the principal setup of an atomic force microscope (AFM) for the investigation of topography and material contrast on the nanometer scale

In the case of the contact mode, the sample is usually scanned perpendicular to the longitudinal axis of the cantilever. As soon as the tip reaches an elevated or depressed surface structure, the deflection of the cantilever is changed and as a result the laser spot is moved vertically between the upper segments A and B and the lower segments C and D of the photodiode. The feedback control then changes the vertical position of the sample by applying a corresponding potential to piezo I (see Fig. 6.2) until the deflection of the cantilever reaches the set point. When scanning the sample in permanent contact with the tip, the cantilever is also twisted due to the friction force between the tip and the surface, resulting in a horizontal movement of the laser spot on the photodiode (see Fig. 6.2). The stronger the frictional force between tip and sample, the larger the torsion of the cantilever and thus the horizontal deflection of the spot on the photodiode. As the friction force is depending on the material, the twisting of the cantilever is more or less pronounced for different materials. This offers the possibility to "visualize" a material contrast and to determine the distribution of different materials on a substrate.

Apart from scanning in the contact mode, the AFM can also be operated in the *intermittent contact mode*. In this case, the AFM tip is vertically oscillating at an average tip-sample distance which is lower than its oscillation amplitude. As a consequence, the tip periodically hits the surface (so-called "tapping"). For this purpose, the cantilever with the AFM tip at its end is mounted on a piezo crystal (piezo II). An oscillating (AC) potential is applied to this piezo. As a consequence, the crystal changes its vertical dimension according to the value of the AC signal, leading to a vertical oscillation of the AFM tip. According to the frequency of the vertical vibration of the cantilever, the laser spot is moving up and down between the two upper segments (A and B) and the two lower segments (C and D) of the photo diode. The photo currents of the four segments can be read out separately.

When the sample is approached to the AFM tip, the oscillation of the tip is damped as soon as the tip meets the surface. As a result the amplitude of the oscillating cantilever – and thus also the amplitude of the oscillation of the laser spot – is reduced. A certain set point for the amplitude is selected before the tip is approached to the surface. After the tip is approached to the surface, the sample is scanned in lateral direction relative to the tip. When the oscillating AFM tip reaches an elevated (depressed) surface feature, the amplitude of the oscillating cantilever is reduced (increased) and is no longer identical to the preselected set point. As a result, an electronic feedback changes the vertical position of the sample relative to the AFM tip until the tip oscillation amplitude reaches the set point. In this way, the piezo (piezo I in Fig. 6.2) follows the topographic profile of the sample, allowing to record three-dimensional topographical images of the sample surface.

Apart from the topographic structure of the surface the lateral distribution of areas with different chemical properties can be determined both in the contact mode and in the intermittent contact mode. As in the case of the intermittent contact mode the phase of the driving AC potential for piezo II ($U_{\text{piezo II}}$ in Fig. 6.2) is known and the phase of the resulting vertical oscillation of the laser spot on the photodiode is measured, the phase difference Δ phase between the two signals can be determined electronically. This phase difference is a function of the tip–sample interaction (including, e.g., effects of surface polarity, adhesion, elasticity, and damping) and thus also a function of the local surface material properties. Different materials, e.g., distributed on the surface of a composite material, result in different tip–surface interactions and thus in different phase shifts.

In the following we will show that atomic force microscopy is an ideal tool to investigate the adhesion properties of nanoparticles both in air and in situ within aquatic environments with a spatial resolution down to the nanometer scale. AFM allows the monitoring of topographic features such as

- particle geometry,
- adsorption-induced particle deformation,
- particle density,
- particle aggregation, and
- particle distribution on the surface

as a function of time and experimental parameters (e.g., temperature, ionic strength), both in air and under liquid. AFM can, therefore, be used, e.g., to monitor the adsorption and desorption processes in real-time and under controlled conditions. Apart from topographic investigations, AFM-based techniques also allow to obtain information about the (nano)mechanical properties of particles and substrate. Using the AFM tip as a calibrated mechanical probe, properties such as the adhesion strength of the particles on their substrates or interaction forces between particles can be studied. Thus, AFM-based techniques are ideal for studying the behavior of nanoparticulate systems in the water cycle with respect to their interaction, e.g., with soil material and for understanding the parameters which influence adhesion.

6.3 Ex Situ AFM Investigations of Anionic and Cationic Polymer Nanoparticles Adsorbed on Mica Substrates

6.3.1 Motivation for Ex Situ AFM Studies

After a defined volume of the suspensions of the anionic and cationic spherical polyelectrolyte brushes (SPB) was dropped on different but chemically identical mica substrates, the topographical investigations of the local organization of the polymer particles after drying showed that both types of nanoparticles arrange in a significantly different manner: While the anionic SPB are highly ordered, the cationic SPB form irregular, network-like patterns on the surface. For a better understanding of theses observed phenomena, ex situ AFM studies were carried out with both kinds of particles.

6.3.2 Ex Situ AFM Studies: Experimental Details

The investigation of the adsorbed particles was carried out with atomic force microscopy (AFM). All AFM images were taken with a commercial AFM operated in intermittent contact mode. For phase contrast imaging, a homebuilt phase imaging electronics was used. Imaging was performed at ambient conditions, using bar-shaped silicon cantilevers with a force constant of 40 N/m and a resonance frequency of 325 kHz. The images represent unfiltered original data, represented in a linear gray scale. For sample preparation, a droplet of 20 μL of a solution containing the respective polymer particles dispersed in pure water was put onto a freshly cleaved mica surface (muscovite, white mica) and dried in air. The concentration of the particles was chosen small enough to avoid multilayer formation. The adsorption experiments and AFM experiments were reproduced several times and with different particle series independently, yielding comparable results. After drying, studies of the long-time behavior were performed to check if the particle structures change with time. No such changes as a function of time were observed in our experiments. All AFM data presented in this chapter show typical structures which are representative for the corresponding sample.

6.3.3 Adsorption of Anionic Polymer Nanoparticles

We first discuss the results obtained by AFM for the anionic spherical polyelectrolyte brushes (SPB). Figure 6.3a displays the AFM topography image of a 6 × 6 μm scan area of a mica substrate which is covered with anionic spherical polyelectrolyte brushes (SPB). Figure 6.3b and d show the AFM topography and the corresponding phase image, respectively, of anionic SPB in a larger magnification. The cross section of two single particles along the white line in Fig. 6.3b is displayed in Fig. 6.3c. Figures 6.3b and c show sharp, regular borders and smooth surfaces of the particles with an average diameter of 200 nm. Taking a closer look at the structure of the particles in Fig. 6.3b, one can see that each particle exhibits a bright center region, which also is obvious as a dark, well-defined area with an average diameter of about 150 nm in the corresponding phase image in Fig. 6.3d. When operating an AFM in intermittent contact mode to image phase contrast, the phase shift between the oscillation of the AC potential driving piezo II (see Fig. 6.2) and thus the cantilever and the vertical oscillation of the laser spot on the photodiode is recorded. In Fig. 6.3d three different phase shifts are detected, corresponding to the substrate,

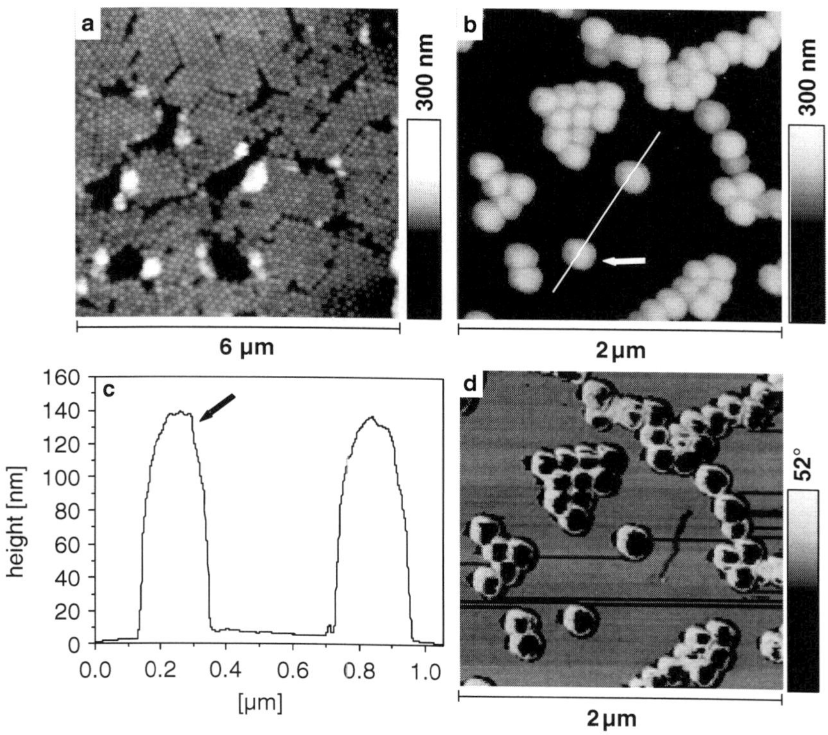

Fig. 6.3 Intermittent contact mode AFM images of topography (**a**, **b**) and phase contrast (**d**) of anionic polymer particles on mica. (**c**) Shows the cross section of two single particles along the *white line* in (**b**); the *arrow* in (**b**, **c**) assigns the particle in (**b**) to the corresponding cross section. It is seen that the particles have sharp regular boundaries and smooth surfaces. A well-defined phase contrast between the particles and the substrate is shown in (**d**). The anionic SPB form a layer with a two-dimensional long-range order on the surface (**a**)

to the edges of the SPB (bright rim) and to the center part of the SPB (dark spots). The phase shift depends on the dissipative tip–sample forces such as local adhesion, capillary forces, and viscoelastic damping. The difference in phase shift detected at the edges of the polymer spheres (bright rim in Fig. 6.3d) compared to the phase shift at the center part of the SPB (black areas in Fig. 6.3d) can be explained as a result of the different tip–sample contact geometries (and thus different dissipative tip–sample forces) when the AFM tip scans across the particle. The tip contacts the rim of the polymer particle with the tip flank whereas the tip apex gets in contact with the center part of the SPB. The spherical shape of the particles that is expected from synthesis is visible even within the aggregates. The comparison of the particle dimensions reveals that the diameters of the particle images in Fig. 6.3b are comparable with the corresponding diameters in Fig. 6.3d but are significantly smaller (50–60%) than the diameter of the cores of the anionic SPB plus twice the contour lengths of the attached polymer chains (cf. Table 6.1). Obviously, the chains of the anionic SPB have contracted during the drying process on the substrate. As intermittent contact mode images from pure polystyrene core particles show a very similar phase contrast as the anionic SPB (not shown here), we can assume that the negatively charged SPB exhibit a dense solid shell as expected. Figure 6.3a shows a two-dimensional long-range order within the layer of the anionic polymer particles. The layer shows small highly ordered arrays of particles divided by cracks or voids. These features have been seen in other studies of negatively charged polymer particles as well (Evers et al., 2000, 2002). The formation of the arrays takes place in a multistep process: After the deposition of a drop of particle dispersion on the mica surface, the water begins to evaporate and the concentration of the particles increases. Gravitational forces on the small anionic SPB particles can be neglected (Kralchewsky and Nagayama, 2000) and the electrosteric stabilization of the particles must lead to a strong mutual repulsion as long as water is present (Guo and Ballauff, 2000, 2001; Borisov et al., 1991). This can be argued from the finding that even high concentrations of monovalent added salt (up to 3 M) have no influence on the colloidal stability of these systems (Guo and Ballauff, 2001). However, when the thickness of the liquid layer of the drying droplet becomes smaller and smaller, the strong lateral capillary force leads to an attractive interaction between the particles. Thus, the film formation and deposition of the particles are expected to take place at the withdrawing rim of the drying drop. The growth of the two-dimensional array is facilitated by a convective liquid flow at the rim of the drop which is responsible for the transport of the particles to the boundary of the drop. As the anionic polymer particles consist of a polystyrene core and a negatively charged polymer brush, only a repulsive interaction between the particles and the negatively charged mica surface is expected. Therefore, particles should move to the boundary of the drop by convective flow without sticking to the surface. Hence, if the evaporation of the water is slow enough, the particles have the possibility to rearrange within the wet particle film. Therefore, this rearrangement leads to the densely packed arrangement of the particles as seen in Fig. 6.3a. This also corresponds to the case of bare polystyrene particles studied by many groups so far (Evers et al., 2000, 2002; Rakers et al., 1997; Wen et al., 2002).

6.3.4 Adsorption of Cationic Polymer Nanoparticles

A totally different ordering behavior is seen when using suspensions of the cationic spherical polyelectrolyte brushes (SPB) on mica. Figure 6.4a shows the large-scale arrangement of the particles resulting from drying a dilute suspension on the mica surface.

The particles are arranged in long chains that form a two-dimensional network on the substrate. Moreover, Fig. 6.4a shows that the drying process of the droplet has generated a gradient in particle concentration that increases from the upper right to the lower left of the image. In situ AFM investigations of freshly cleaved mica immersed in a suspension of cationic polymer particles showed that the polymer particles seem to stick to the surface as soon as they get into contact with it. It is, therefore, obvious that the attractive interaction between the cationic particles and the negatively charged mica surface is much stronger than between anionic spheres and mica. As a consequence of this, capillary forces that lead to a dense packing of the anionic spheres are now partially balanced by the adsorption of the cationic spheres onto the mica surface. Moreover, the adsorbed particles obviously act as nucleation centers for other polymer spheres. This becomes more evident when zooming into Fig. 6.4a. Figure 6.4b shows that only small disordered aggregates of particles are formed. Therefore, the aggregates seen in Fig. 6.4a consist of few particles only that form a network-like structure. Zooming into Fig. 6.4b, Fig. 6.4d and e corroborate this explanation for the totally different behavior of the cationic particles as compared to the anionic particles: First of all, Fig. 6.4d demonstrates that the adsorbed cationic SPB show different shapes in the topography image as compared to the corresponding phase image in Fig. 6.4e. This is in contrast to the behavior observed for the anionic polymer particles (see Fig. 6.3b and 6.3d). The topography cross section of the cationic SPB in Fig. 6.4c shows a regular shape of the particles as in the case of the anionic SPB (Fig. 6.3c). The phase contrast image of the cationic particles in Fig. 6.4e, however, is not as smooth and well defined as in the case of the anionic spheres (Fig. 6.3d). Most importantly, in Fig. 6.4e a corona is observed which is surrounding the positively charged SPB. This corona is clearly visible in the phase image of Fig. 6.4e but can hardly be detected in the topography image in Fig. 6.4d. The size of the particles (without corona) is roughly the same when it is determined from the topography image (Fig. 6.4d) and from the phase image (Fig. 6.4e). The particle diameter thus determined is considerably lower than the diameter of the particles corresponding to fully extended chains (ca. 320 nm; diameter of core plus twice the contour lengths of the attached chains, cf. Table 6.1). From these finding it must be concluded that the cationic particles strongly interact with mica. The corona visible in Fig. 6.4e is due to the positively charged polymer chains on the surface of the cationic particles that are strongly attached to the negatively charged mica surface. The phase-sensitive detection in the AFM analysis is able to visualize these attached chains directly. Moreover, Fig. 6.4e demonstrates that the chains have not retracted during the drying process. This indicates that the interaction between the polyelectrolyte chains and the surface already exists when water is still present. For geometrical reasons the particle diameters (including the

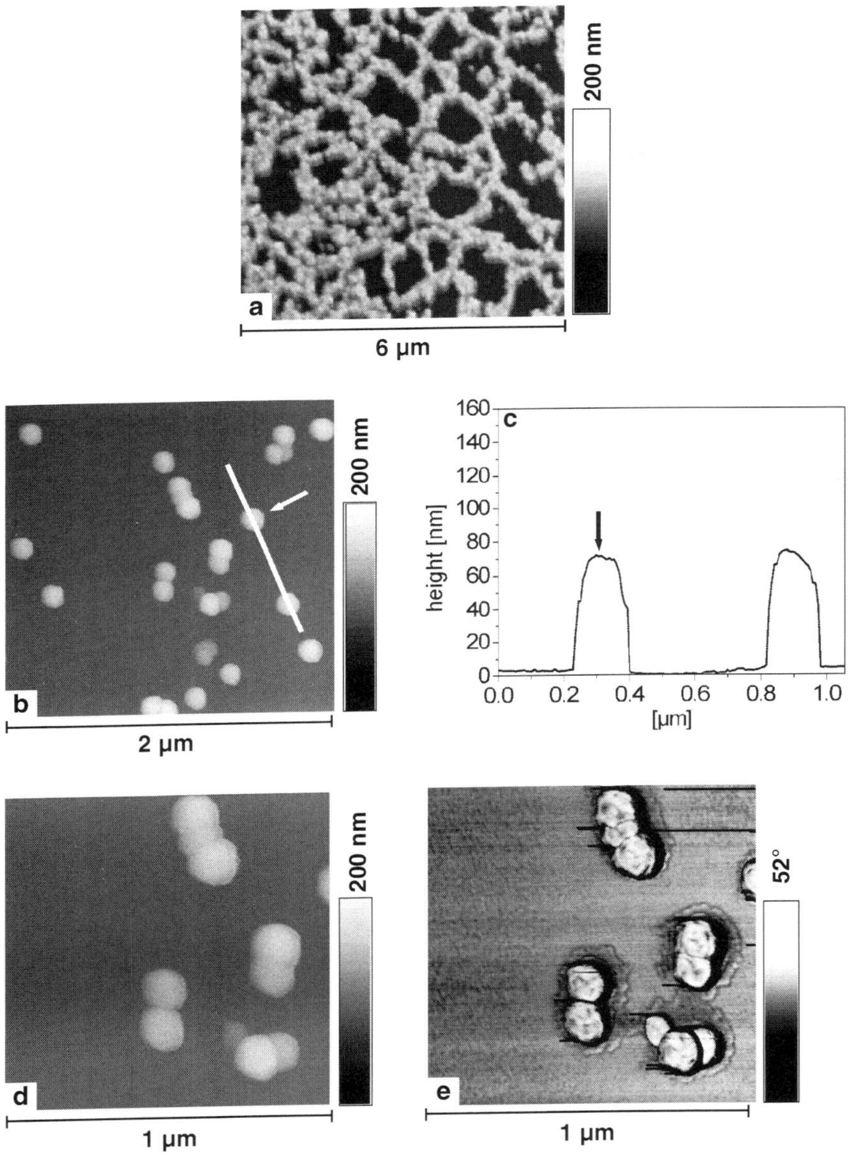

Fig. 6.4 Intermittent contact AFM images of topography (**a**, **b**, **d**) and phase contrast (**e**) of cationic polymer particles (see Table 6.1) having positively charged chains attached to their cores. The solid substrate is mica. (**c**) Shows the cross section of two single particles along the *white line* in (**b**) and the *arrow* in (**c**) marks the corresponding particle in (**b**). In (**e**) the particles show an irregular phase distribution on their surfaces. They are surrounded by a corona consisting of the polyelectrolyte chains that are attached to the solid mica surface. As shown in (**a**), the particles tend to form network-like aggregates on the surface without long-range two-dimensional order

corona) determined from Fig. 6.4e must still be smaller than the maximal diameter of the particles (cf. Table 6.1). This is found indeed when comparing the overall particle diameters including the corona detected by AFM (ca. 220 nm) to the theoretical maximum size derived from the core diameter and the contour length of the chains (ca. 320 nm; see above). A schematic rendition of the process of drying is given on the right-hand side of Fig. 6.5: The substrate becomes the counterion for the positively charged polyelectrolyte chains and balances a part of the charge of the shell of the cationic SPB. The sodium counterions as well as the cations balancing the charge of the mica surface are released in this way. Thus, the entropy of the entire system is increased. This "counterion release force" has recently been the subject of a thorough theoretical study (Fleck and von Grünberg, 2001). It is seen experimentally in investigations of the interaction of DNA with appropriate solid substrates (Wagner et al., 2000; Maier and Rädler, 2000). The strong interaction of the positive polyelectrolyte chains and the negative surface thus effected must lead to a partial spreading of the chains which is detected as the corona around the SPB in Fig. 6.4e. Moreover, this attachment counterbalances partially the strong capillary forces that tend to aggregate the particles during the drying process.

As a consequence of the sticking of the particles to the surface, a loose network of particles is formed upon drying (see Fig. 6.4a), but no close packing is observed as in the case of the anionic SPB particles (see Fig. 6.3a). It is now obvious that the anionic particles (see Fig. 6.3) must interact in a totally different fashion as shown schematically on the left-hand side of Fig. 6.5: The particles are at first repelled from the negatively charged surface. The counterions are kept within the shell upon drying, and a smooth surface results. After evaporation of water the particles are attracted to the surface only by conventional van der Waals attraction if their distance to the surface is small enough (Russel et al., 1989). The chains of the particles keep the cores sufficiently above the surface, however, unless all water has evaporated. Hence, the van der Waals interaction of the particles with the surface remains small, and their surface diffusion remains unhampered up to the point where virtually all water has been evaporated. This results in a long-range two-dimensional order of the particles on the surface as seen in Fig. 6.3a.

6.4 In Situ AFM Investigations of Adsorption and Desorption of Cationic Polymer Nanoparticles on Mica Substrates

6.4.1 Motivation for In Situ AFM Studies

It was shown in Sect. 6.2 that the adsorption of *cationic* spherical polyelectrolyte brushes (SPB) on mica is driven by the counterion release force which is much more efficient than the adsorption of the *anionic* polymer brushes predominated by the weak attractive van der Waals forces. Although these results derived from dried particles on mica give valuable information about the different driving forces of particle adsorption as a function of the particle properties, it is necessary to get detailed

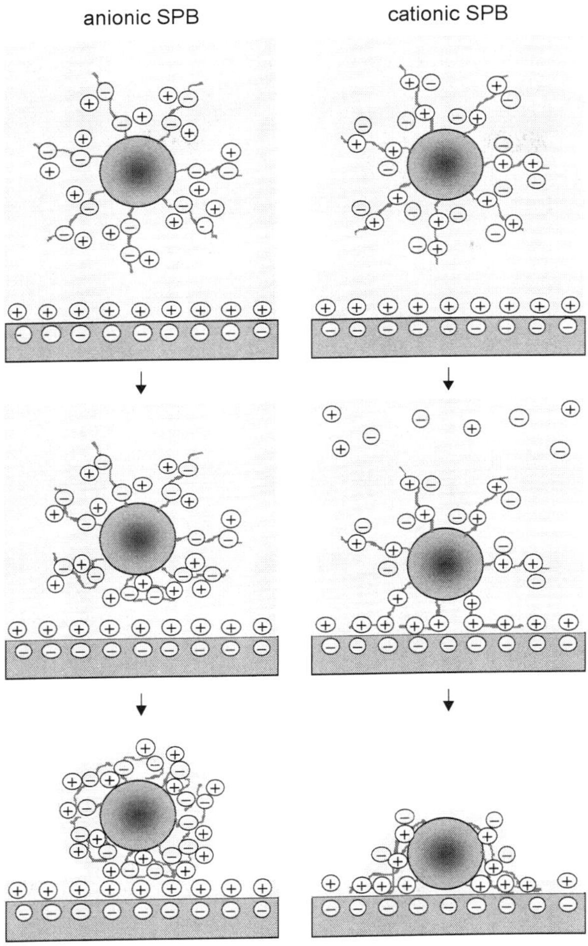

Fig. 6.5 Scheme of the interaction of the anionic and the cationic polymer particles with negatively charged surfaces. The *left-hand side* displays schematically the interaction of the negatively charged particles (see Table 6.1) with the solid mica substrate, whereas the *right-hand side* displays the interaction of the positively charged spherical polyelectrolyte brushes. The solid substrate which bears a negative charge becomes the counterion of the positive chains attached to the cationic particles. The respective number of counterions both of the particles and of the substrate is, therefore, released. This leads to a strong attractive force between the particles and the substrate ("counterion release force"; see Fleck and von Grünberg, 2001; Maier and Rädler, 2000; Wagner et al., 2000 for further details). The negatively charged particles (*left-hand side*), on the other hand, do not exhibit this interaction with the mica surface. Their spatial structure obtained upon drying resembles the result found for uncoated polystyrene particles in previous investigations (Evers et al., 2000, 2002)

information on the in situ processes during the adsorption and how the adsorption can be influenced. Particularly for the adsorption processes of nanoparticulate systems on mineral surfaces in wet soil material the ionic strength is relevant for the adsorption and desorption processes and thus for the mobility of the particles in the soil. Therefore, as the counterion release force should depend on the ionic strength in solution, the influence of the amount of added electrolyte on the adsorption behavior of cationic SPB was investigated. It is demonstrated that the efficiency of particle adsorption on mica based on the counterion release force can indeed be tuned by the ionic strength of the solution: By an increase of the ionic strength, a desorption of the cationic particles from the mineral surface could be induced and observed in situ and in real time by atomic force microscopy (AFM).

6.4.2 In Situ AFM Studies: Experimental Details

For the in situ AFM investigation of the adsorption behavior of the particles in liquid a piece of mica was freshly cleaved and a circle line was drawn on the mica surface with a water-resistant marker (Fig. 6.6). This line acts as a hydrophobic barrier which avoids the spreading of the suspension over the entire surface area. A drop of diluted suspension of the particles (300 μL) was deposited on the inner part of the circle. To avoid the adsorption of the particles on the back of the cantilever during the AFM investigation which would result in a loss of laser light intensity reaching the photodiode of the AFM, the suspension was subsequently substituted by a drop of bi-distilled water after exposure to the suspensions. This washing procedure was repeated five times and removes most of the suspended polymer particles. Then the substrate was fixed on the sample holder of the AFM and the AFM tip on the cantilever, which is mounted underneath a glass piece, is approached to the surface of the mica substrate. A stable meniscus is formed between the substrate and the glass piece and the hydrophobic barrier avoids the spreading of the liquid (Fig. 6.6). For the investigations in liquid the AFM was operated in the intermittent contact mode with bar-shaped silicon cantilevers having an average force constant of 1.75 N/m. The operation frequency of the cantilevers was approx. 30 kHz for all experiments done in situ. For the intermittent contact mode investigations the tip amplitude is damped to 80% of the free off-surface value before the feedback loop is triggered and the sample is withdrawn.

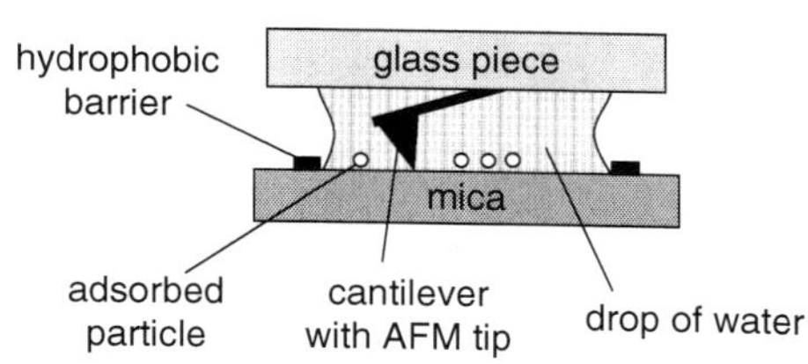

Fig. 6.6 Experimental setup for the in situ AFM studies of the adsorption and desorption of cationic spherical polyelectrolyte brushes

6.4.3 *In Situ AFM Investigation of the Adsorption of Cationic Polymer Nanoparticles*

Figure 6.7 shows the liquid-AFM amplitude images (intermittent contact mode) of cationic SPB adsorbed on a mica substrate 5 min (Fig. 6.7a) and 1 h (Fig. 6.7b) after the suspension was dropped onto the mica substrate. For the analysis of the AFM results obtained in the liquid the amplitude images are used, because they show a much better contrast between the particles and the substrate than the corresponding topography images. Figure 6.7b shows a zoom-out image of Fig. 6.7a. It was obtained after 12 AFM image scans of the surface area displayed in Fig. 6.7a. The number of particles on the 10×10 μm surface area in Fig. 6.7b is about 220.

All particles which can be observed in Fig. 6.7a can also be found in Fig. 6.7b. This indicates that no particles were removed during the 12 AFM image scans which took place before the zoom-out displayed in Fig. 6.7b was carried out. The zoom-out in Fig. 6.7b also shows that there are no particles deposited along the border of the scan area shown in Fig. 6.7a. This is an important indicator that no particles were removed from their positions during the scan 5 min after the suspension was dropped onto the mica surface.

For anionic spherical polyelectrolyte brushes a high mobility of the particles on the mica surface during the evaporation process of the water was found (see Sect. 6.3.3). As a result a densely packed, well-ordered particle array was formed after drying the sample (Fig. 6.3a). This behavior was explained by strong capillary forces between the particles caused by the receding water droplet so that the adhesion force between the anionic particles and the surface is overcompensated. Therefore, the effect of the drying process on the particle position on the surface can be used as

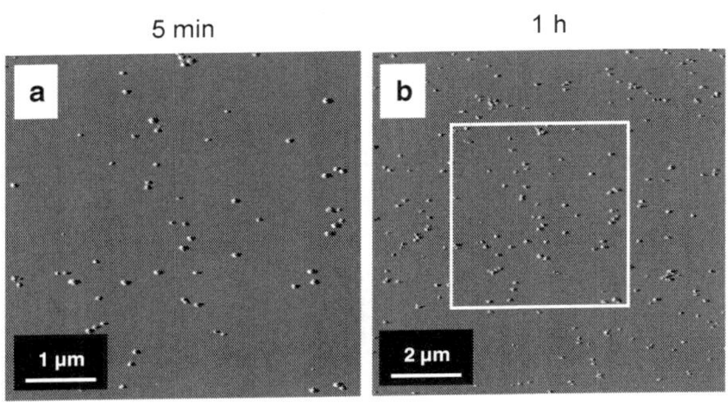

Fig. 6.7 Liquid-AFM amplitude images (intermittent contact mode) of cationic spherical polyelectrolyte brushes on mica 5 min (**a**) and 1 h (**b**) after the suspension was dropped onto the mica surface and after 12 subsequent AFM scans. (**a**) and (**b**) show different magnifications of the identical location on the sample surface. The white frame in (**b**) marks the scan area of (**a**); the particle density for (**b**) is about 220 particles/100 μm^2; all particles observable in (**a**) can also be found in (**b**) and no scan window can be detected in (**b**), indicating that the particles are not (re)moved by the AFM tip during the scan; scan area in (**a**) 5×5 μm, scan area in (**b**) 10×10 μm

a first qualitative indicator for the force with which a particle is adsorbed on the surface.

Here we investigated the effect of the drying process on the adsorption of the cationic SPB in more detail. AFM studies of identical sample areas before and after drying the surface were carried out. Figure 6.8A shows the liquid-AFM amplitude image (10 × 10 μm) of cationic SPB on mica and Fig. 6.8B shows the image of the identical sample area after drying the suspension. It is obvious that the particle density has increased during the drying process. This can be explained to the fact that during the washing process described in the experimental part (Sect. 6.4.2) not all dispersed particles are removed from the liquid above the surface. During the drying process these particles reach the surface and adsorb. The images 6.8a and b in Fig. 6.8 show AFM zoom-ins of the surface areas displayed in Fig. 6.8A and B (white frames), respectively. They give a more detailed information about the influence of the drying process on the adsorbed particles. To compare images 6.8a and b, all particles of image 6.8a (before drying) were marked with a circle. In a second step this arrangement of the circles and the AFM image of the corresponding surface area after drying were superposed and image 6.8b was obtained. As the center of each circle is occupied by a particle in Fig. 6.8b, it can be concluded that no particle which can be detected in Fig. 6.8a was removed during the drying process.

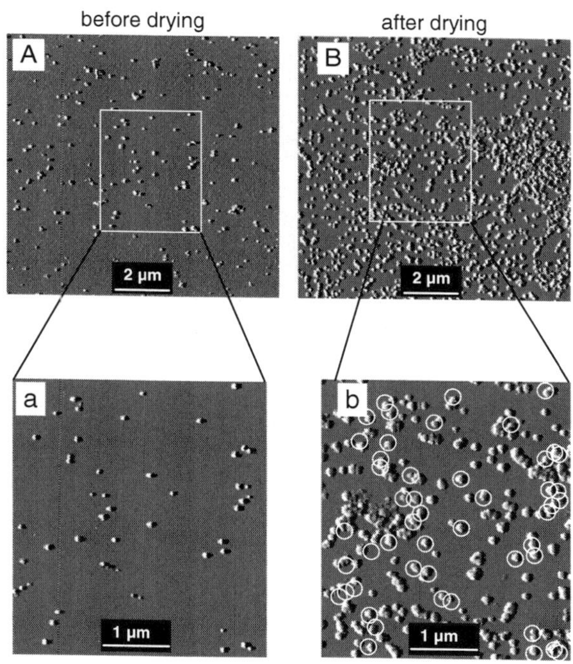

Fig. 6.8 AFM amplitude images (intermittent contact mode) of cationic polymer particles adsorbed on mica; (**A**) shows an image obtained in liquid and (**B**) an image of the identical surface area after drying the substrate. (**a**) and (**b**) were obtained by zooming into (**A**) and (**B**) (*white frames*), respectively. The *white rings* in (**b**) mark those particle positions which are occupied by a particle in (**a**). As a result all particle positions from (**a**) remain occupied in (**b**) without exception, which means that no particles were removed from their original positions in spite of the strong capillary forces acting during the drying process

This indicates a strong interaction between the particles and the surface. In the case of a weak particle–surface interaction the capillary forces – operative during the drying process – would remove the particles from their original positions. This was shown for anionic particles on mica in Sect. 6.3.3. However, it is not observed in the case of the cationic SPB studied here in situ.

To exclude possible changes of the mica surface due to cleavage in air and the surface exposure to the particle suspension, additional experiments were carried out where mica was cleaved within the suspension. There were no significant differences in particle density between mica substrates which were cleaved in liquid or in air.

All findings are easily explained in terms of the counterion release force: As soon as the particles come into contact with the substrate, the negative counterions of the positively charged SPB chains are substituted by the negative charges of the freshly cleaved mica substrate. Concomitantly, the positive counterions of the mica are substituted by the positive charges of the SPB chains. As a result of the process of adsorption, a corresponding number of counterions are released and the entropy of the entire solution is increased. This creates a strong adhesion of the particles.

6.4.4 In Situ AFM Investigations of the Desorption of Cationic Polymer Nanoparticles

We now discuss the influence of the ionic strength onto the counterion release force in solution. Evidently, a sufficiently high concentration of added salt is expected to diminish this force significantly and no adsorption should take place any more. Figure 6.9a shows a liquid-AFM amplitude image ($5.3 \times 4.7\ \mu$m) of cationic polymer particles adsorbed on the mica surface. As about 5% of the counterions of the SPB polyelectrolyte chains leave the shell material the ionic strength within the suspension is about 10^{-4} mol/L. No additional electrolyte was added to the solution. The particle density in Fig. 6.9a is about 210/100 μm^2. The ionic strength was then increased within the meniscus between the mica substrate and the glass piece (Fig. 6.6) by lifting up the AFM tip and adding one drop (30 μL) of an aqueous 0.1 M NaCl solution to the suspension. In this way the ionic strength of the volume was increased to ca. 10^{-2} mol/L. The first amplitude image of the surface 9 min after the electrolyte was added is shown in Fig. 6.9b, which represents the identical surface area as Fig. 6.9a.

From the first scan after increasing the ionic strength we see that about one quarter of the particles were removed from the surface. Figure 6.9c shows that after the first scan about 70% of the particle positions occupied in Fig. 6.9a remain occupied while about 20% of the particles were removed from their original positions. The relatively high amount of particles (about 12%) appearing on new positions in Fig. 6.9b may be the result of the particle displacement by the AFM tip and/or an additional adsorption due to the increased ionic strength.

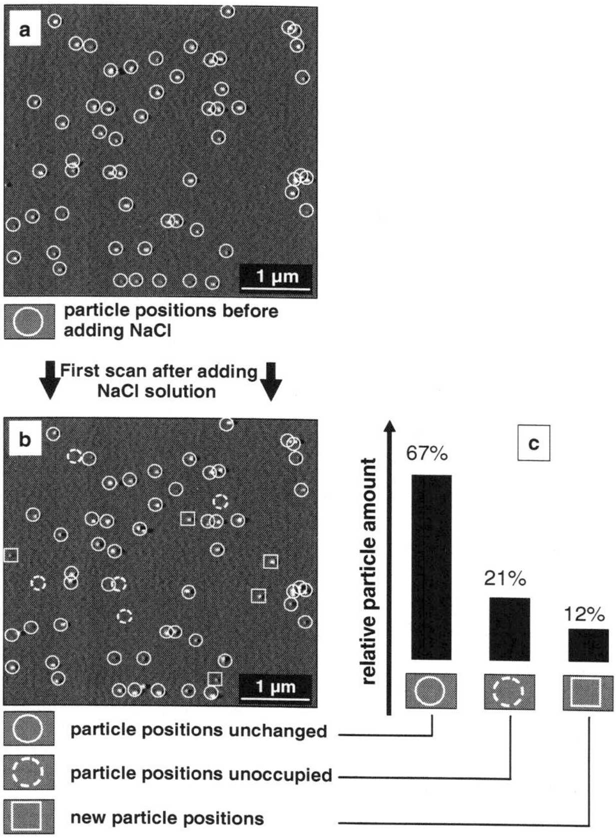

Fig. 6.9 Liquid-AFM amplitude images (scan area 5.3 × 4.7 μm) of cationic polymer particles adsorbed on freshly cleaved mica. In (**a**) the ionic strength is ca. 10^{-4} mol/L. (**b**) The first AFM scan after increasing the ionic strength to 10^{-2} mol/L by adding aqueous solution of NaCl is shown. The statistics in (**c**) summarizes the influence of the increase in ionic strength on the particle density and on the particle distribution on the surface

The results of the AFM investigation presented in Fig. 6.9 show that the adhesion force of the particles is significantly reduced due to the reduced activation barrier for the particle desorption which is the result of the increased ionic strength.

This has also been proved directly by repeated AFM scans after increasing the ionic strength of the solution. Figure 6.10a–f shows a sequence of liquid-AFM amplitude images of the identical mica surface area *before* (Fig. 6.10a) the addition of the aqueous solution of NaCl and *after* the increase of the ionic strength of the suspension. The parameter of the different graphs in Fig. 6.10 is the number of AFM scans. Each scan removes more and more particles from the surface. Figure 6.10f shows that after 90 min (corresponding to eight intermittent contact mode image scans) no more particles can be detected on the substrate surface any

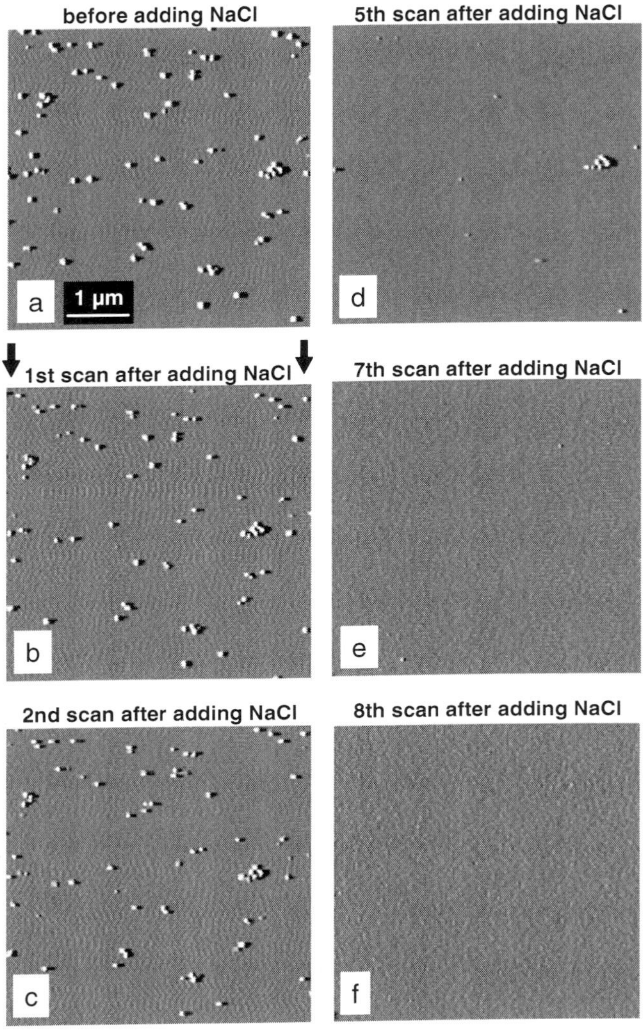

Fig. 6.10 Liquid-AFM amplitude images (5 × 5 μm) of cationic polymer particles adsorbed on freshly cleaved mica without adding aqueous solution of NaCl to the suspension (**a**), immediately after the increase of the ionic strength to 10^{-2} mo/L (**b**), and after different numbers of scans after NaCl addition (**c**–**f**); all images represent the identical surface area; after the addition of NaCl no particles can be detected anymore after scanning the identical surface area eight times

more. This result must be compared to the corresponding results of the in situ AFM studies in pure water mentioned earlier, where at low ionic strength no particle was removed even after 12 scans (Fig. 6.7). This demonstrates clearly a significantly reduced adhesion force between the particles and the surface if the ionic strength is sufficiently high.

6.5 Conclusions

An investigation of the interaction of anionic and cationic spherical polyelectrolyte brushes with negatively charged mica surfaces by AFM in intermittent contact mode was presented. It was demonstrated on dried samples that the negatively charged spherical polyelectrolyte brushes (SPB) are repelled from the mica surface and that the particle–particle interaction is dominating as compared to the interaction of the particles with the substrate. The positively charged SPB, however, interact strongly with the mica surface as expected and network-like structures without long-range two-dimensional order are formed on the surface. The shell of these particles containing positively charged polyelectrolyte chains spreads over the negative surface and anchors the particles. Hence, a shell of polyelectrolyte chains is a highly efficient means of adjusting the interaction of colloidal particles with solid, mineral substrates.

In the case of cationic spherical polyelectrolyte brushes the adsorption on mica substrates was investigated in situ, i.e., *in suspension,* by AFM. The results show that the adsorption of the particles on negatively charged mica substrates is influenced by the particle–surface interaction. Strong adhesive interaction between the cationic SPB and the mica surface is present within suspensions in pure water. Increasing the ionic strength in the system leads to a weakening of the particle–substrate interaction. The results demonstrate that the counterion release force is responsible for the strong interaction between the particles and the surface in the case of cationic SPB suspensions. From a practical point of view this means that it is possible (i) to control the adhesion force of the cationic SPB on the substrate by varying the ionic strength within the suspension and (ii) to remove particles from the substrate which are already adsorbed by increasing the ionic strength. The findings presented here not only open up new perspectives for technological application but also help to understand the basics of the adsorption and desorption processes of nanoparticles on mineral surfaces and thus to understand the transport of nanoparticulate pollutions in soils.

Acknowledgment This work was financially supported within the DFG-Center for Functional Nanostructures (CFN), by the Research Prize for Applied Science ("Landesforschungspreis") Baden-Württemberg, by the Deutsche Forschungsgemeinschaft (SFB 481, Bayreuth), by the European Community (Project "POLYAMPHI"), by the Fonds der Chemischen Industrie, and by the Landesstiftung Baden-Württemberg within the Research Network of Excellence "Functional Nanostructures".

References

Almeida AT, Gliemann H, Schimmel Th, Petri DFS (2003) Adsorption kinetic of enolase on silicon studied by in-situ ellipsometry and wet AFM. Acta Microsc A 12: 41–44.
Almeida AT, Gliemann H, Schimmel Th, Petri DFS (2005) Characterisation of PMMA/PVB blend films by means of AFM. Microsc Microanal 11: 122–125.
Ballauff M (2007) Spherical polyelectrolyte brushes. Progr Polymer Sci 32: 1135–1151.
Baumann T, Fruhstorfer P, Klein T, Niessner R (2006) Colloid and heavy metal transport at landfill sites in direct contact with groundwater. Water Res 40: 2776–2786.

Berlinger A, Gliemann H, Barczewski M, Durigon PER, Petri DFS, Schimmel Th (2001) Influence of sulphonating on polymer and polymer blend surfaces studied by atomic force microscopy. Surf Interface Anal 32: 144–147.

Binnig G, Quate CF, Gerber C (1986) Atomic force microscope. Phys Rev Lett 56: 930–933.

Binnig G, Rohrer H (1982) Scanning tunneling microscopy. Helvetica Physica Acta. 55: 726–735.

Borisov OV, Birshtein TM, Zhulina EB (1991) Collapse of grafted polyelectrolyte layers. J Phys II 1: 521.

Das B, Guo X, Ballauff M (2002) The osmotic coefficient of spherical polyelectrolyte brushes in aqueous salt-free solution. Prog Colloid Polym Sci 121: 34.

Distler D (1999) Wässrige Polymerdispersionen. Wiley-VCH, New York.

Evers M, Palberg T, Dingenouts N, Ballauff M, Richter H, Schimmel Th (2000) Vitrification in restricted geometry: dry films of colloidal particles. Progr Colloid Polym Sci 115: 307–314.

Evers M, Schöpe HJ, Palberg Th, Dingenouts N, Ballauff M (2002) Residual order in amorphous dry films of polymer latices: indications of an influence of particle interaction. J Non-Cryst Sol 307–310: 579–583.

Fleck C, von Grünberg HH (2001) Counterion evaporation. Phys Rev E 63: 061804.

Fleer GJ, Cohen Stuart MA, Scheutjens JMHM, Cosgrove T, Vincent B (1993) Polymers at Interfaces. Chapman and Hall, London.

Gliemann H, Almeida AT, Petri DFS, Schimmel Th (2007) Nanostructure formation in polymer thin films influenced by humidity. Surf Interface Anal 39: 1–8.

Gliemann H, Mei Y, Ballauff M, Schimmel Th (2006) Adhesion of spherical polyelectrolyte brushes on mica: an in-situ AFM investigation. Langmuir 22: 7254–7259.

Guo X, Ballauff M (2000) Spatial dimensions of colloidal polyelectrolyte brushes as determined by dynamic light scattering. Langmuir 16: 8719–8726.

Guo X, Ballauff M (2001) Spherical polyelectrolyte brushes: comparison between annealed and quenched brushes. Phys Rev E 64: 051406.

Guo X, Weiss A, Ballauff M (1999) Synthesis of spherical polyelectrolyte brushes by photoemulsion polymerization. Macromolecules 32: 6043–6046.

Israelachvili JN (1992) Intermolecular and Surface Forces. Academic Press, London.

Jusufi A, Likos CN, Löwen H (2002a) Conformations and interactions of star-branched polyelectrolytes. Phys Rev Lett 88: 018301.

Jusufi A, Likos CN, Löwen H (2002b) Counter ion-induced entrophic interactions in solutions of strongly stretched, osmotic polyelectrolyte stars. J Chem Phys. 116: 11011.

Kaufhold S, Kaufhold A, Jahn R, Brito S, Dohrmann R, Hoffmann R, Gliemann H, Weidler PG, Frechen M (2009) A new massive deposit of Allophane raw material in Ecuador. Clays Clay Miner 57: 72–81.

Keller AA, Auset M (2007) A review of visualization techniques of biocolloid transport processes at the pore scale under saturated and unsaturated conditions. Adv Water Res 30: 1392–1407.

Kralchewsky PA, Nagayama K (2000) Capillary interactions between particles bound to interfaces, liquid films and biomembranes. Adv Colloid Interface Sci 85: 145–192.

Linke-Schaetzel M, Bhise AD, Gliemann H, Koch Th, Schimmel Th, Balaban TS (2004) Self-assembled chromophores for hybrid solar cells. Thin Solid Films 451–452: 16–21.

Maier B, Rädler JO (2000) DNA on fluid membranes: a model polymer in two dimensions. Macromolecules 33: 7185–7194.

Marra A, Pleuvrel-Disdier E, Wittemann A, Guo X, Ballauff M (2003) Rheology of dilute and semidilute suspensions of spherical polyelectrolyte brushes. Colloid Polym Sci 281: 491–496.

Mei Y, Wittemann A, Sharma G, Ballauff M, Koch Th, Gliemann H, Horbach J, Shimmel Th (2003) Engineering the interaction of latex spheres with charged surfaces: AFM investigation of spherical polyelectrolyte brushes on mica. Macromolecules 36: 3452–3456.

Montero-Pancera S, Gliemann H, Petri DFS, Schimmel Th (2005) Adsorption behavior of creatine phosphokinase onto silicon wafers: comparison between ellipsometric and atomic force microscopy investigation. Microsc Microanal 11: 56–59.

Montero-Pancera S, Gliemann H, Schimmel Th, Petri DFS (2006a) Adsorption behavior and activity of hexokinase. J Colloid Interface Sci 302: 417–423.

Montero-Pancera S, Gliemann H, Schimmel Th, Petri DFS (2006b) Effect of pH on the immobilization and activity of creatine phosphokinase. J Phys Chem B 110: 2674–2680.
Müller M, Schimmel Th, Häußler P, Fettig H, Müller O, Albers A (2006) Finite element analysis of V-shaped cantilevers for atomic force microscopy under normal and lateral force loads. Surf Interface Anal 38: 1090–1095.
Napper DH (1983) Polymeric Stabilization of Colloidal Dispersions. Academic Press, London.
Pang GKH, Baba-Kishi KZ, Patel A (2000) Topographic and phase-contrast imaging in atomic force microscopy. Ultramicroscopy 81: 35.
Pfrang A, Reznik B, Schimmel Th, Gerthsen D (2003) Comparative study of differently textured pyrolytic carbon layers by atomic force, transmission electron and polarized light microscopy. Carbon 41: 181–185.
Pincus P (1991) Colloid stabilization with grafted polyelectrolytes. Macromolecules 24: 2912–2919.
Rakers S, Chi LF, Fuchs H (1997) Influence of the evaporation rate on the packing order of polydisperse latex monofilms. Langmuir 13: 7121–7124.
Russel WB, Saville DA, Schowalter WR (1989) Colloidal Dispersions. Cambridge University Press, Cambridge.
Wagner K, Harries D, May S, Kahl V, Rädler JO, Ben-Shaul A (2000) Direct evidence for counterion release upon cationic lipid–DNA condensation. Langmuir 33: 303–306.
Wen L, Wu RC, Eschenazi E, Papadopoulos K (2002) AFM of amidine latex particles attachment on mica. Colloids Sur A 197: 157–165.
Zhong Q, Inniss D, Kjolle K, Elings VB (1993) Fractured polymer/silica fiber surface studied by tapping mode atomic force microscopy. Surf Sci Lett 290: L688.

Chapter 7
X-Ray Spectromicroscopy Studies of Nanoparticles in Aqueous Media

Jürgen Thieme, Sophie-Charlotte Gleber, Julia Sedlmair, Jens Rieger, Jürgen Niemeyer and John Coates

7.1 X-Ray Spectromicroscopy

In 1952 Horst Wolter wrote that X-radiation in the energy range between the K-absorption edge of carbon and of oxygen is very well suited for microscopy purposes. For this energy range he created the phrase “water window” (Wolter, 1952). The reason for this namegiving becomes clear when looking at the linear absorption coefficients plotted in Fig. 7.1.

The intensity I_1 of X-rays transmitted through an object can be calculated using the Lambert–Beer equation (Attwood, 2000):

$$I_1 = I_0 \exp(-\mu_l d)$$

Here I_0 is the incident radiation, μ_l is the linear absorption coefficient, and d is the thickness of the object along the path of the X-rays. A good measure for calculating the penetration of X-rays through a sample is to determine at what thickness the transmitted radiation I_1 is weakened to $1/e$ of the incident intensity I_0. The resulting penetration depth $1/\mu_l$ for water ranges from 2 μm at the carbon K-edge up to 10 μm at the oxygen K-edge (Kirz et al., 1995).

In Fig. 7.1 the linear absorption coefficients μ_l of water, of the mineral smectite, and of the organic substance phenol are shown as a function of the incident X-ray energy. Smectite is a clay mineral abundant in soils, and phenolic systems can be found in a large set of organic molecules in the environment. Both can be seen as representatives for inorganic or organic matter. The linear absorption coefficient μ_l of water is, compared to μ_l of organic or inorganic matter between the K-absorption edge of carbon and of oxygen, significantly lower. This absorption difference is the source for a natural amplitude contrast when imaging specimen with X-rays

---

J. Thieme (✉)
NSLS-II Project, Brookhaven National Laboratory, Upton, NY 11973, USA
e-mail: jthieme@bnl.gov

F.H. Frimmel, R. Niessner (eds.), *Nanoparticles in the Water Cycle*,
DOI 10.1007/978-3-642-10318-6_7, © Springer-Verlag Berlin Heidelberg 2010

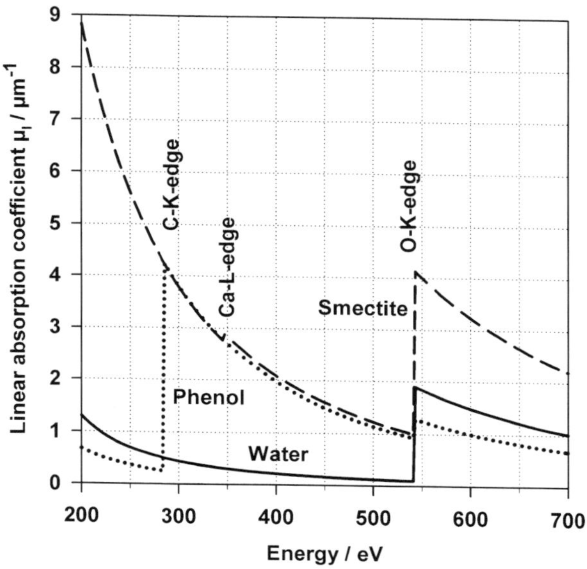

Fig. 7.1 Linear absorption cross section μ_l of a clay mineral (smectite), an organic molecule (phenol) and of water as a function of X-ray energy. The absorption edges of oxygen, calcium, and carbon are marked

within this energy range. Therefore, samples can be studied directly in aqueous media without fixation or staining.

This is true for the following reason, too: In the X-ray energy region of the water window the complex refractive index is, as everywhere else

$$n = 1 - \delta - i\beta$$

The values for δ and β are very small leading to a refractive index very close to unity (Henke et al., 1993). As a result scattered X-ray light will not be reflected from inner surfaces in inhomogeneous media (Pohl, 1967). Clear images can therefore be expected even when studying thick and heterogeneous specimen.

The spatial resolution achievable when looking with a microscope at a sample is directly related to the energy of the radiation. Increasing the energy implies increasing the resolution capabilities, i.e., the higher the energy the smaller the structures visible. Compared to visible light, the energy of the X-rays used for experiments in the water window is much higher. This means that with an X-ray microscope a much better resolution can be achieved than with a conventional microscope using visible light. At present, the smallest structures visible with an X-ray microscope are less than 15 nm in size (Chao et al., 2005).

A light source usable for spectromicroscopy experiments has to be highly brilliant and must supply radiation in a wide energy range as well. Up to now, only electron storage rings, i.e., large-scale facilities providing synchrotron radiation, are capable of meeting these needs. Therefore, all spectromicroscopy stations are set up at these light sources. Using a monochromator it is possible to reduce the bandwidth of the radiation to a very small value and thus to tune the energy of this radiation in very fine steps. This allows for mapping elements and for the determination of binding states within the sample under investigation. For obtaining a distribution map of a single element within a sample it is sufficient to take two images with X-ray energies above and below the absorption edge of this element. Dividing the spatially resolved absorption values gives rise to a map of the distribution of this element in the sample.

To examine binding states of the chemical elements within the sample it is possible to excite and detect near-edge resonances for NEXAFS- (= XANES-) studies (Stöhr, 1992). When the used X-ray energy is tuned across the absorption edge, resonances appear reflecting the chemical bonding state of the element. These resonances superimpose the step-like rise in absorption due to the element. Important absorption edges within the energy range of the water window are listed in Table 7.1. Knowing the intensity I_0 impinging on the sample and measuring the intensity I transmitted through the sample, spectra can be created showing the optical density $\mu_l d$ as a function of the X-ray energy. From these spectra the binding states within the sample can be derived as shown below.

In summary, X-ray microscopy is a tool capable of imaging particles in the colloidal size range directly in their aqueous environment with high spatial resolution and furthermore it is possible to combine this with high spectral resolution for spectromicroscopy studies.

Fresnel zone plates, i.e., circular diffraction gratings that work in X-radiation like thin lenses in visible light, are used as high resolution optical elements in X-ray microscopy, both in the transmission X-ray microscope and in the scanning transmission X-ray microscope (Schmahl and Rudolph, 1969).

The optical setup of a transmission X-ray microscope consists of a monochromator, a condenser, and a micro-zone plate. The monochromator reduces the bandwidth of the synchrotron radiation coming from the electron storage ring to at least $E/\Delta E$ of several hundreds. This reduction of the bandwidth is necessary as the micro-zone plate only acts as a high resolution optical element when supplied with quasi-monochromatic radiation. The condenser focuses the radiation on the object. The micro-zone plate is placed behind the object. It creates an enlarged image of the object in the image plane, which is recorded by a CCD camera with exposure times in the range of seconds. The Institute for X-ray Physics in the University

Table 7.1 K- and L-absorption edges within the energy range of the water window

Element	Absorption edge	Energy (eV)
O	K	543
N	K	410
Ca	L_2, L_3	350, 346
K	L_2, L_3	297, 295
C	K	284

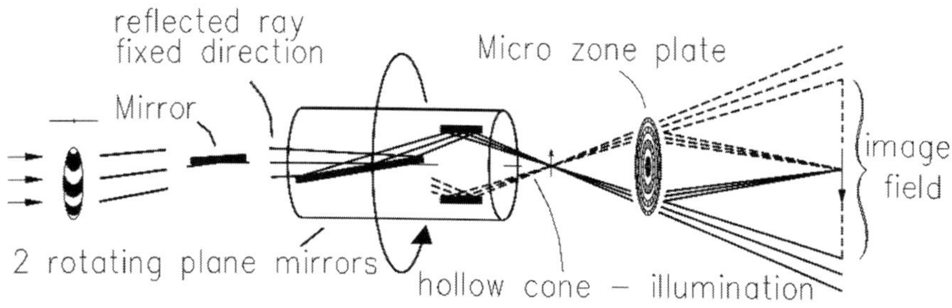

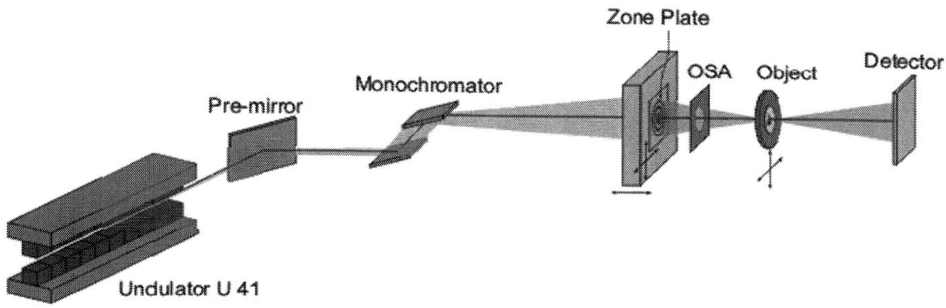

Fig. 7.2 Sketches of the optical setups of the transmission X-ray microscope (*top*) and the scanning transmission X-ray microscope (*bottom*) at the electron storage ring BESSY II in Berlin

of Göttingen has developed several generations of these X-ray microscopes. The X-ray images presented here have been taken with the X-ray microscope designed for and operated with the undulator U41 at the electron storage ring BESSY II in Berlin (Guttmann et al., 2003). A sketch of the optical setup is shown in the upper drawing of Fig. 7.2. The X-radiation coming from the undulator U-41 is reduced in bandwidth and focused onto the object by an off-axis transmission zone plate. A mirror set, consisting of one fixed and two rotating mirrors, provides for aperture matching with the micro-zone plate and for incoherent illumination of the sample. Using this rotating mirror condenser the bandwidth reduction of the radiation is $E/\Delta E = 2000$, leading to a very high clarity of the images (Niemann et al., 2003). A backside illuminated thinned CCD camera records the image generated by the micro-zone plate. Other transmission X-ray microscopes are operated as well at the Advanced Light Source, Berkeley, USA (Meyer-Ilse et al., 2000,Fischer et al., 2006), the storage ring Astrid in Arhus, Denmark (Medenwaldt and Uggerhøj, 1998), and in a dual microscopy setup (TwinMic) at ELETTRA, Trieste, Italy (Kaulich et al., 2003).

For a scanning transmission X-ray microscope the bandwidth of the polychromatic synchrotron radiation is reduced by a grating monochromator to $E/\Delta E$ of several thousands. The micro-zone plate focuses this radiation into a spot of diffraction limited size onto the object. A detector records the radiation transmitted through the object and the signal is stored in a computer. By scanning either the focus spot over the object or the object across this spot and recording the transmitted signal as a function of scan position, an image can be generated in the computer. Due to the scanning process, the total time needed for the generation of an image is longer than with a transmission X-ray microscope. Tuning the energy of the X-radiation with the monochromator and keeping the position of the light spot on the object the transmitted signal can be measured as a spectrum from this specific object point. By using a segmented detector it is possible to detect the signal not only in bright field mode but also in dark field and in differential phase contrast mode simultaneously. In the lower drawing of Fig. 7.2 the setup of the scanning transmission X-ray microscope at BESSY II is shown (Wiesemann et al., 2003). It is located side by side with the transmission X-ray microscope. The monochromator consists of a plane mirror and a plane grating with variable line density (Wiesemann et al., 2001). The latter provides for reduction of the bandwidth of the radiation and for a coherent illumination of the micro-zone plate. Thus, the zone plate creates the desired diffraction limited small spot. The zone plate, and so the spot, is moved by a piezo scanning device with an accuracy of a few nanometers across the object. To choose beforehand regions of interest, the object can be looked at using a visible light microscope, which allows as well for pre-focusing. Subsequently, the object can be scanned coarsely by using stepper motors for gaining a more detailed overview. A pn-CCD camera behind the object acts as a configured detector. Additional scanning transmission X-ray microscopes for the energy range discussed here are set up at the electron storage rings NSLS (Jacobsen et al., 1991) and ALS (Warwick et al., 1998, Ade et al., 2003) in the USA, the Canadian Light Source (Kaznatcheev et al., 2007), and the Swiss Light Source (Flechsig et al., 2006).

In general, transmission X-ray microscopes are used to take high resolution images from an object with exposure times of a few seconds or less, whereas with the slower scanning transmission X-ray microscope spectromicroscopy analysis can be performed. An intrinsic advantage of a scanning transmission X-ray microscope is that the micro-zone plate is located in front of the sample. Therefore, all radiation penetrating the sample is collected by the detector. Zone plates show diffraction efficiencies of around 10%, meaning that only this part of the radiation is used in a transmission X-ray microscope for image generation. This results in a much lower dose and thus lower radiation damage when studying a sample with a scanning transmission X-ray microscope.

7.2 Visualization of Nanoparticles

Dispersed clay and soil samples, and dispersions with zinc containing particles, all in aqueous media, have been imaged with the transmission X-ray microscope. The morphology of clusters built up by these particles is clearly imaged.

Two clay samples, Na^{+}-montmorillonite and nontronite (Fig. 7.3), and colloidal structures within a chernozem soil (Fig. 7.4) have been imaged. The images clearly show the expected morphology of clusters of colloids or nanoparticles. A micro-zone plate with $dr_n = 25$ nm has been used as high resolution X-ray

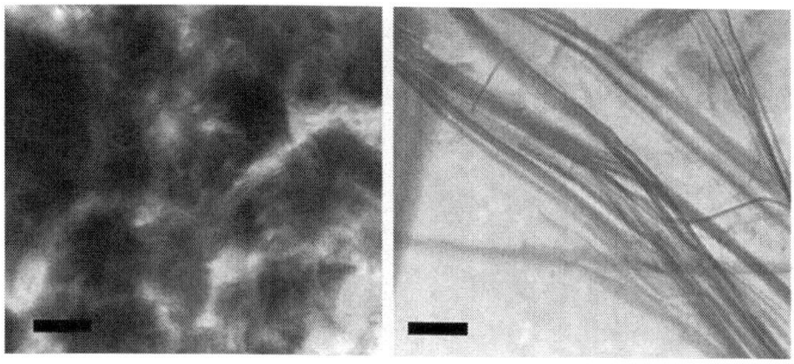

Fig. 7.3 Transmission X-ray microscopy images of two clay types, Na^+-montmorillonite (*left*) and nontronite (*right*), imaged in aqueous media with the X-ray microscope at BESSY II with a resolution of about 25 nm at an X-ray energy of $E = 520$ eV. The scale bar indicates 2 μm

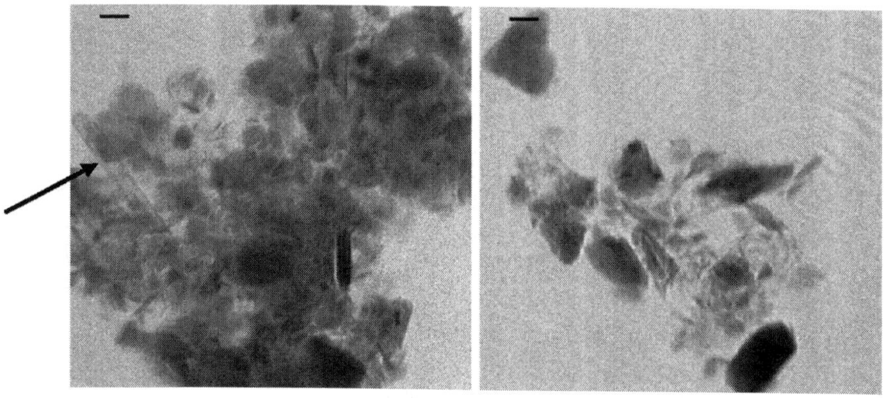

Fig. 7.4 Transmission X-ray microscopy images of colloids in an chernozem soil, imaged in aqueous media with the X-ray microscope at BESSY II with a resolution of about 25 nm at an X-ray energy of $E = 520$ eV. The scale bar indicates 1 μm

objective, resulting in a resolution in the images of about the same value. Thus, single particles within the clusters can easily be identified.

The clay samples have been chosen to demonstrate the variety in shapes of clay particles. Na^+-montmorillonite consists of small particles stacked together whereas nontronite particles are of much larger size.

Samples have been taken from the A_H horizon of an otherwise well-characterized chernozem soil from an area close to Göttingen (Ahl et al., 1985). This chernozem is an alkaline soil (pH = 8.2–8.4) with a high organic content (4.1%). The X-ray images show that in the colloidal size range clusters of particles feature many different appearances. The trained eye, however, quickly identifies in each cluster particles of similar shapes. In this size range all clusters mainly consist of clay particles making these similarities understandable. Microorganisms are very often attached to the particles. They can be found in the left image of Fig. 7.4 as rod-like shapes with a distinct membrane attached to other soil particles, indicated by the arrow.

X-ray microscopy images usually are two-dimensional projections, therefore it is not possible to clarify an exact spatial arrangement of colloidal structures. All particles are clearly visible, but the question remains whether they are attached to each other or whether there is a distance between them not visible in that particular projection. X-ray stereo microscopy can probe the proximity relation between particles with just two images (Gleber et al., 2003a). Tomography based on X-ray microscopy images will reveal the full three-dimensional morphology of the sample. For these experiments the transmission X-ray microscope of the Center for X-ray Optics at beamline 6.1.2 of the Advanced Light Source, Lawrence Berkeley Laboratory, Berkeley, is very well suited (Thieme et al., 2003). This can be seen in the following example. To obtain information about the morphological changes of clusters of humic substances as a result of a redox change, the three-dimensional structure has been imaged with cryo-tomography based on X-ray microscopy images. For this the samples were filled into capillaries with diameters around 8–10 μm, then shock-frozen and kept at liquid nitrogen temperatures to prevent changes in the sample due to radiation damage during recording the tomographic data set. The sample is a microbially reduced humic substance from Suwannee River, a well-characterized humic substance from the International Humic Substances Society. Simply exposing the test tube to air has reoxidized the reduced sample. In Fig. 7.5 two images of a set taken for tomographic reconstruction are shown revealing already the changes in the spatial arrangement of the humic substances. The resolution of the tomographic data set has been determined to be 45 nm. Using tomography it can be verified that humic substances, appearing as a sponge-like structure in a reduced state, change

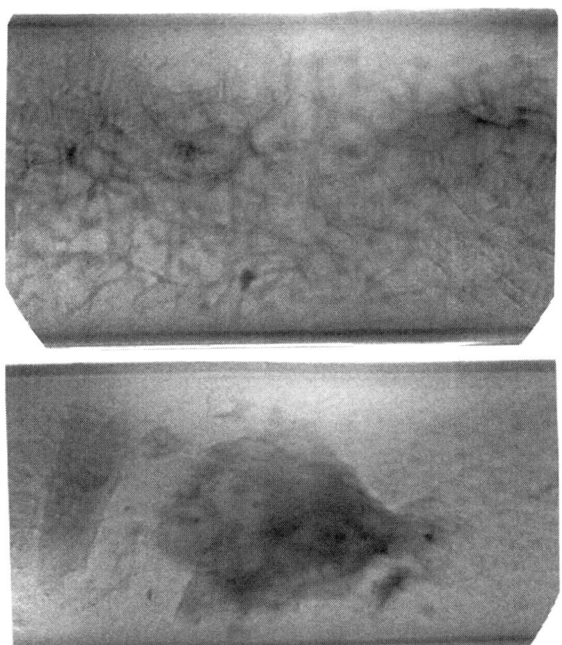

Fig. 7.5 X-ray micrographs of humic substances in a reduced (*top image*) and a reoxidized (*botttom image*) state. The samples were filled in capillaries with diameters of 8–10 μm and kept at cryogenic temperatures

dramatically their morphology to isolated clusters when reoxidized (Coates et al., 2000,Thieme et al., 2007).

Precipitation reactions, i.e., the formation of colloidal solids or nanoparticles from supersaturated solutions, are still far from being fully understood even after more than 100 years of research. Processes occurring at a molecular and nanoscale level seem to be much more complex than usually assumed in the past. It has become apparent that solid-phase conversions, aggregation processes, and recrystallization phenomena during precipitation have to be taken into account. A fundamental understanding of structure evolution in precipitation reactions is only possible if the processes can be followed from commencement of supersaturation on which the precipitation initiates. The ability of X-ray microscopy to study samples under ambient conditions with virtually no sample preparation allows for the study of these aqueous systems (Rieger et al., 2000). In addition, time-resolved studies can be carried out (Rieger et al., 2007).

Therefore, X-ray microscopy is a powerful tool for the investigation of precipitation reactions. As an example, the precipitation of Zn containing particles, mainly $Zn(OH)_2$, is shown in Fig. 7.6, where a $ZnCl_2$ solution and NaOH solution have

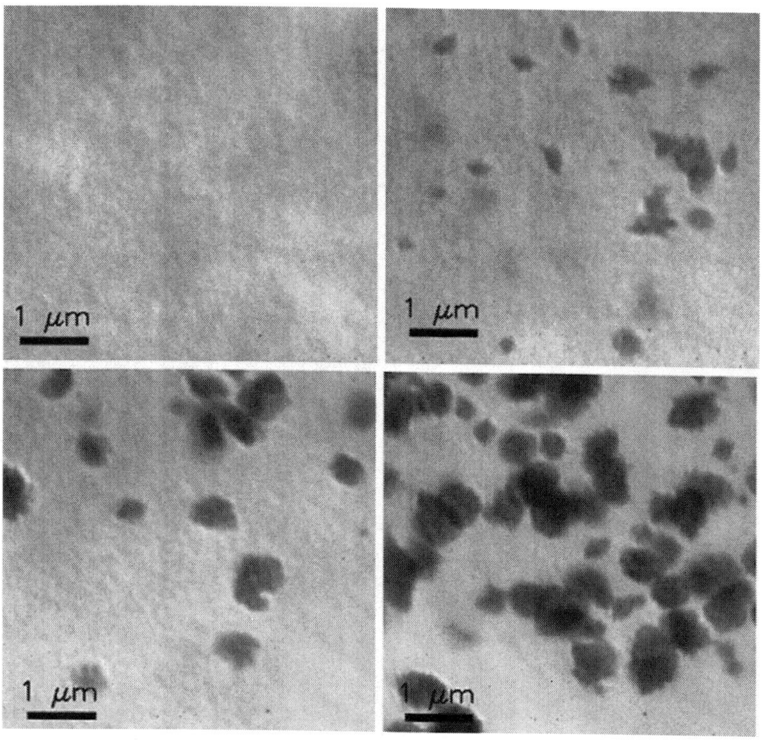

Fig. 7.6 Zinc containing particles in an aqueous dispersion imaged as a time sequence with the X-ray microscope at BESSY II with a resolution of approx. 25 nm at an X-ray energy of E = 520 eV with exposure times of a few seconds. The images span over 50 min after the first mixing of $ZnCl_2$ and NaOH solutions

been mixed. In the beginning, the dispersion did not show particles on the resolution scale of the X-ray microscope as can be seen in the upper left image of Fig. 7.6. After 30 min, particles a few hundred nanometers in size are visible on X-ray images, shown in the upper right of Fig. 7.6. After another 20 min the particles have become larger and more particles appear as can be seen in the lower images of Fig. 7.6.

7.3 Morphology and Chemistry

To combine morphological information of a sample with knowledge of the chemical composition within the sample, a very useful method is to take a set of images at closely spaced energies throughout the region of interest near to the absorption edge of the desired element (Jacobsen et al., 2000). This approach is visualized in Fig. 7.7, where some of a stack of images from a soil sample taken with the scanning transmission X-ray microscope at BESSY II are shown. Here it was measured around the K-absorption edge of carbon. The full set spans a three-dimensional coordinate system of x, y, and the energy. The energy ranges here from 280 to 295 eV with steps of 0.5 eV. Due to possible drifts or movements within the instrumentation during recording, the images have to be aligned to each other. Afterward, a spectrum can be obtained by selecting an object region and plotting the optical density of that region as a function of energy. Spectra can be obtained from discrete image pixels as well as from larger regions by integrating the optical density of all pixels within the chosen area. This experiment is more time consuming than simply taking a spectrum from a fixed position, but spatially and spectrally resolved data of an object are available at the same time (Mitrea et al., 2008). Thus, spectromicroscopy provides information about the chemistry within a sample in combination with high spatial resolution. The radiation dose applied to a single point within the sample is the same for point spectra as for stacks.

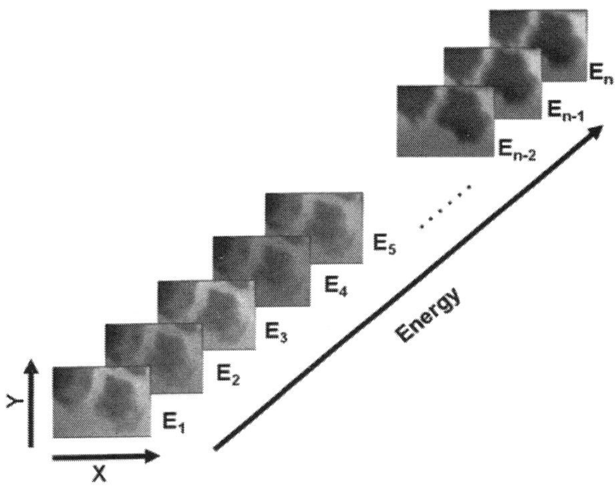

Fig. 7.7 A stack of images as a function of energy taken with the scanning transmission X-ray microscope at BESSY II, showing colloidal structures within a chernozem soil. The image size is $8 \times 6\ \mu m^2$ with 160×120 pixel^2 and the dwell time is 12 ms/pixel, the energy range for the full stack is 280–295 eV, step size $\Delta E = 0.5$ eV

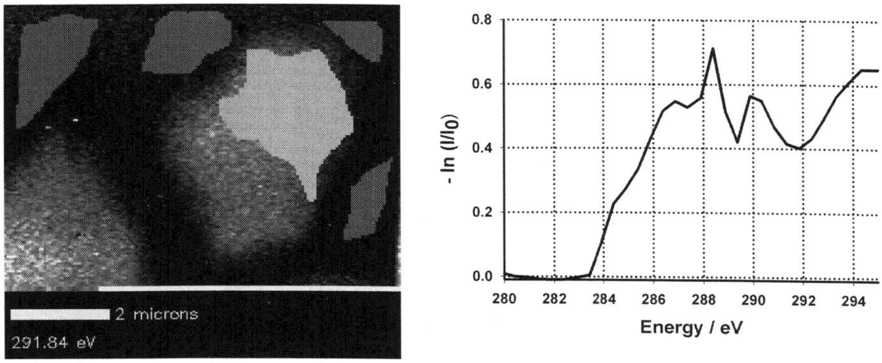

Fig. 7.8 Selected object region marked in *light gray* from the stack of images in the previous figure and the corresponding spectrum at the carbon K-absorption edge. The regions used for extracting I_0 are marked in *dark gray*. The stack was taken with the scanning transmission X-ray microscope at BESSY II, the image size is $8 \times 6\ \mu m^2$ with 160×120 pixel^2 and with a dwell time of 12 ms/pixel, the energy range for the full stack is 280–295 eV, with an energy step size of $\Delta E = 0.5$ eV

A sample of the chernozem soil mentioned above has been taken as a sample for demonstration. A small drop of a 1% (w/w) dispersion in deionized water has been placed on a 100-nm thick Si_3N_4 membrane and let dry. As a result, small clusters of soil colloids were found on the membrane.

A stack of images of this chernozem sample has been taken at energies around the K-absorption edge of carbon. Figure 7.8 shows on the left side one image of the whole stack, grayscale inverted. Four regions are marked in dark gray and one region marked in light gray. Integrating the transmission signal I_1 of the light gray region as a function of energy and using the regions marked in dark gray in the same way to obtain I_0 the spectrum below each image could be obtained as $-\ln(I_1/I_0)$. This yields the optical density. The derived spectrum in the right image clearly reveals resonance features of the content of organic carbon of the sample. To extract chemical information from the spectra, the program SpecFit has been developed (Gleber et al., 2003b). With this tool it is possible to assemble the spectrum using combinations of Gaussians and Lorentzians for the resonances and arctan functions for the corresponding edge jumps. In Fig. 7.9 the achieved result for the spectrum in Fig. 7.8 is shown. The aromatic, phenolic, and carboxylic content of the sample can be extracted from the spectrum, revealing the nature of the main organic components in that soil sample.

7.4 Summary

X-ray microscopy is a very suitable tool for imaging nanoparticles and colloidal structures in aqueous media. Two-dimensional imaging and three-dimensional tomographic reconstruction are very powerful methods for examining the morphology of clusters of nanoparticles and colloids. The analysis of samples from

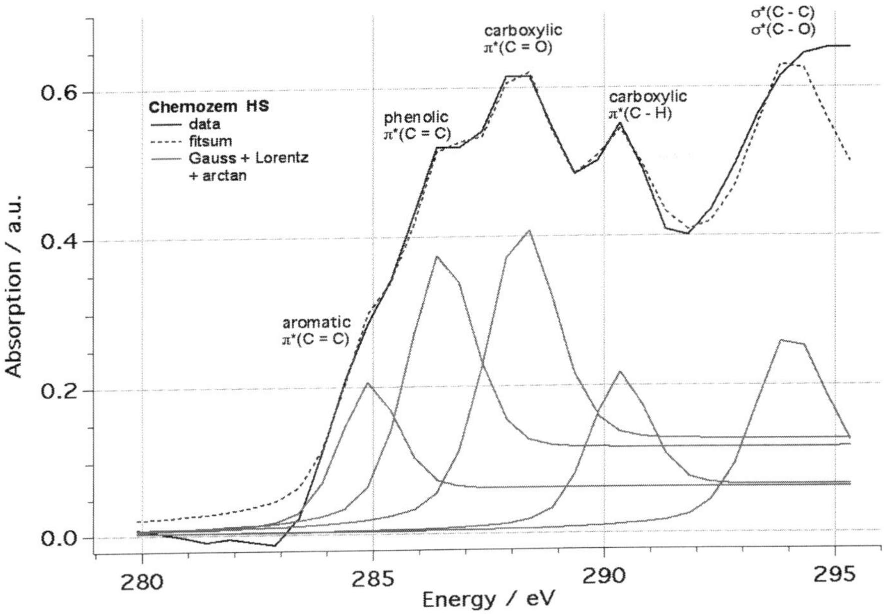

Fig. 7.9 XANES spectrum of the A_H horizon of a chernozem soil obtained from an image stack at the K-absorption edge of carbon (see Fig. 7.8). Functional groups are fitted as described in the text, indicating contributions from aromatic, phenolic, and carboxylic groups to the spectrum

environmental science, which are very heterogeneous physically as well as chemically, is a real challenge for X-ray spectromicroscopy but possible. Spatially resolved spectromicroscopy studies allow for the characterization of structure and chemistry of nanoparticles and colloidal structures in relation to other elements. The results presented here document the possibilities for studies not only comparing spectra obtained from bulk samples but looking in detail into the nanometer to micrometer size range within a sample.

Acknowledgments The work presented here has been supported by the Federal Minister of Education and Research (BMBF) under the contract numbers 05KS1MG1/9 and 02WU9893/0, by the Office of Naval Research under contract number N00014-99-10371, and by DFG under contract numbers Th 445/8-1 and Th 445/8-2 and within the SFB755 "Nanoscale Photonic Imaging." We would like to thank the staff of ALS and BESSY II for providing excellent working conditions when performing experiments.

References

1. Ade H, Kilcoyne A, Tyliszczak T, Hitchcock P, Anderson E, Harteneck B, Rightor E, Mitchell G, Hitchcock A, Warwick T (2003) Scanning transmission X-ray microscopy at a bending magnet beamline at the advanced light source. In Susini J, Joyeux D, Polack F (eds.) 7th International Conference on X-ray Microscopy, pp. 3–8, J. Phys. IV France, 104.

2. Ahl C, Frede H-G, Gäth S, Meyer B (1985) Böden aus Löß des Leinetalgrabens und seiner Hochflächen-Umrandung. Mitteilungen der Deutschen bodenkundlichen Gesellschaft 42: 359–434.
3. Attwood D (2000) Soft X-rays and Extreme Ultraviolet Radiation – Principles and Applications. Cambridge University Press, Cambridge.
4. Chao W, Harteneck B, Liddle J, Anderson E, Attwood D (2005) Soft X-ray microscopy at a spatial resolution better than 15 nm. Nature 435: 1210–1213.
5. Coates J, Chakraborty R, O'Connor S, Schmidt C, Thieme J (2000) The geochemical effects of microbial humic substances reduction. Acta Hydrochim Hydrobiol 28(7): 420–427.
6. Fischer P, Kim D-H, Chao W, Liddle J, Anderson E, Attwood D (2006) Soft X-ray microscopy of nanomagnetism. Materials Today 9(1–2): 26–33.
7. Flechsig U, Quitmann U, Raabe J, Böge M, Fink R, Ade H (2006) The PolLux microspectroscopy beamline at the swiss light source. In: Choi J-Y, Rah S (eds.) Proceedings of Ninth International Conference on Synchrotron Radiation Instrumentation, 2006 pp. 505–508, AIP Conference Proceedings 879.
8. Gleber S, Knöchel C, Thieme J, Rudolph D, Schmahl G (2003a) 3-D computer reconstruction of X-ray microscopy objects from stereo images. In: Susini J, Joyeux D, Polack F (eds.) 7th International Conference on X-ray Microscopy, pp. 639–642, J. Phys. IV France, 104.
9. Gleber G, Thieme J, Niemeyer J, Feser M (2003b) Interaction of organic substances with iron studied by 01s spectroscopy – Development of an analysis program. In: Susini J, Joyeux D, Polack F (eds.) 7th International Conference on X-ray Microscopy, pp. 429–432, J. Phys. IV France, 104.
10. Guttmann P, Niemann B, Rehbein S, Knöchel C, Rudolph D, Schmahl G (2003) The transmission X-ray microscope at BESSY II. In: Susini J, Joyeux D, Polack F (eds.) 7th International Conference on X-ray Microscopy, pp. 85–90, J. Phys. IV France, 104.
11. Henke B, Gullikson E, Davis J (1993) X-ray interactions: photoabsorption, scattering, transmission, and reflection at E = 50–30000 eV, Z = 1–92. Atomic Data and Nuclear Data Tables 54: 181–342.
12. Jacobsen C, Flynn G, Wirick S, Zimba C (2000) Soft x-ray spectroscopy from image sequences with sub-100 nm spatial resolution. J Microsc 197: 173–184.
13. Jacobsen C, Williams S, Anderson E, Browne M, Buckley C, Kern D, Kirz J, Rivers M, Zhang X (1991) Diffraction-limited imaging in a scanning transmission X-ray microscope. Optics Comm 86: 351–364.
14. Kaulich B, Bacescu D, Cocco D, Susini J, Salome M, Dhez O, David C, Weitkamp T, Di Fabrizio E, Cabrini S, Morrison G, Charalambous P, Thieme J, Wilhein T, Kovac J, Podnar M, Kiskinova M (2003) Twinmic: a European twin microscope station combining full-field imaging and scanning microscopy. In: Susini J, Joyeux D, Polack F (eds.) 7th International Conference on X-ray Microscopy, pp. 103–107, J. Phys. IV France, 104.
15. Kaznatcheev K, Karunakarana C, Lanke U, Urquhart S, Obst M, Hitchcock A (2007) Soft X-ray spectromicroscopy beamline at the CLS: Commissioning results. NIM A 582: 96–99.
16. Kirz J, Jacobsen C, Howells M (1995) Soft X-ray microscopes and their biological applications. Q Rev Biophys 28(1): 33–130.
17. Medenwaldt R, Uggerhøj E (1998) Description of an X-ray microscope with 30 nm resolution. Rev Sci Instr 69(8): 2974–2977.
18. Meyer-Ilse W, Denbeaux G, Johnson L, Bates W, Lucero A, Anderson E (2000) The high resolution X-ray microscope, XM-1. In: Meyer-Ilse W, Warwick T, Attwood D (eds.) X-ray Microscopy, Proceedings of the 6th International Conference on X-ray Microscopy, Berkeley, California, August 1999, pp. 129–134, AIP Conference Proceedings, American Institute of Physics, Melville, 507.
19. Mitrea G, Thieme J, Guttmann P, Heim S, Gleber S (2008) X-ray spectromicroscopy with the scanning transmission X-ray microscope at BESSY II. J Synchrotron Rad 15: 26–35.
20. Niemann B, Guttmann P, Rehbein S, Knöchel C (2003) Concept and realization of the novel rotating condenser-monochromator at the Göttingen TXM at BESSY II. In: Susini J, Joyeux

D, Polack F (eds.) 7th International Conference on X-ray Microscopy, pp. 273–276, J. Phys. IV France, 104.

21. Pohl R (1967) Einführung in die Physik, 3. Teil – Optik und Atomphysik, 180, 12th ed. Springer, Berlin.
22. Rieger J, Frechen T, Cox G, Heckmann G, Schmidt C, Thieme J (2007) Precursor structures in the crystallization/precipitation processes of $CaCO_3$ and control of particle formation by polyelectrolytes. Faraday Discuss 136: 265–277.
23. Rieger J, Thieme J, Schmidt C (2000) Study of precipitation reactions by x-ray microscopy: $CaCO_3$ precipitation and the effect of polycarboxylates. Langmuir 16: 8300–8305.
24. Schmahl G, Rudolph D (1969) Lichtstarke Zonenplatten als abbildende Systeme für weiche Röntgenstrahlung. Optik 29: 577–585.
25. Stöhr J: (1992) NEXAFS Spectroscopy. Springer Series in Surface Sciences. Springer-Verlag, Berlin.
26. Thieme J, McNulty I, Vogt S, Paterson D (2007) X-ray spectromicroscopy – a tool for environmental sciences. Env Sci Technol Oct 15: 6885–6889.
27. Thieme J, Schneider G, Knöchel C (2003) X-ray tomography of a microhabitat of bacteria and other soil colloids with sub-100 nm resolution. Micron 34: 339–344.
28. Warwick T, Franck K, Kortright J, Meigs G, Moronne M, Myneni S, Rotenberg E, Seal S, Steele W, Ade H, Garcia A, Cerasari S, Denlinger J, Hayakawa S, Hitchcock A, Tyliszczak T, Kikuma J, Rightor E, Shin H-J, Tonner B (1998) A scanning transmission x-ray microscope for materials science spectromicroscopy at the advanced light source. Rev Sci Instr 69(8): 2964–2973.
29. Wiesemann U, Thieme J, Guttmann P, Früke R, Niemann B, Rudolph D, Schmahl G (2001) Construction of a scanning transmission X-ray microscope at the undulator U-41 at BESSY II. NIM A 467–568: 430–434.
30. Wiesemann U, Thieme J, Guttmann P, Früke R, Rehbein S, Niemann B, Rudolph D, Schmahl G (2003) First results of the new scanning transmission X-ray microscope at BESSY-II. In: Susini J, Joyeux D, Polack F (eds.) 7th International Conference on X-ray Microscopy, pp. 95–98, J. Phys. IV France, 104.
31. Wolter H (1952) Spiegelsysteme streifenden Einfalls als abbildende Optiken für Röntgenstrahlen. Ann Phys 6(10): 94–114.

Chapter 8
In Situ Measurements on Suspended Nanoparticles with Visible Laser Light, Infrared Light, and X-Rays

Harald Zänker

8.1 Introduction

The presence of engineered nanoparticles (ENPs) in the water cycle is a subject of discussion among toxicologists, producers of nanomaterials, environmentalists, and politicians. Currently, it is more a vision (or an apprehension) to become exposed to such particles by drinking tap water, having a wash, or going for a swim because the release of ENPs to the environment is still marginal. But this may change. The exposure of men, animals, and plants to ENPs might become a major environmental challenge of the future. Nevertheless, at this stage the influence of ENPs on the environment is still minimal and there is also hardly any experience with measuring such artificial nanoparticles within the complex matrices of environmental samples.

But there is experience with measuring natural nanoparticles in environmental waters. Over decades, naturally occurring nanoparticles (colloids) such as iron-containing, silicate-containing, or organic particles have been studied with numerous methods (cf. Buffle and van Leeuwen, 1992; 1993; Frimmel et al., 2007; Wilkinson and Lead, 2007). Because of the lack of knowledge concerning the behavior and the lack of experience concerning the measurement of artificial nanoparticles under environmental conditions, this chapter will (with few exceptions) focus on measurements referring to natural nanoparticles. It is expected that future measurements on artificial nanoparticles will benefit a lot from the existing measuring experience on natural nanoparticles.

What are the measurands that are of interest (for both ENPs and natural nanoparticles)? It is obvious that information on the absence or presence of a particle under discussion is a primary goal. Thus, it is important to recognize if matter in the colloidal form is present, and if so, what its concentration is. But this is not enough if predictions on the toxicity, the transport behavior in water bodies, the transport behavior in biological systems, the ability to carry environmental pollutants, the

---

H. Zänker (✉)
Institute of Radiochemistry, Forschungszentrum Dresden-Rossendorf, D-01314 Dresden, Germany
e-mail: h.zaenker@fzd.de

F.H. Frimmel, R. Niessner (eds.), *Nanoparticles in the Water Cycle*,
DOI 10.1007/978-3-642-10318-6_8,

aggregation behavior, the aging behavior, etc., are needed. For such complex predictions the knowledge of further particle properties is inevitable. These are particle size, size distribution, shape, surface area, chemical composition, crystal structure, surface functionality, affinity to environmental contaminants, surface charge, and others.

8.2 Strategies of Particle Characterization

8.2.1 Invasive Methods

Any method of nanoparticle characterization, no matter which of the above-mentioned "measurands" are to be determined with it, as well as any method of nanoparticle separation, can be assigned to one of two main categories: invasive techniques and "non-invasive" techniques. Filtration/ultrafiltration (Buffle et al., 1992; Lead et al., 1997), scanning electron microscopy (Buffle and Leppard, 1995), energy dispersive X-ray analysis (Kaplan et al., 1994), transmission electron microscopy (Nomizu et al., 1988), atomic force microscopy in air (Namjesnik-Dejanovic and Maurice, 1997), X-ray photoelectron spectroscopy (Ding et al., 2000), secondary ion mass spectrometry (Novikov et al., 2006), centrifugation/ultracentrifugation (Jones and Bryan, 1998; Zänker et al., 2006), size exclusion chromatography (von Wandruszka et al., 1999), hydrodynamic chromatography, split-flow-lateral-transport-thin-cell-fractionation (George et al., 2000), or field-flow fractionation (Klein and Niessner, 1998; Lyven et al., 2003; Prestel et al., 2006; von der Kammer et al., 2005) belong to the invasive techniques. They have been applied extensively on natural environmental nanoparticles. It is easily conceivable that invasive methods have their disadvantages (cf. Johnston et al., 1993) since they are more plagued by artifacts than non-invasive ones. Nevertheless, for certain measuring problems their application is justified; in order to get certain information, invasive techniques cannot be avoided. Moreover, note that the "degree of invasivity" differs greatly between the methods mentioned. Scanning electron microscopy, for instance, is a very invasive method. It requires separation of the particles from the solution (filtration), drying, possibly coating with gold or carbon, possibly application of vacuum. Conversely, ultracentrifugation and, even more so, field-flow fractionation are only very moderately invasive (flow field-flow fractionation only causes dilution of the sample which normally does not have significant influence on the nanoparticles) making it hard to assign these latter methods to one of the two categories.

8.2.2 Non-invasive Methods

There is no in situ or non-invasive technique in the strictest sense of the term. Measuring the properties of an object of matter such as a colloidal particle is always getting in interaction with this object. However, as already mentioned, there are ways to minimize this interaction; it is understood in this chapter that "non-invasive" methods are in reality always little-invasive methods or "minimum invasive methods." There are also ways to corroborate results by using methods of particle characterization that are based on different physical principles and have, therefore, different types of interactions with the measured object (i.e., they are associated with different types of artifacts).

The so-called non-invasive techniques of particle characterization very often use photons such as the photons of visible laser light, infrared light, or X-rays as probes. They are able to work in the presence of the original water and, thus, their influence on the colloidal system is little to none. Such photon-based methods are

- *photon correlation spectroscopy* (Filella et al., 1997; Schurtenberger and Newman, 1993) and *laser-induced breakdown detection* (Kim and Walther, 2007) for particle size determination,
- *laser Doppler velocimetry* (Fourest et al., 1994; Riley, 2005) for determining electrophoretic mobility and zeta potential,
- *time-resolved laser fluorescence spectroscopy* (Geipel et al., 1996), *attenuated total reflection Fourier transform infrared spectroscopy* (Johnston et al., 1993; Müller et al., 2009), and *X-ray absorption spectroscopy (XAS) with synchrotron radiation* (Charlet and Manceau, 1993; Gaillard, 2007) for identifying the binding states of contaminants onto nanoparticles.

Examples of using these six methods for the characterization of environmentally relevant nanoparticles are given in Sect. 8.3. Other minimum invasive methods of determining particle size are static light scattering (Schurtenberger and Newman, 1993) and single particle counting (Knollenberg and Veal, 1992; Walther et al., 2006). A new very gentle method for the characterization and separation of particles of a size from several hundred nanometers is photophoretic velocimetry in a flowing solvent (Helmbrecht et al., 2007). It allows properties such as the size or the refractive index of the single particles to be analyzed. Additional minimum invasive methods applicable for identifying binding states of adsorbants onto colloids are nuclear magnetic resonance (NMR) spectroscopy (Johnston et al., 1993), electron spin resonance (ESR) spectroscopy (Boughriet et al., 1995), Raman spectroscopy (Hedderman et al., 2006), and coherent anti-Stokes Raman scattering (CARS) spectroscopy (Druet and Teran, 1981). Apart from these "bulk methods," little-invasive techniques of microscopy have evolved that image the shape of nanoparticles such as liquid-cell atomic force microscopy (Plaschke et al., 2002) or that give "chemical images" such as X-ray spectromicroscopy (Thieme et al., 2007), laser-scanning CARS microscopy (Cheng et al., 2002), surface-enhanced Raman scattering (SERS) microscopy (Alvarez-Puebla et al., 2004), or tip-enhanced Raman spectroscopic (TERS) microscopy (Schmid et al., 2008).

8.2.3 Invasive vs. Non-invasive

Many colloidal systems are in fragile equilibrium. In particular, hydrophobic colloids are thermodynamically unstable and exist thanks to electrostatic or steric stabilization only. They tend to reduce their specific surface area by aggregation (and sedimentation), i.e., the slightest perturbations from outside may change them. In general, methods of sampling, sample preparation, and measurement which exert as little influence on them as possible should therefore be preferred to others. Such "non-invasive" methods are in the focus of this chapter. However, non-invasive methods have also disadvantages over invasive ones. These are primarily

- the measuring results of non-invasive methods are, as a matter of fact, more representative of the real colloidal system; but they are often also less unambiguous,
- non-invasive methods do normally not fractionate the nanoparticles which prevents subsequent experiments on individual particle fractions being performed,
- less invasive methods are often more cumbersome than more invasive ones (e.g., liquid-cell atomic force microscopy vs. atomic force microscopy in air).

The best choice between invasive and non-invasive methods of nanoparticle characterization normally depends on the desired information. Some of the measuring values are very sensitive to solution conditions. This refers, for instance, to particle size. This also refers to measuring values characterizing the binding form of contaminants onto nanoparticles. Such measuring values should be determined in situ, i.e., with a non-invasive method. On the other hand, the chemical composition of the particles is often rather robust to influences from the liquid phase and can in such cases be determined by filtration through filters of pore size as small as possible and

- either determining the difference of the element concentrations before and after filtration (e.g., by ICP-MS)
- or analyzing the filter cake after washing and digesting (e.g., by ICP-MS)

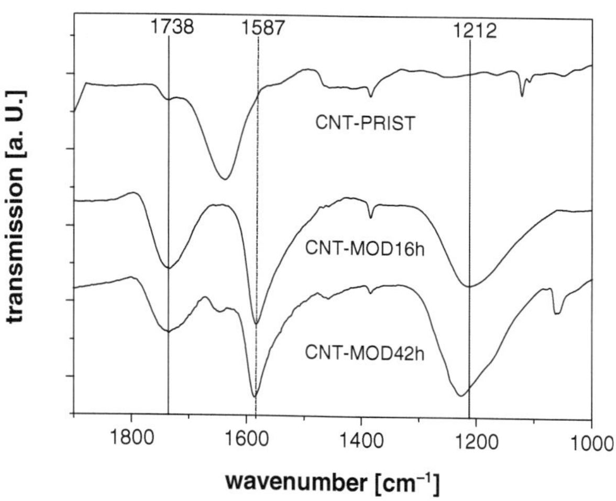

Fig. 8.1 FTIR spectra (KBr pellet) of pristine (CNT-PRIST) and modified CNTs (CNT-MOD16h and CNT-MOD42h) (from Schierz and Zänker, 2009; with permission)

which is a fairly invasive way of determination. Another example of an analysis relatively insensitive to undesired influences and undefined falsifications which can well be carried out with an invasive technique may be the detection of the functional groups on the surface of nanoparticles. In an oxidation experiment carboxyl groups were produced on carbon nanotubes (CNTs) by heating them in a mixture of concentrated HNO_3 and H_2SO_4 under reflux for 3 h at 100°C and subsequent stirring under sonication for two different periods of time (16 and 42 h) at 70°C (Schierz and Zänker, 2009). The modified CNTs were removed, washed, dried, and mixed with KBr. Pellets were pressed of the mixture for Fourier transform infrared spectroscopy (FTIR). Figure 8.1 shows that the IR absorption bands at 1738 cm^{-1} (stretching vibration of C=O) and at 1212 cm^{-1} (stretching vibration of C–O) are missing in the pristine CNTs, but appear in the modified CNTs. These two absorption bands together represent the carboxyl groups, i.e., the carboxyl groups are reliably detected also with this invasive technique.

8.3 In situ Techniques

8.3.1 General

It would be beyond the scope of this chapter to give a comprehensive overview of particle characterization by non-invasive methods. The intention is to give several examples of the application of such methods, to show the general strategies, to indicate typical problems and difficulties of their application, and to demonstrate how such difficulties can be overcome.

8.3.2 Visible Laser Light

Particle size of very small nanoparticles at high concentration by photon correlation spectroscopy (PCS). PCS is based on the analysis of the time-dependent fluctuations of the light scattered by particles when illuminated by a laser beam. Since the fluctuations are due to the particle movement, they contain information on the particles' diffusion coefficient(s). The primary information obtained by a PCS measurement is the autocorrelation function $C(\tau)$ of the scattered light intensity fluctuations which is defined by (cf. Ford, 1985; Weiner, 1984)

$$C(\tau) = \lim_{N\to\infty} \frac{1}{N} \sum_{j=1}^{N} n_j n_{j-m}, \quad m = 1,2,3,...M \tag{8.1}$$

$C(\tau)$ = autocorrelation function
n = number of pulses registered during the sampling time $\Delta t = t - \tau$,
with t = time, $t = j\Delta t$, τ = delay time, $\tau = m\Delta t$, Δt = time increment into which the time axis has been divided (sampling time),

N = total number of samples $T/\Delta t$,
with T = measuring time,
M = number of correlator channels,
j, m = running terms.

More information on the principles of PCS can be found in articles by Ford (1985), Tscharnuter (2000), or Weiner (1984).

Figure 8.2 shows a puddle of acid rock drainage (ARD) solution in an abandoned Zn–Pb–Ag mine at Freiberg, Germany. Such highly mineralized, red-colored ARD solutions are formed in ore or coal mines due the sulfide oxidation process. The figure also shows precipitates of ochre which usually accompany ARD solutions in mines. The solutions can be very acidic (see, for instance, Nordstrom et al., 2000) and there is general lack of knowledge as to the chemical speciation of their constituents. In particular the colloid chemistry of such waters has scarcely been studied.

ARD from the puddle depicted in Fig. 8.2 was investigated by PCS (BI-go, Brookhaven Instr., USA) using a laser of a wavelength λ of 514.5 nm (Zänker et al., 2002). The samples were of reddish color but visibly clear. The pH was 2.7, the sulfate concentration was 411 mmol/L. Table 8.1 gives the chemical analysis. The goal was to find out if nanoparticles play a role in the chemistry of this solution.

Parallel filtration of aliquots of the ARD solution with Nuclepore filters of decreasing pore size (i.e., of a series of subsamples of the original solution) and centrifugation of aliquots (subsamples of the original solution again) with increasing centrifugal force were performed. Figure 8.3 shows the autocorrelation functions of the raw sample and the filtered sample fractions. The shape of the autocorrelation functions is controlled by the time-dependent relaxation of particle concentration fluctuations. In our case, these concentration fluctuations are primarily caused by the

Fig. 8.2 Acid rock drainage solution in the "Himmelfahrt Fundgrube" mine at Freiberg, Germany (length of the puddle: ca. 1 m)

Table 8.1 Chemical data of the ARD solution of pH 2.7 (for additional data cf. Zänker et al., 2002)

Component	Concentration/mmol L^{-1}
Fe	94.3
Zn	82.2
Mg	93.7
Al	54.4
As	6.73
Pb	0.09
Cd	0.26
Sulfate	411

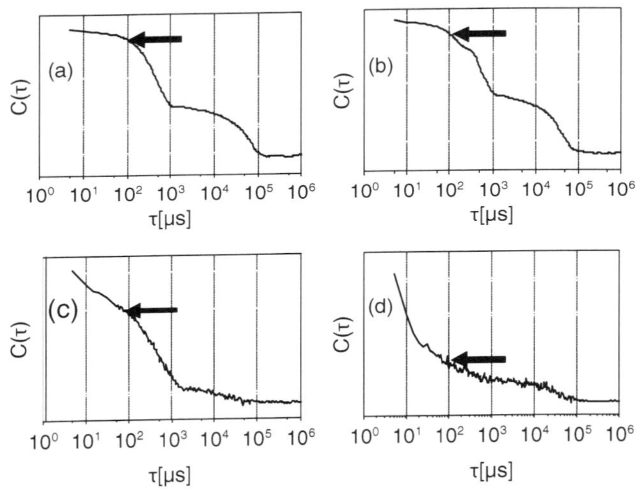

Fig. 8.3 Autocorrelation functions of the light scattered from the ARD sample: (**a**) raw sample, (**b**) 5-μm filtrate, (**c**) 400-nm filtrate, and (**d**) 50-nm filtrate (from Zänker et al., 2002; with permission)

translational diffusion of rigid particles, i.e., the autocorrelation functions provide the distribution of the diffusion coefficients and the particle size (equivalent spherical diameter) distributions can be calculated with the aid of the Stokes–Einstein equation. Monodisperse spherical particles of 10 nm, for instance, reach practically full relaxation (decay of the autocorrelation function to the baseline) after about 10^2 μs. The autocorrelation function of a polydisperse real sample is the sum of different decaying contributions, normally the sum of exponentials (Grabowski and Morrison, 1983; Stock and Ray, 1985; Schurtenberger and Newman, 1993). As one can see, the first three autocorrelation functions in Fig. 8.3 consist primarily of components of relaxation times larger than 10^2 μs (the point in time of 10^2 μs is highlighted by an arrow). However, the autocorrelation function in Fig. 8.3d (50-nm filtrate) is different from the others. It shows its major decay already before 10^2 μs. These features of the scattered light fluctuations are also reflected in the CONTIN deconvolutions of the autocorrelation functions (Provencher, 1982), i.e., the particle size distributions (Fig. 8.4). The figure shows that the raw sample and the

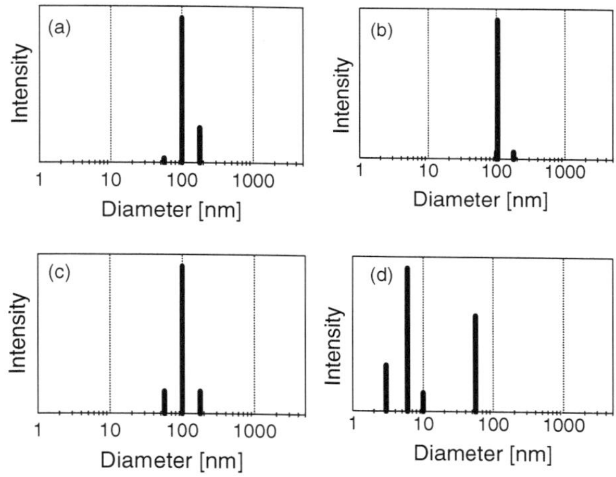

Fig. 8.4 Light-intensity-weighted particle size distribution in the ARD sample according to the CONTIN deconvolution of the autocorrelation functions. (**a**) Raw sample, (**b**) 5-μm filtrate, (**c**) 400-nm filtrate, and (**d**) 50-nm filtrate. Removing the larger submicron particles results in the appearance of the weakly scattering ultrafine particles (<10 nm) (from Zänker et al., 2002; with permission)

first two filtrates are dominated by particles of about 100 nm (this peak corresponds to the hump between 10^3 and 10^5 μs in the autocorrelation function) whereas most of the larger submicron particles have been removed by the 50-nm filtration which makes ultrafine particles of <10 nm detectable by PCS (the detection of extremely small colloid particles can be prevented in the presence of larger submicron particles due to optical masking which results from the d^6 dependence of the scattered light intensity on the particle diameter d in the particle size range of Rayleigh scattering, i.e., for particles of a size of $d \leq \lambda/10$). Similar results were obtained by the centrifugation experiments. In this case the ultrafine particles appeared in the PCS spectra after a 1-h centrifugation at $\geq 15{,}000\times g$.

ICP-MS/AAS on the retentates of the 50-nm filters revealed that the strongly scattering 100-nm particles are a trace component of only about 20 mg/L consisting primarily of Fe and As compounds. On the other hand, ultrafiltration with 3-kD filters showed that at least 680 mg/L Fe, 230 mg/L As, and 20 mg/L Pb occurred in the form of the ultrafine nanoparticles of <10 nm in this ARD solution which means that at least 15% of the Fe, 50% of the As, and 80% of the Pb were colloidal heavy metal/metalloid.

Particle size of nanoparticles at very low concentration by laser-induced breakdown detection (LIBD). A method of a lower particle concentration detection limit is LIBD. Figure 8.5 shows the LIBD setup according to Opel et al. (2007). A pulsed Nd:YAG laser is used as the light source; the laser beam reaches the cuvette via beam adjustment and beam diagnostics units and a lens system for focusing (laser wavelengths: 532 nm). The laser pulse energy is adjusted in a way that

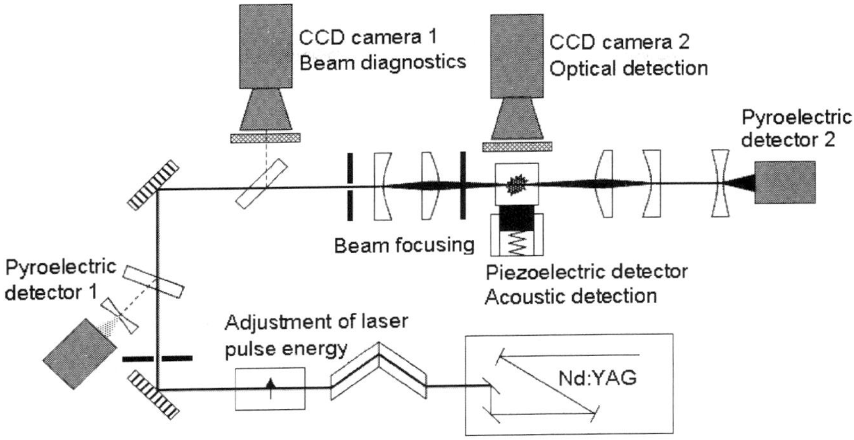

Fig. 8.5 Setup of LIBD (from Opel et al., 2007; with permission)

- no breakdown events occur in the liquid and
- breakdown events occur on solids.

Thus the laser pulses may cause a breakdown event as soon as a particle enters the laser focus. The events are detected acoustically by a piezoelectric detector and optically by a CCD camera. From the breakdown probability (the number of breakdowns per number of laser pulses) information on the particle concentration can be derived. The so-called ignition length is obtained by the optical detection and denotes the length of the focal area of the laser beam in propagation direction where colloids of certain size cause breakdown events. The ignition length is proportional to the average particle size (Bundschuh et al., 2001; Scherbaum et al., 1996); this dependence can be quantified by a calibration of the system with reference particles. Here, suspensions of polystyrene latex particles in ultrapure water were used for calibrating the apparatus. Note that LIBD is less troubled by masking of small particles by larger ones than PCS. The intensity of scattered light, which is used in PCS, is proportional to d^6 for particles of the size range of Rayleigh scattering whereas it is proportional to roughly d^2 for particle diameters larger than about $d = \lambda/6$. This makes PCS relatively insensitive to small particles in the presence of larger particles which scatter much more light than the small particles. LIBD, on the other hand, shows a dependency of about d^2 also for very small particles (Plaschke et al., 2001) making it more sensitive for the detection of such small particles and less susceptible to masking problems by larger ones. Nevertheless, a d^2 dependence is still considerable and a spatiotemporal separation of the particles before detection is often desirable also for LIBD.

The LIBD apparatus depicted in Fig. 8.5 was applied to determine the particle concentrations and sizes of several commercial mineral waters (Opel et al., 2006). Mineral waters are very poor in particles since they are de-ironed which also

Table 8.2 Mean diameter d_P and number concentration c_P of particles in four mineral waters determined by LIBD (p_{BD}: breakdown probability, z_F: focal length)

Water	p_{BD}	$z_F/\mu m$	d_P/nm	$c_P/N\,L^{-1}$
Hunyadi Janos	0.181	1438	20	3.7×10^9
Margon medium	0.381	1540	30	1.6×10^{10}
Margon still	0.133	1514	25	2.1×10^9
San Pellegrino	0.052	1459	20	9.2×10^9

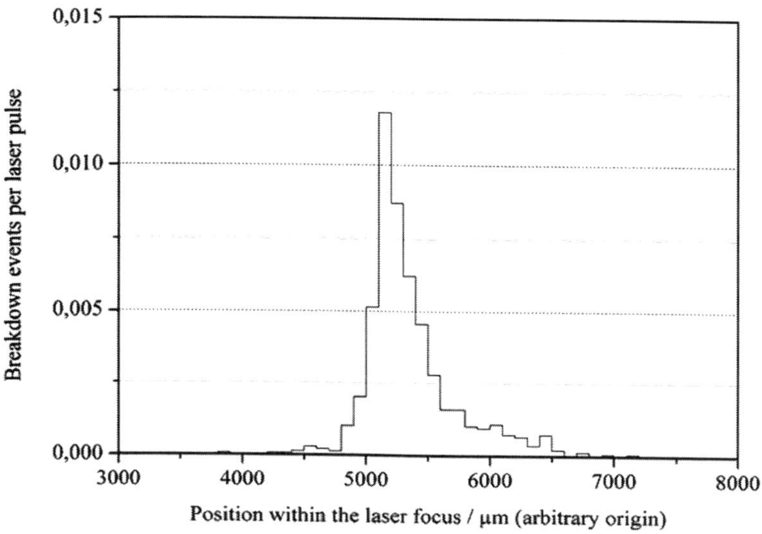

Fig. 8.6 Frequency distribution of breakdown events in the focal area of the laser beam for San Pellegrino mineral water (from Opel et al., 2006)

removes most of the naturally occurring particles. The breakdown probabilities and ignition lengths for the investigated mineral waters are given in Table 8.2. Table 8.2 also gives the resulting particle sizes and particle concentrations. Note that PCS as a classical characterization method for colloidal solutions was not able to determine the particle size in any of the mineral water samples. Figure 8.6 exhibits the longitudinal distribution of the breakdown events within the focal area of the laser beam for the San Pellegrino mineral water from which the mean ignition length and the mean particle diameter are derived. Monomodal particle size distributions would yield one nearly Gaussian peak. The distribution in Fig. 8.6 shows a significant deviation from the Gaussian shape. This suggests that there is a more complicated particle size distribution such as a bimodal distribution in the mineral water.

A method of determining particle size distributions of aquatic colloids by LIBD was suggested by Walther et al. (2004). It is intended to be used for bulk samples without any sample preparation. Another way of determining particle size distributions by LIBD is the above-mentioned detection after spatiotemporal particle separation. Such a separation is ideally done by field-flow fractionation (cf. Thang et al. 2000). Field-flow fractionation overcomes the limits of LIBD in particle sizing resolution whereas the high sensitivity of LIBD counteracts the detection problems often caused by field-flow fractionation due to sample dilution.

Zeta potential of carbon nanotubes (CNTs) by laser Doppler velocimetry. The zeta potential of a colloidal particle is the potential at the shear plane between the ions which are "stuck" to the particle when it moves through the solution and the ions which are stripped off from the particle due to particle motion. The zeta potential can be measured by electrokinetic methods such as laser Doppler velocimetry (cf. Riley, 2005) which is similar to PCS. There are various technical versions of laser Doppler velocimetry, however, the basic principle is always the same. The particles are exposed to an electrical field causing them to move between the electrodes. By splitting and recombining an illuminating laser beam, interference fringes can be produced in the cell. As particles move through the fringes under the influence of the electric field, intensity fluctuations are produced whose frequencies are directly related to the velocity of the particles. The intensity fluctuations are detected by a detector such as a photomultiplier and analyzed in a similar way as in PCS. Consequently, the primary information obtained is the electrophoretic mobility of the particles. The zeta potential is derived from the electrophoretic mobility using the Smoluchowski equation or the Hückel equation. The information obtained is information on the effective particle charge in an electric field. There may be contradictions to charge information from titration experiments. For instance, the isoelectric point (laser Doppler velocimetry) may deviate from the point of zero charge (titration) which is due to possible shielding of charged surface sites that are accessible to titrants but do not influence the electrostatic properties of the particle in the electric field. Limitations of the method arise from problems such as too strong electrode polarization or too strong Joule heating. Such problems are encountered with samples of high electrical conductivity (high electrolyte concentration), high viscosity, or low particle charge (low zeta potential). Particle sedimentation can also be a problem.

In Fig. 8.1 the modification of CNTs by surface oxidation (generation of carboxyl groups) is demonstrated. It is obvious that such surface modifications also influence the surface charge and, thus, the zeta potential of nanoparticles. Figure 8.7 shows the zeta potential for the above-discussed pristine and modified CNTs depending on pH as determined by laser Doppler velocimetry (Schierz and Zänker, 2009). A significant shift of the zeta potential toward more negative values caused by the formation of the deprotonable COOH groups is to be seen.

Binding states of uranium(VI) onto lipopolysaccharide macromolecules by time-resolved laser fluorescence spectroscopy (TRLFS). There are many interactions between nanoparticles and environmental contaminants such as hydrophobic organic toxicants or heavy metals in natural waters. The reaction of the uranyl cation, UO_2^{2+}, with lipopolysaccharide (LPS) from *Pseudomonas aeruginosa* (commercially available as a powder prepared by trichloroacetic acid extraction) was studied with the aid of both potentiometric titration and TRLFS (Barkleit et al., 2008). The TRLFS apparatus applied is described in an article by Geipel et al.

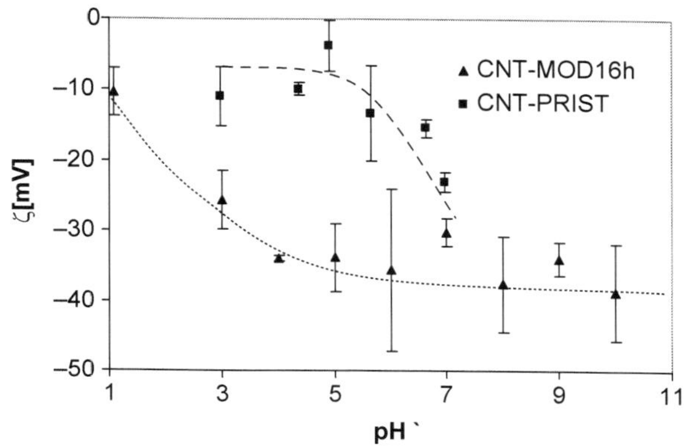

Fig. 8.7 Zeta potential of pristine and modified CNTs depending on pH ($I = 10$ mM NaCl, error bars: $\pm 2\sigma$) (from Schierz and Zänker, 2009; with permission)

(1996). A Nd:YAG laser system with a fast pulse generator and a digital delay generator was used. The excitation wavelength of the uranyl fluorescence was 266 nm; the pulse energy was 0.2–0.5 mJ. The time-resolved fluorescence emission from the bulk of the sample was detected using a detector interface and a spectrograph. For Gram-negative bacteria, the LPS unit in the outer membrane provides the major binding sites for metal ions. These potential binding sites are carboxyl, phosphoryl, hydroxyl, and/or amino groups. The objective was to identify the major binding sites for the uranium on the LPS macromolecules and to assess the relative stability of the different surface complexes formed.

The LPS under study has a molecular weight of 50–100 kD, i.e., the LPS molecules are much bigger than the uranyl ions or inorganic uranyl complexes. Because of this difference the complexation of uranyl with LPS can be regarded as a surface complexation reaction between a nanoparticle (the macromolecule) and the uranyl. The uranyl adopts the transport behavior of the LPS macromolecules in an aquatic system and will no longer behave as an individual small metal ion (or, in a "real-world" system, it even adopts the transport behavior of a bacterium since the LPS is part of a bacterial membrane if it is not separated from it).

Figure 8.8 gives a summary of the measured fluorescence (photoluminescence) spectra of the UO_2^{2+}-LPS system at different pHs. For comparison, the uranyl spectra at pH 2.5 (100% UO_2^{2+}(aq)) and pH 6.0 (main species $(UO_2)_3(OH)_5^+$) are included. A strong increase in fluorescence intensity, connected with a red shift of the peak maxima of about 8 nm at pH 2.5 and up to 11 nm at neutral pH, compared to the free uranyl ion UO_2^{2+}(aq), was observed. At very low LPS concentrations (≤ 0.05 g/L) uranyl hydroxide species appear at pH 7, but at higher LPS concentration uranyl hydroxide formation was not observed until pH 8. Thus, the experiment takes advantage of the high stability of the UO_2^{2+}-LPS complex species. Due to this high stability of the complex the presence of free UO_2^{2+}(aq) or dissolved uranyl hydroxid species was negligible for most of the samples, and for the few low-LPS samples the disturbance by such species could be quantified from the different fluorescence properties of the inorganic uranyl species and be taken into consideration.

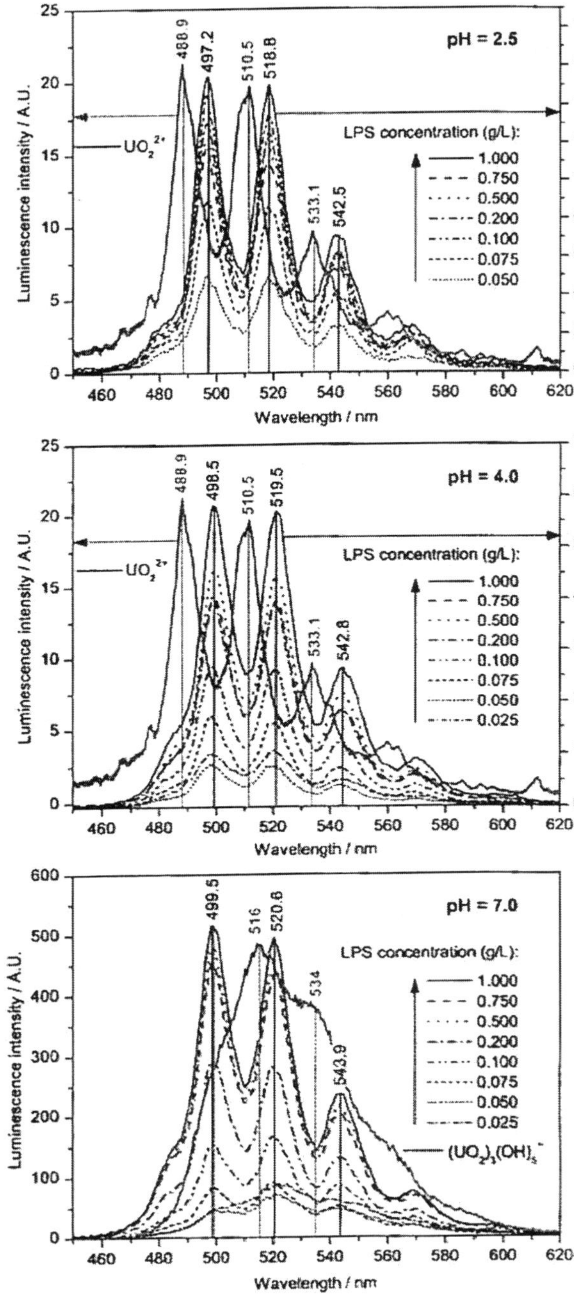

Fig. 8.8 Fluorescence spectra at fixed uranyl concentration (10^{-5} M) and various LPS concentrations at different pH values (from Barkleit et al., 2008; with permission)

A red shift of about 8–11 nm is very typical of uranyl–phosphoryl complexes (Barkleit et al., 2008 and literature therein). On the other hand, most of the carboxylate complexes do, like inorganic uranyl carbonates, not show photoluminescence at room temperature, i.e., they would not be recognized by fluorescence spectroscopy. Inorganic phosphate causes a still stronger red shift (about 10–14 nm), i.e., it should be ruled out as an important reaction partner of the uranyl. Thus, uranyl–phosphoryl complexes are the most probable fluorescent entities in these samples according to the location of the fluorescence peaks.

Further information comes from the time-resolved measurements with the help of which the lifetimes of the excited states can be determined. Lifetime measurements allow the estimation of the number of fluorescent species in the sample. They can also help to recognize whether or not inorganic fluorescent species such as UO_2^{2+}(aq) or uranyl hydroxide species are present. The lifetime studies support the conclusions that can be drawn from the red shift of the peak maxima in Fig. 8.8.

It follows from this study based on potentiometric titration (cf. Barkleit et al., 2008) and TRLFS that there is a coordination of the uranyl ion with both the phosphoryl and the carboxyl groups of the LPS. The phosphoryl complexes are detectable by fluorescence spectroscopy, the carboxylate complexes are not. Under conditions of ligand excess, coordination via phosphoryl groups prevails, whereas under conditions of a slight deficit of phosphoryl groups (deficit of LPS) also carboxyl groups play an important part for uranyl coordination.

The study is an example of using a minimum invasive technique for the elucidation of binding states at the surface of nanoparticles under well-defined laboratory conditions, i.e., of an in situ study under idealized, not "real-world," conditions.

8.3.3 Infrared Light

Binding states of uranium(VI) and carbonate on ferrihydrite nanoparticles by attenuated total reflection Fourier transform infrared spectroscopy (ATR-FTIR). Near-neutral or slightly acidic floodwaters of abandoned mines very often contain significant amounts of iron-rich nanoparticles (cf. Zänker et al., 2003). However, these nanoparticles are different from the ultrafine particles found in strongly acidic ARD solutions as the solution depicted in Fig. 8.2. In laboratory simulation experiments with ferrihydrite at near-neutral or slightly acidic pH values the binding states of the uranyl ion on freshly precipitated ferrihydrite and the influence of carbonate on these binding states were investigated by ATR-FTIR (Ulrich et al., 2006). Figure 8.9 shows the principle of ATR-FTIR. ATR-FTIR enables water-containing

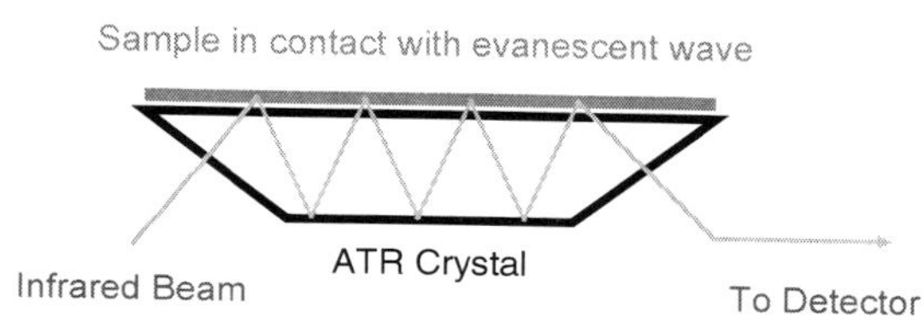

Fig. 8.9 Principle of ATR-FTIR

samples (i.e., the bulk of the samples, not single particles) to be investigated in spite of the strong IR absorption by water. The infrared beam is guided through the ATR crystal by total reflections. However, at every reflection point the infrared light (the evanescent wave) slightly penetrates the sample which produces the IR spectrum.

The ferrihydrite nanoparticles precipitated at varying uranyl and carbonate concentrations were separated from the solution as a wet paste by ultracentrifugation without any further treatment. Note that more than 98% of the uranyl added was adsorbed to the ferrihydrite (information from ultrafiltration plus ICP-MS) making dissolved U species negligible for the following considerations. An aliquot of the wet ferrihydrite precipitate was spread over the optical crystal surface as a thin layer (<1 mm). The spectrum of a U-free ferrihydrite precipitate prepared at pH 5.5 and pCO_2 < 0.2 Pa, i.e., without adsorption of U(VI) and carbonate, shows a broad absorption band in the region of 937 cm^{-1} representing the δ(OH) bending vibration of the ferrihydrite (Fig. 8.10, curve A right). The spectra of ferrihydrite samples containing UO_2^{2+} ions exhibit an additional band at ~903 cm^{-1}, the intensity of which correlates with the increasing U(VI) concentration (Fig. 8.10, curves C-E right). This band represents the antisymmetric stretching vibration (ν_3) of the UO_2^{2+} cation adsorbed to the Fe hydroxide phase. Contrary to the samples

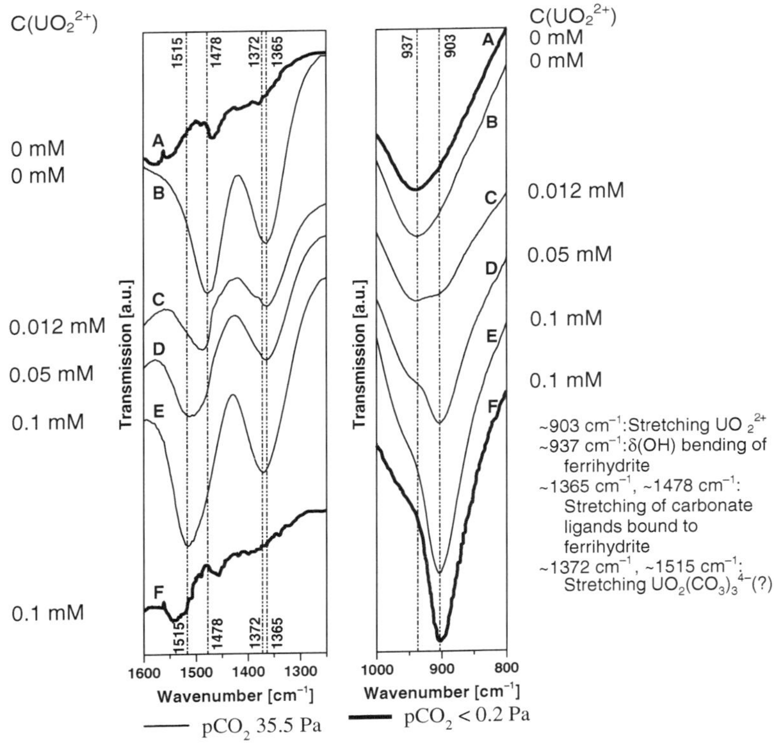

Fig. 8.10 ATR-FTIR spectra of U(VI) adsorbed onto ferrihydrite. Samples prepared at pH 5.5 and at different UO_2^{2+} concentrations (from Ulrich et al., 2006; with permission)

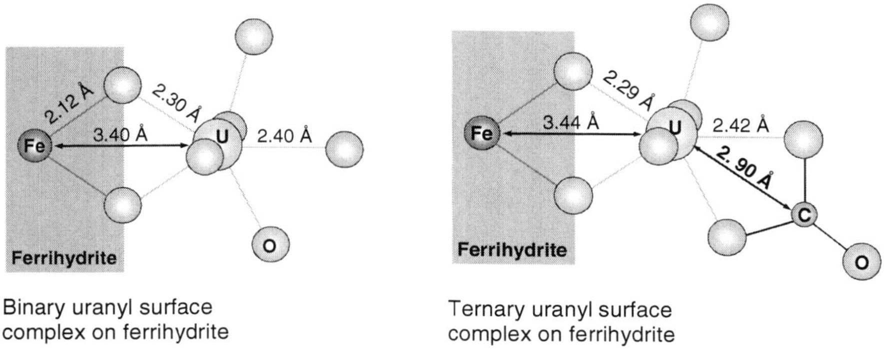

Fig. 8.11 Structure of a binary and of a ternary surface complex of uranyl on ferrihydrite colloids as derived from ATR-FTIR (and EXAFS) spectroscopy (from Ulrich et al., 2006; with permission)

prepared at atmospheric pCO_2 of 35.5 Pa, the spectra of both samples prepared while excluding CO_2 from the system have no absorption bands in the range of 1,250–1,600 cm^{-1}, thus verifying the absence of adsorbed carbonate (Fig. 8.10, curves A-F left). The other sample spectra show additional bands around 1,365 and 1,500 cm^{-1} (Fig. 8.10, curves B-E left) which are due to the symmetric (ν_s) and antisymmetric (ν_{as}) stretching vibration of carbonate ligands bound to the ferrihydrite phase. These bands show a steady shift to higher wavenumbers with increasing U(VI) concentration. In particular the frequency of the $\nu_{as}(CO_3{}^{2-})$ vibration shifts from 1,478 to 1,515 cm^{-1}, the latter being reached by the sample prepared with initially 0.1 mM $UO_2{}^{2+}$. A similar, but smaller shift from 1,365 to 1,372 cm^{-1} was observed for the $\nu_s(CO_3{}^{2-})$ vibration ($\Delta\nu_s \sim 7$ cm^{-1}; Fig. 8.10, curve E left). The experiment elucidates the interrelation of the different interactions in the system which are in competition with each other: the ferrihydrite–uranyl interaction, the ferrihydrite–carbonate interaction, and the interaction of adsorbed uranyl with carbonate (dissolved uranyl-carbonato complexes play a minor role because, as mentioned above, almost all the uranium is adsorbed to the ferrihydrite under the conditions chosen). The resulting surface complexes of uranyl on ferrihydrite are depicted in Fig. 8.11 (for details see Foerstendorf and Heim, 2009; Ulrich et al., 2006). An alternative ternary surface complex in which a carbonate ligand acts as a bridge between uranyl and ferrihydrite was suggested by Rossberg et al. 2009. The formation of the binary surface complex is promoted by low carbonate concentration and low pH. The formation of the ternary surface complex is supported by higher carbonate concentration and higher pH. These interrelations directly influence the transport behavior of U(VI) in mine waters such as floodwaters of abandoned uranium mines.

8.3.4 X-Rays

Binding states of arsenic(V) on acid rock drainage colloids by extended X-ray absorption (EXAFS) spectroscopy. As was shown in Sect. 8.3.2, the acid rock drainage (ARD) solution depicted in Fig. 8.2 and characterized by Table 8.1 contained a high concentration of ultrafine nanoparticles of less than 10 nm in size. Furthermore, a more aggregated ochre-like long-term precipitate evolved in this solution within months. The nanoparticles seem to be an intermediate of this ochre precipitation process (cf. Zänker et al., 2002). The ARD solution was studied by EXAFS spectroscopy, a version of X-ray absorption spectroscopy (XAS) which belongs to the few methods capable of getting information on the binding states of adsorbed trace elements at the surface of solids in the original aquatic environment. For the principles of EXAFS see Charlet and Manceau (1993). The raw ARD solution, two filtrates from filters of different pore size (5 μm, 1 kD), the 1-kD retentate, the long-term precipitate (all in the presence of the original Water), and two reference minerals (as dry samples) were investigated (Moll et al., 2000; Zänker et al., 2002).

The EXAFS data were recorded at the Rossendorf Beamline (ROBL) at the ESRF in Grenoble. For details concerning this beamline see Matz et al. (1999). The data were treated using the EXAFSPAK software developed by George and Pickering (1995). Theoretical backscattering phase and amplitude functions, $\delta(k)$ and $F(k)$, used in data analysis, were calculated using the FEFF7 (Zabinsky et al., 1995) program. The EXAFS oscillations were isolated using standard procedures for pre-edge subtraction, spline removal, and data normalization (Koningsberger and Prins, 1988). EXAFS spectra provide mainly two parameters:

- the distance between metal (or metalloid) atoms and
- the number of metal/metalloid neighbors of the element studied (coordination number).

They are defined by the type of linkage and the ionic radii of the neighboring atoms. Their knowledge allows the determination of the type of linkage and, therefore, the type of adsorption between a metal and an adsorbent.

Figure 8.12 shows the measured arsenic K-edge EXAFS spectra and the corresponding Fourier transforms (FTs) obtained from the model compounds, the ARD raw sample, the ARD filtration fractions, and the ochre-like long-term precipitate. As can be seen, the FTs of the ARD fractions are quite different from that of the pure arsenate solution. They indicate a pronounced As–Fe contribution at about 3.28 Å. On the other hand, the As–Fe distance in the ARD samples was by 0.04–0.08 Å shorter than that of the minerals scorodite ($FeAsO_4 \cdot 2H_2O$) and bukowskyite ($Fe_2AsO_4SO_4OHA \cdot 7H_2O$).

It follows from the EXAFS results that the arsenate is bound to the colloid particles by surface complexation (formation of a bidentate binuclear inner-sphere complex). However, the transformation of the colloidal material into the more aggregated ochre-like long-term precipitate results in a change of the arsenate binding in this material. Obviously, the incorporation of the arsenate into the interior of the iron hydroxy sulfate crystal structures is more pronounced in the precipitate which was indicated by an increase in the Fe–As coordination number. The mechanisms of this incorporation are either the substitution of arsenate for sulfate in the iron hydroxy

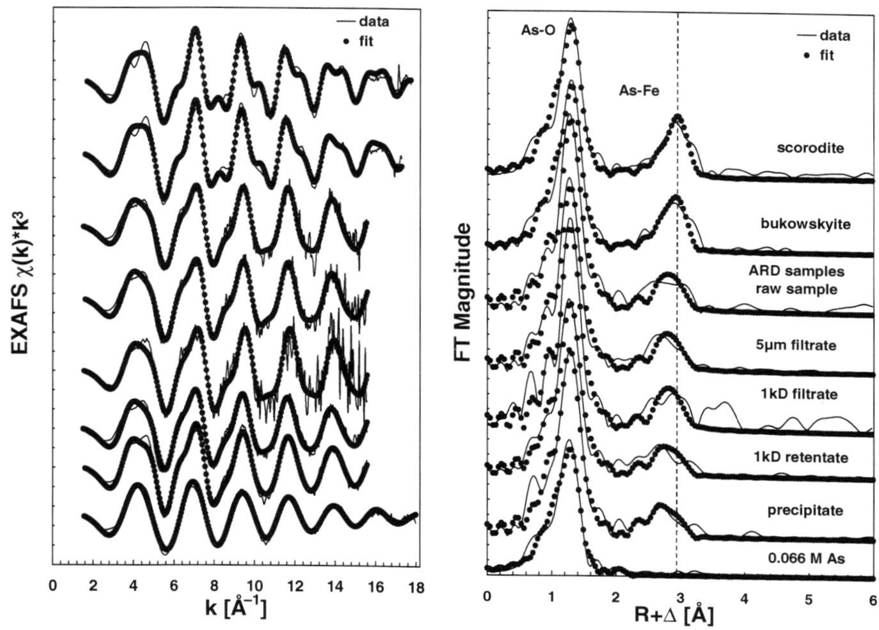

Fig. 8.12 As K-edge k^3-weighted EXAFS spectra (*left*) and the corresponding Fourier transforms, FTs (*right*), of model compounds and ARD fractions. The FTs are not corrected for EXAFS phase shifts. The peaks appear at shorter distances ($R+\Delta$) relative to the true near-neighbor distances (R) (from Zänker et al., 2002; with permission)

sulfate structure or the formation of extremely small scorodite clusters as occlusions within iron hydroxy sulfate (for details see Zänker et al., 2002).

8.4 Conclusions

For six examples of nanoparticle systems the application of "in situ" methods of particle characterization which use photons as probes were demonstrated. The six examples are *selected* examples. Some of them refer to relatively "ideal" laboratory systems, some represent first steps toward real-world systems. The examples have in common that the methods applied worked satisfyingly under the conditions given in the experiments.

It is understood that "in situ" or "non-invasive" methods are in reality always little-invasive methods. There is no in situ or non-invasive technique in the strictest sense of the term since measuring the properties of an object of matter such as a colloidal particle is always getting in interaction with this object. However, there are ways to minimize these interactions.

All the methods applied in the examples of this chapter have the potential to be used also for the characterization of ENPs. However, even for the use for natural nanoparticles, all the methods may fail if the nanoparticle system differs from the system presented here. TRLFS, for instance, normally fails if iron-rich particles are to be investigated (Fe(III) quenches the fluorescent light). LIBD is hardly applicable on humic macromolecules since their breakdown threshold does not differ enough from that of water. PCS fails if there is a mixture of large and very small particles because of masking problems. Sometimes it is possible to counteract these problems. For instance, the masking problems in PCS can sometimes be overcome by invasive pretreatment of the sample as demonstrated for ARD solution in the first example, i.e., by filtration or centrifugation. Sometimes there are no remedies.

Similar problems as with the natural nanoparticles are to be expected for ENPs. The next step for ENP characterization should be to test the in situ methods on "ideal" ENP laboratory systems. Note that ENPs are not a uniform category. Although "ENPs" are subsumed under a common name, they represent entirely different chemical systems (TiO_2, ZnO, Ag(0), Fe(0), CNTs, spherical fullerenes, quantum dots). This great variety of ENPs will not ease the development of characterization techniques for such particles.

However, the real difficulty for ENP characterization will be the change from laboratory to true real-world systems. The most serious challenges concerning natural systems are probably the low ENP concentration to be expected in natural samples and the need to differentiate between (i) ENPs, (ii) natural nanoparticles, and (iii) associates of ENPs with natural nanoparticles. The efforts to tackle these challenges are only at the beginning worldwide.

References

Alvarez-Puebla RA, Garrido JJ, Aroca RF (2004) Surface-enhanced vibrational microspectroscopy of fulvic acid micelles. Analytical Chem 76: 7118–7125.

Barkleit A, Moll, H, Bernhard, G (2008) Interaction of uranium(VI) with lipopolysaccharide. Dalton Transact 21: 2879–2886.

Boughriet A, Cordier C, Deram L, Ouddane B, Chamley H, Wartel M (1995) Coprecipitation/accumulation/distribution of manganese and iron, and electrochemical characteristics of Mn in calcareous seawater. Fresenius J Anal Chem 352: 341–353.

Buffle J, Perret D, Newman M (1992) The use of filtration and ultrafiltration for size fractionation of aquatic particels, colloids and macromolecules. In: Buffle J, van Leeuwen HP (eds.) Environmental Particles. Vol. 1 Environmental Analytical and Physical Chemistry Series (IUPAC), Lewis Publishers, Boca Raton, pp. 171–230.

Buffle J, van Leeuwen HP (1992) Environmental Particles. Vol. 1 Environmental Analytical and Physical Chemistry Series (IUPAC), Lewis Publishers, Boca Raton.

Buffle J, van Leeuwen HP (1993) Environmental Particles. Vol. 2. Environmental Analytical and Physical Chemistry Series (IUPAC), Lewis Publishers, Boca Raton.

Buffle J, Leppard GG (1995) Characterization of Aquatic Colloids and Macromolecules. 2. Key Role of Physical Structures on Analytical Results. Environ Sci Technol 29: 2176–2184.

Bundschuh T, Hauser W, Kim JI, Knopp R, Scherbaum FJ (2001) Determination of colloid size by 2-D optical detection of laser induced plasma. Coll Surf A: Physicochem Eng Asp 180: 285–293.

Charlet L, Manceau A (1993) Structure, formation, and reactivity of hydrous oxide particles: insights from x-ray absorption spectroscopy. In: Buffle J, van Leeuwen HP (eds.)

Environmental Particles. Vol. 2. Environmental Analytical and Physical Chemistry Series (IUPAC). Lewis Publishers, Boca Raton, pp. 117–164.
Cheng JX, Jia YK, Zheng G, Xie XS (2002) Laser-scanning coherent anti-stokes Raman scattering microscopy and applications to cell biology. Biophys J 83: 502–509.
Ding M, De Jong BHWS, Roosendaal SJ, Vredenberg A (2000) XPS studies on the electronic structure of bonding between solid and solutes: adsorption of arsenate, chromate, phosphate, Pb^{2+}, and Zn^{2+} ions on amorphous black ferric oxyhydroxide. Geochim Cosmochim Acta 64: 1209–1219.
Druet SAJ, Teran JPE (1981) CARS spectroscopy. Prog Quantum Electronics 7: 1–72.
Filella M, Zhang J, Newman ME, Buffle J (1997) Analytical applications of photon correlation spectroscopy for size distribution measurements of natural colloidal suspensions: capabilities and limitations. Coll Surf A: Physicochem Eng Asp 120: 27–46.
Foerstendorf H. Heim K (2009) Spectroscopic identification of ternary carbonate complexes upon U(VI) sorption onto ferrihydrite. Geochim Cosmochim Acta 73: A 386.
Ford NC (1985) Light scattering apparatus. In: Pecora, R (ed.) Dynamic Light Scattering: Applications of PCS. Plenum Press, New York, pp. 7–58.
Fourest B, Hakem N, Guillaumont R (1994) Characterization of colloids by measurement of their mobilities. Radiochim Acta 66/67: 173–179.
Frimmel F, von der Kammer F, Flemming KC (2007) Colloidal Transport in Porous Media. Springer-Verlag, Berlin, Heidelberg.
Gaillard JF (2007) Probing environmental colloids and particles with x-rays. In: Wilkinson KJ, Lead JR (eds.) Environmental Colloids and Particles. IUPAC Series on Analytical and Physical Chemistry of Environmental Systems. Wiley Interscience, New York, pp. 613–666.
Geipel G, Brachmann A, Brendler V, Bernhard G Nitsche H (1996) Uranium(VI) sulfate complexation studied by time-resolved laser-induced fluorescence spectroscopy (TRLFS). Radiochim Acta 75: 199–204.
George GN, Pickering IJ (1995) EXAFSPAK, A suite of computer programs for analysis of X-ray absorption spectra. Stanford Synchrotron Radiation Laboratory, Stanford, USA.
George S, Steinberg SM, Hodge V (2000) The concentration, apparent molecular weight and chemical reactivity of silica from groundwater in Southern Nevada. Chemosphere 40: 57–63.
Grabowski EE, Morrison ID (1983) Particle size distribution from analysis of quasi-elastic light-scattering data. In: Dahneke, BE (ed.) Measurement of Suspended Particles by Quasi-Elastic Light Scattering. Wiley-Interscience, New York, pp. 199–236.
Hedderman TG, Keogh SM, Chambers G, Byrne HJ (2006) In-depth study into the interaction of single walled carbon nanotubes with anthracene and p-terphenyl. J Phys Chem B 110: 3895–3901.
Helmbrecht C, Niessner R, Haisch C (2007) Photophoretic velocimetry for colloid characterization and separation in a cross-flow setup. Anal Chem 79: 7097–7103.
Johnston CT, Sposito G, Earl WL (1993) Surface spectroscopy of environmental particles by Fourier-transform infrared and nuclear magnetic resonance spectroscopy. In: Buffle J, van Leeuwen HP (eds.) Environmental Particles. Vol. 2. Environmental Analytical and Physical Chemistry Series (IUPAC), Lewis Publishers, Boca Raton, pp. 1–36.
Jones MN, Bryan ND (1998) Colloidal properties of humic substances. Adv Coll Interface Sci 78: 1–48.
Kaplan DI, Hunter DB, Bertsch PM, Bajt S, Adriano DC (1994) Application of synchrotron x-ray fluorescence spectroscopy and energy dispersive x-ray analysis to identify contaminant metals on groundwater colloids. Environ Sci Technol 28: 1186–1189.
Kim JI, Walther C (2007) Laser-induced breakdown detection. In: Wilkinson KJ, Lead JR (eds.) Environmental Colloids and Particles. IUPAC Series on Analytical and Physical Chemistry of Environmental Systems. Wiley Interscience, New York, pp. 555–612.
Klein, T., Niessner, R. (1998) Characterization of heavy-metal-containing seepage water colloids by flow FFF, ultrafiltration, ELISA and AAS. Mikrochim Acta 129: 47–55.
Knollenberg RG, Veal DL (1992) Optical monitors, counters, and spectrometers: performance characterization, comparison, and use. J IES 35: 64–81.

Koningsberger D E, Prins R (1988) X-ray absorption: principles, applications, techniques for EXAFS, SEXAFS, and XANES. Wiley-Interscience, New York.

Lead, J R, Davison W, Hamilton-Taylor J, Buffle J (1997) Characterizing colloidal material in natural waters. Aquatic Geochem 3: 213–232.

Lyven B, Hassellöv M, Turner DR, Haraldsson C, Andersson K (2003) Competition between iron- and carbon-based colloidal carriers for trace metals in a freshwater assessed using flow field-flow fractionation coupled to ICPMS. Geochim Cosmochim Acta 67: 3791–3802.

Matz W, Schell N, Bernhard G, Prokert F, Reich T, Claußner J, Oehme W, Schlenk R, Dienel S, Funke H, Eichhorn F, Betzl M, Pröhl D, Strauch U, Hüttig G, Krug H, Neumann W, Brendler V, Reichel P, Denecke MA, Nitsche H (1999) ROBL – a CRG beamline for radiochemistry and material research at the ESFR. J Synchrotron Rad 6: 1076–1085.

Moll H, Zänker H, Richter W, Brendler V, Reich T, Hennig C, Roßberg A, Funke H, Kluge A (2000) XAS Study of Acid Rock Drainage Samples From an Abandoned Zn-Pb-Ag Mine at Freiberg, Germany. 2nd Euroconference and NEA Workshop on Speciation, Techniques, and Facilities for Radioactive Materials at Synchrotron Light Sources. Actinide-XAS-2000, Grenoble, France.

Müller K, Foerstendorf H, Tsushima S, Brendler V, Bernhard G (2009) Direct spectroscopic characterization of aqueous Actinyl(VI) species: a comparative study of Np and U. J Phys Chem A 113: 6626–6632.

Namjesnik-Dejanovic K, Maurice PA (1997) Atomic force microscopy of soil and stream fulvic acids. Coll Surf A: Physicochem Eng Asp 120: 77–86.

Nomizu T, Goto K, Mizuike A (1988) Electron microscopy of nanometer particles in freshwater. Anal Chem 60: 2653–2656.

Nordstrom DK, Alpers CN, Ptacek CJ, Blowes DW (2000) Negative pH and extremely acidic mine waters from iron mountain, California. Environ Sci Technol 34: 254–258.

Novikov AP, Kalmykov SN, Utsunomiya S, Ewing RC, Horreard F, Merkulov A, Clark SB, Tkachev VV, Myasoedov BF (2006) Colloid transport of plutonium in the far-field of the Mayak production association, Russia. Science 314: 638–641.

Opel K, Zänker H, Bernhard G (2006) Characterization of the colloidal inventory of mineral waters and physiological salines by laser-induced breakdown detection (LIBD). Report FZR-443, Forschungszentrum Rossendorf, p. 19.

Opel K, Weiß S, Hübener S, Zänker H, Bernhard G (2007) Study of the solubility of amorphous and crystalline uranium dioxide by combined spectroscopic methods. Radiochim Acta 95: 143–149.

Plaschke M, Schäfer T, Bundschuh T, Ngo Manh T, Knopp R, Geckeis H, Kim, JI (2001) Size characterization of Bentonite colloids by different methods. Anal Chem 73: 4338–4347.

Plaschke M, Römer J, Kim, JI (2002) Characterization of Gorleben groundwater colloids by atomic force microscopy, Environ Sci Technol 36: 4483–4488.

Prestel H, Niessner R, Panne U (2006) Increasing the sensitivity of asymmetrical flow field-flow fractionation: slot outlet technique. Anal Chem 78: 6664–6669.

Provencher W (1982) A constrained regularization method for inverting data presented by linear algebraic or integral equations. Comput Phys Commun 27: 213–227.

Riley J (2005) Charge in colloidal systems. In: Cosgrove T (ed.) Colloid Science. Principles, Methods and Applications. Blackwell Publishing, Oxford, pp. 14–35.

Rossberg A, Ulrich KU, Weiss S, Tsushima S, Hiemstra T, Scheinostt AC (2009) Identification of Uranyl Surface Complexes an Ferrihydrite: Advanced EXAFS Data Analysis and CD-MUSIC Modeling. Environ Sci Technol 43: 1400–1406.

Scherbaum FJ, Knopp R, Kim JI (1996) Counting of particles in aqueous solutions by laser-induced photoacoustic breakdown detection. Appl Phys B 63: 299–306.

Schierz A, Zänker H (2009) Aqueous suspensions of carbon nanotubes: surface oxidation, colloidal stability and uranium sorption. Environ Pollut 157: 1088–1094.

Schmid T, Messmer A, Yeo BS, Zhang WH, Zenobi R (2008) Towards chemical analysis of nanostructures in biofilms II: tip-enhanced Raman spectroscopy of alginates. Anal Bioanal Chem 391: 1907–1916.

Schurtenberger P, Newman ME (1993) Characterization of biological and environmental particles using static and dynamic light scattering. In: Buffle J, van Leeuwen HP (eds.) Environmental Particles. Vol. 2. Environmental Analytical and Physical Chemistry Series (IUPAC), Lewis Publishers, Boca Raton, pp. 37–115.

Stock RS, Ray WH (1985) Interpretation of photon correlation spectroscopy data: a comparison of analysis methods. J Polym Sci Polym Phys Ed 23: 1393–1447.

Thang NM, Knopp R, Geckeis H, Kim JI, Beck HP (2000) Detection of nanocolloids with flow-field flow fractionation and laser-induced breakdown detection. Anal Chem 72: 1–5.

Thieme J, McNulty I, Vogt S, Paterson D (2007) X-ray spectromicroscopy – A tool for environmental sciences. Environ Sci Technol 41: 6885–6889.

Tscharnuter W (2000) Photon correlation spectroscopy in particle sizing. In: Meyers RA (ed.) Encyclopedia of Analytical Chemistry. John Wiley & Sons, Chichester, pp. 5469–5485.

Ulrich KU, Rossberg A, Foerstendorf H, Zänker H, Scheinost AC (2006) Molecular characterization of uranium(VI) sorption complexes on iron(III)-rich acid mine water colloids. Geochim Cosmochim Acta 70: 5469–5487.

von der Kammer F, Baborowski M, Friese K (2005) Application of a high-performance liquid chromatography fluorescence detector as a nephelometric turbidity detector following Field-Flow Fractionation to analyse size distributions of environmental colloids. J Chromatography A 1100: 81–89.

von Wandruszka R, Schimpf M, Hill M, Engebretson R (1999) Characterization of humic acid size fractions by SEC and MALS. Organic Geochemistry 30: 229–235.

Walther C, Cho HR, Fanghänel T (2004) Measuring multimodal size distribution of aquatic colloids at trace concentrations. Appl Phys Lett 85: 6329–6331.

Walther C, Buchner S, Filella M, Chanudet V (2006) Probing particle size distributions in natural surface waters from 15 nm to 2 μm by a combination of LIBD and single-particle counting. J Colloid Interface Sci 301: 532–537.

Weiner BB (1984) Particle sizing using photon correlation spectroscopy. In: Barth, HG (ed.) Modern Methods of Particle Size Analysis. John Wiley & Sons, New York, pp. 93–116.

Wilkinson KJ, Lead JR (2007) Environmental Colloids and Particles. IUPAC Series on Analytical and Physical Chemistry of Environmental Systems. Series Editors: Buffle J, van Leeuwen HP, Wiley Interscience, New York.

Zabinsky SI, Rehr JJ, Ankudinov A, Albers RC, Eller MJ (1995) Multiple-scattering calculations of x-ray absorption spectra. Phys Rev B 52: 2995–3008.

Zänker H, Moll H, Richter W, Brendler V, Hennig C, Reich T, Kluge A, Hüttig G (2002) The colloid chemistry of acid rock drainage solution from an abandoned Zn-Pb-Ag mine. Appl Geochem 17: 633–648.

Zänker H, Richter W, Hüttig G (2003) Scavenging and immobilization of trace contaminants by colloids in the waters of abandoned ore mines. Coll Surf A: Physicochem Eng Asp 217: 21–31.

Zänker H, Hüttig G, Arnold T, Nitsche H (2006) Formation of iron-containing colloids by the weathering of Phyllite. Aquatic Geochem 12: 299–325.

Chapter 9
Coupling Techniques to Quantify Nanoparticles and to Characterize Their Interactions with Water Constituents

Markus Delay, Luis A. Tercero Espinoza, George Metreveli, and Fritz H. Frimmel

9.1 Introduction

Nanotechnology has become a promising branch of scientific research and industrial production. Due to the rapid progress and production of nanosized materials, nanotechnology has developed to be a key technology for the twenty-first century (Luther, 2004). Typically, nanotechnology considers structures which are smaller than 100 nm in at least one dimension (Paschen et al., 2004). The properties of nanosized structures and particles are thoroughly different from those of bulk materials (macro scale), mainly concerning the chemical reactivity or electrical conductivity (Duphny Guzmán et al., 2006; Grüne et al., 2005; Krug et al., 2004; Long et al., 2006; Roco, 2005). The properties of engineered nanoparticles (ENP) and hence the effects on environmental processes and human health are strongly related to the nanometer scale of the substance, to the large surface to mass ratio, and/or to the surface properties of the components (Krug et al., 2004; Duphny Guzmán et al., 2006; Long et al., 2006). As a consequence, nanoparticles show outstanding potential benefits for various industrial, medical, pharmaceutical, cosmetic, and life science applications (Betorelle et al., 2006; Paschen et al., 2004; Wagner and Wechsler, 2004). Progress in the nanosciences enables the specific engineering of nanoparticles with defined material properties such that a wide spectrum of ENP can be produced. Some typical and widely used nanoparticles were described by Fahrner (2005), Fleischer (2002), Krug (2005), Löchtefeld et al. (2005), Masciangioli and Zhang (2003), Nowack and Bucheli (2007), Theodore and Kunz (2005), and Wiesner et al. (2006).

Examples of nanomaterials and their applications are the following:

- Metal oxides: TiO_2 (pigment, photocatalyst, UV absorber), Fe_2O_3, Fe_3O_4 (pigment, pharmaceutical products, medicine), ZnO and ZrO_2 (surface hardener), SiO_2 (additive for polymers),

F.H. Frimmel (✉)
Engler-Bunte-Institut, Chair of Water Chemistry, Karlsruhe Institute of Technology (KIT), Engler-Bunte-Ring 1, D-76131 Karlsruhe, Germany
e-mail: fritz.frimmel@kit.edu

F.H. Frimmel, R. Niessner (eds.), *Nanoparticles in the Water Cycle*,
DOI 10.1007/978-3-642-10318-6_9, © Springer-Verlag Berlin Heidelberg 2010

- Metals: gold and silver nanoparticles (catalysts, electronic devices),
- Quantum dots: CdTe, GaAs (semiconductors, electronic devices),
- Carbon based nanomaterials (CBN): black carbon (ink for printers, additive in tires), fullerenes (additive in grease), carbon nanotubes (additive in polymers, accumulators, liquid fuel cells).

Nanotechnology shows high potential to improve environmental quality and sustainability through applications in treatment processes and remediation techniques (Joo and Cheng, 2006; Masciangioli and Zhang, 2003). Reactive nanoparticles are useful remediation tools (Obare and Meyer, 2005; Wiesner et al., 2006): Zero valent iron nanoparticles, for example, are highly reactive and are a promising technology for the in situ remediation of trichlorethene (TCE) (Liu et al., 2005; Wang and Zhang, 1997). Polymer-supported hydrated Fe(III) nanoparticles find use for the removal of arsenic (Cumbal and Sengupta, 2005). Furthermore, titanium dioxide is applied for in situ anion immobilization (Mattigod et al., 2005) or for photocatalytic degradation of organic pollutants like pharmaceuticals (Doll and Frimmel, 2005; Kutsuna et al., 1999).

However, there is a conflict between the power of the applicability of ENP and their benefits on the one hand and the possible risk of undesirable environmental impacts after application on the other hand. Due to their widespread use (e.g., nanosilver as an antimicrobial agent in socks) and their increasing production, emission of ENP into aquatic environmental systems via municipal or industrial waste waters and atmospheric emissions is likely (Benn and Westerhoff, 2008; Nowack and Bucheli, 2007; Oberdörster et al., 2005). Nevertheless, little is known about the fate and environmental behavior of ENP in aquatic systems. Although there is currently a focus on atmospheric nanoparticles in the literature (Biswas and Wu, 2005), the behavior of ENP in aqueous environmental systems is of high relevance as water quality parameters might be influenced by ENP.

The fundamental understanding of the behavior of ENP and their stability in aquatic systems is a key issue regarding human health and environmental risk assessment. There is experimental and clinical evidence for toxic effects of nanometer sized materials (Nel et al., 2006), and nanoparticles have been shown to stimulate the formation of oxyradicals (Oberdörster et al., 2005). Titanium dioxide can induce the production of reactive oxygen species (ROS) and lead to neurotoxic effects by interacting with brain microglia (Long et al., 2006). The antibacterial activity of fullerene/water suspensions (Lyon et al., 2006) and silver nanoparticles (Pal et al., 2007) also demonstrates the potential of ENP to interact and react with organisms. In vitro cytotoxicity of oxide nanoparticles and their uptake into human lung cells were found by Brunner et al. (2006) and Limbach et al. (2005, 2007). Rothen-Rutishauser et al. (2006) described the uptake of nanoparticles into red blood cells. Uptake kinetics of nanoparticles by cells is highly influenced by the size and the shape of the nanoparticles (Chithrani et al., 2006), and definitely by the phase they are in.

There is a severe deficiency of data on the occurrence and environmental concentrations of ENP (Mueller and Nowack, 2008) due to the lack of adequate standardized methods rather than to missing analytical instruments. There are promising approaches for the characterization of natural colloids and ENP by analytical coupling techniques such as the coupling of asymmetrical flow field-flow fractionation (AF^4) (separation unit) or transport columns with inductively coupled plasma mass spectrometry or multi-angle laser light scattering (detection unit) (Metreveli et al., 2005; Prestel et al., 2005; von der Kammer et al., 2005; Wilkinson and Lead, 2006). It is known that natural aquatic colloids and some nanoparticles can adsorb or bind organic or inorganic contaminants, that they can be transported in aqueous systems over large distances due to their high mobility, and that they can reach the drinking water supply systems (Kretzschmar et al., 1999; Lead and Wilkinson, 2007; Lecoanet and Wiesner, 2004; Lecoanet et al., 2004; Metreveli et al., 2005; Zhou et al., 2007). Findings for natural colloids have shown to be also valid for ENP like synthetic three-layer clay minerals (Metreveli et al., 2005). In this context, the interaction of natural organic matter (NOM) with ENP is crucial as NOM is ubiquitous in natural water and has the potential to change the physical and chemical properties of pristine ENP. As several recent publications have shown, partial NOM adsorption onto ENP or coating of ENP by NOM can strongly influence the agglomeration behavior and stability, the surface charge, the morphology, and thus, the transport behavior of ENP and their interaction with pollutants in environmental systems (Diegoli et al., 2008; Hyung et al., 2007; Hyung and Kim, 2008; Pelley and Tufenkji, 2008; Tercero Espinoza et al., 2008; Xie et al., 2008).

The objective of this chapter is to give examples for coupling techniques for the characterization of nanoparticles and their interactions with heavy metals and NOM. The application of these techniques contributes to a deeper understanding of the fate, stability, and reactions of ENP in aqueous systems.

9.2 Principle of Analytical Coupling Techniques

Assessing the environmental impact of ENP demands powerful analytic methods allowing

- the quantification and basic characterization of ENP (concentration, size distribution, surface charge),
- the characterization of their interactions (agglomeration, interactions with organic and inorganic pollutants, interactions with other water constituents, sorption behavior), and
- the characterization of their transport behavior.

The determination and quantification of ENP at concentration ranges typical of environmental systems demand the separation of ENP from complex matrices and mixtures, an adequate sample preparation including pre-concentration, and a subsequent highly sensitive detection of the particles. For these purposes, the application of analytical coupling techniques has turned out to be adequate and powerful.

Generally, analytical coupling techniques are widely used, for example, for element speciation (Dunemann and Begerow, 1995). Coupling of ICP-MS or inductively coupled plasma optical emission spectrometry (ICP-OES) with different chromatographic or chromatography-like systems such as size exclusion chromatography (SEC) (Schmitt et al., 2000, 2003), ion chromatography (IC) (Bissen and Frimmel, 2000), flow field-flow fractionation (FFFF) (Hasselöv et al., 1999a; Baalousha et al. 2006a, b), or sedimentation field-flow fractionation (SdFFF) (Schmitt et al., 2002; Hasselöv et al., 1999b) have been applied successfully to investigate element speciation and interactions between elements (mostly heavy metals) and inorganic nanoparticles or colloids.

The general principle of the setup of analytical coupling techniques is the combination of a separation unit and at least one detection instrument as illustrated in Fig. 9.1. Principally, it is also possible to run the separation unit and the following detection in an *off-line* operation mode. This includes the collection of sample fractions after the separation step and the subsequent characterization of the individual fractions (*off-line*). However, concerning sample preparation, sample stability, analytical resolution, and sensitivity of the detection, the *on-line* mode is more advantageous.

The following units are frequently applied for the analytical separation (and preconcentration) step:

- SEC: separation of organic (macro) molecules,
- AF^4: separation of organic and inorganic colloids and nanoparticles,
- Transport columns: experimental simulation of transport processes in porous media.

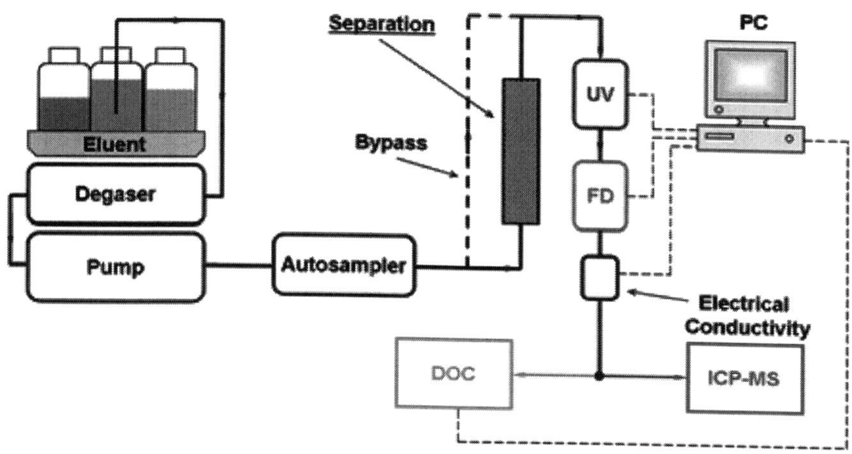

Fig. 9.1 General setup of analytical coupling experiments. Further details are given in the text

Depending on the object of investigation, the subsequent highly sensitive instruments can be used to detect nanoparticles and (in)organic compounds:

- UV and fluorescence detectors (FD), multi-angle laser light scattering (MALLS),
- Nanoparticle analyzer based on laser-induced breakdown detection (NPA-LIBD),
- Detection of dissolved organic carbon (DOC),
- ICP-MS and ICP-OES.

Generally, coupling of instruments is abbreviated by a "/". As an example, coupling of AF^4 and ICP-MS is given as "AF^4/ICP-MS." For the determination of recovery rates of the separation step (interaction of the analyte and the separation system), the coupling unit should include the opportunity of a direct bypass sample stream to the detectors.

In the following sections, examples of applications and further details on the coupling techniques will be given.

9.3 Examples of Applications

9.3.1 Size Exclusion Chromatography with Dissolved Organic Carbon Detection (SEC/DOC)

9.3.1.1 Background

As a first example, we explore the use of size exclusion chromatography to fractionate NOM primarily according to molecular size (Tercero Espinoza et al., 2008). With SEC/DOC, we examine the adsorption of NOM from a bog lake (Lake Hohloh, Black Forest, Germany) onto a metal oxide surface (TiO_2 nanoparticle agglomerates) and the changes in molecular size which occur upon ultraviolet irradiation in the presence of TiO_2 as a photocatalyst. The detection in this case was performed by both ultraviolet absorption ($\lambda = 254$ nm) and dissolved organic carbon (DOC), but the emphasis in this discussion will be on the SEC/DOC results. A critical evaluation of the SEC/DOC method is given by Lankes et al. (2009).

Before considering the two topics mentioned above, we introduce some concepts that will be useful in (1) understanding the underlying process (photocatalytic degradation) and (2) reading the SEC/DOC chromatograms presented throughout this section.

Heterogeneous Photocatalysis with Titanium Dioxide

Heterogeneous photocatalysis with titanium dioxide as the photocatalyst (HP-TiO_2) is one of the so-called advanced oxidation processes (AOP). These rely on the oxidative power of hydroxyl radicals (•OH), which can be produced in water by various methods. One of these methods is the excitation of titanium dioxide nanoparticles by irradiation with $\lambda \leq 388$ nm (Braun et al., 1991; Oppenländer, 2003) and is commonly referred to as heterogeneous photocatalysis, though other wide band-gap semiconductors may be used instead of TiO_2.

Because of the high reactivity of •OH, reactions take place only at or in the immediate vicinity of the photocatalyst surface. Thus, in a simplified way, the surface of the irradiated photocatalyst may be regarded as a highly reactive wall.

SEC/DOC Chromatograms

A sample SEC/DOC chromatogram of the brown water used for this example is presented in Fig. 9.2. This chromatogram reflects the chromatographable part of the DOC prior to any treatment steps. To a first approximation, larger molecules elute at lower retention time (t_r) than smaller molecules. The chromatograms were divided into three fractions (F1–F3) for ease of analysis, as seen in Fig. 9.2.

F1 was defined to lie in the interval 28.0 min $< t_r <$ 45.8 min, where t_r is the retention time (in this study numerically equal to the elution volume in mL). It appears to be composed of two overlapping peaks, the higher one with maximum at $t_r \approx$ 41.9 min and the shorter one (shoulder) with maximum around $t_r \approx$ 43.5 min. These two peaks are commonly attributed to humic material (Huber et al. 1994). F2 (45.8 min $< t_r <$ 50.7 min, maximum of the underlying peak around $t_r \approx$ 46.8 min) constitutes a second shoulder in the chromatogram of the non-irradiated water samples. This fraction becomes important upon irradiation and thus deserves separate consideration. The DOC contained in F2 is sometimes referred to as "building blocks" and is assumed to contain aromatic and polyfunctional organic acids (Frimmel, 1998; Huber et al., 1994). The third fraction, F3 (50.7 min $< t_r$ $<$ 56.5 min), appears clearly as a peak with maximum at $t_r \approx$ 52.7 min in the chromatograms of both non-irradiated and irradiated samples. Because the samples were

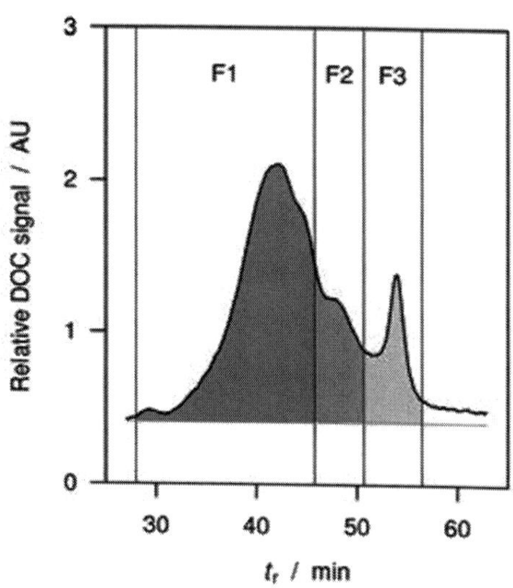

Fig. 9.2 SEC/DOC chromatogram showing the fractions selected to characterize adsorption and degradation of the DOC. The *vertical axis* is directly proportional to the amount of dissolved organic carbon. The *horizontal axis* is the retention time (t_r) in the SEC column and is related but not directly proportional to molecular size

not buffered prior to injection into the SEC/DOC system, this last peak also includes molecules that elute prematurely due to the difference in ionic strength between sample and mobile phase (Huber et al., 1994; Specht and Frimmel, 2000).

9.3.1.2 Adsorption

When the water sample comes into contact with the TiO_2 particles, a part of the DOC adsorbs onto the surface of the particles. In this case, close to 40% of the DOC present initially adsorbs onto the TiO_2 surface after $\approx$ 1 h of contact time, as measured by the difference in DOC before and after adsorption. The concentration of the TiO_2 was 0.5 g/L in all cases. Because of the heterogeneous nature of NOM (Fig. 9.2), it is to be expected that not all DOC fractions adsorb onto the surface of TiO_2 to the same extent. SEC/DOC is a very good tool for investigating this, and relevant chromatograms are shown in Fig. 9.3.

As expected, not all DOC fractions adsorb to an equal extent onto the surface of the photocatalyst. In fact, as seen in Fig. 9.3, most of the adsorbed DOC comes from the larger molecular weight fraction (F1), with very little material adsorbing

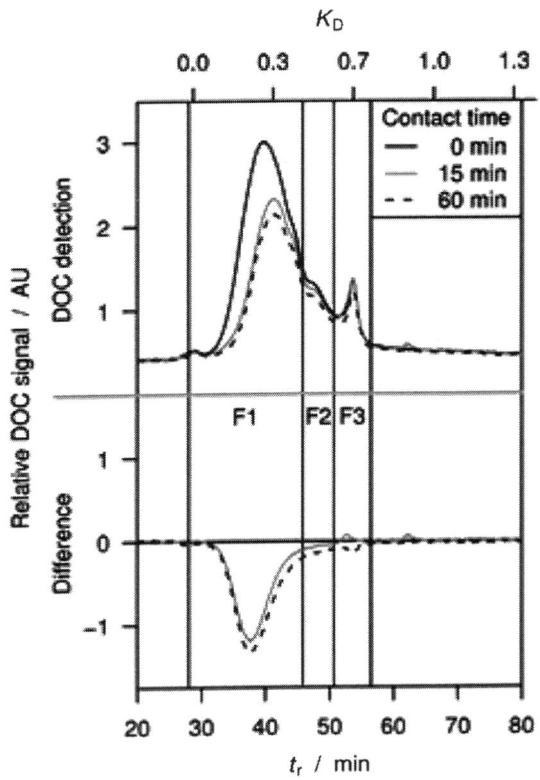

Fig. 9.3 SEC/DOC chromatograms showing the adsorption of NOM onto suspended TiO_2 particles (P25, Evonik) at different contact times. There was no significant change in samples equilibrated for more than 1 h (Tercero Espinoza et al., 2008). © 2008 Elsevier B.V. Reproduced with permission

from F2 and F3. This has one very important consequence: when the TiO_2 surface is irradiated ($\lambda < 388$ nm), F1 is "closest" to the reactive surface. Thus, we may expect to see some preferential degradation of the DOC because proximity to the surface is an important factor in heterogeneous photocatalysis.

9.3.1.3 Degradation

Upon UV irradiation using a solar simulator (Oriel Corp., Stratford, CT, USA), a rapid decline in F1 with an accompanying increase in both F2 and F3 was observed. A chromatogram series is shown in Fig. 9.3 showing the progression from the original sample until almost complete disappearance of DOC in solution. Note that the adsorbed portion of the DOC is not accessible for analysis. Inspection of Fig. 9.4 reveals that the DOC disappeared sequentially, starting with the higher molecular weight material and proceeding to lower molecular weight material. The increase in F2 and F3 is attributed to the metabolites of F1 falling in the size and property range corresponding to F2 and F3, and the order of disappearance is likely directly related to the differences in adsorbed amounts of F1–F3 onto the photocatalyst surface, as discussed in Sect. 9.3.1.2. Thus, a largely unselective process (due to the high reactivity of •OH) shows selectivity due to adsorption effects, and this effect can be studied by examining the DOC not only as a whole but also after fractionation.

To further compare the overall changes in DOC with the changes in the individual fractions, Fig. 9.5 presents integrals of F1–F3 along with the sum of F1–F3. This figure in particular conveys a good sense of the information gained by the fractionation prior to DOC detection: while the shift from higher to lower molecular weight

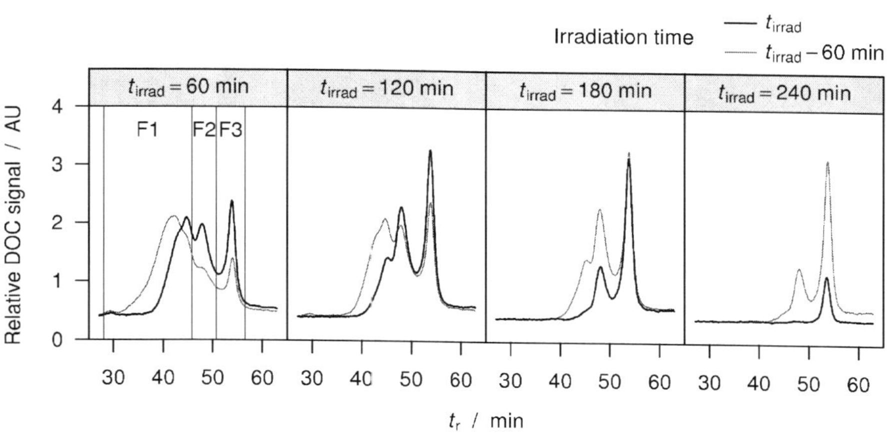

Fig. 9.4 SEC-DOC chromatograms showing the defined fractions (F1–F3) and their time evolution under UV irradiation in the presence of a photocatalyst (suspended TiO_2 nanoparticle agglomerates). Each panel compares the chromatogram after a given irradiation time (t_{irrad}) with the chromatogram at $t_{irrad} = 60$ min (modified from Tercero Espinoza et al., 2008)

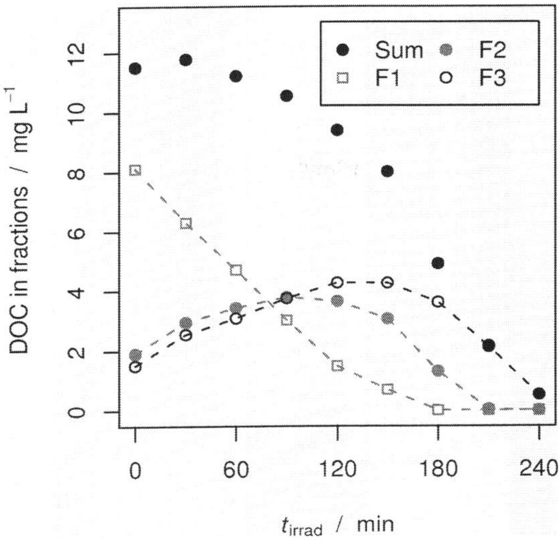

Fig. 9.5 DOC content of each fraction as a function of irradiation time (t_{irrad}) in a TiO_2 (P25) suspension. The dashed lines are intended as a guide to the eye. The filled black dots correspond to the sum of the DOC contained in F1, F2, and F3 (Tercero Espinoza et al., 2008). © 2008 Elsevier B.V. Reproduced with permission

material and the dynamics of this process are very clearly shown in the plots of the individual fractions, this information is missing from the summary view.

9.3.2 *Asymmetrical Flow Field-Flow Fractionation (AF^4) with UV and ICP-MS Detection*

9.3.2.1 Background and Experimental Details

Field-flow fractionation (FFF) is an analytical method similar to chromatography allowing the separation of colloids and nanoparticles in the size range between 1 and 50 μm (Schimpf et al., 2000). In 1966, the method was described for the first time by Giddings (Giddings, 1966). In FFF, particles are separated in a thin channel with a laminar flow field governed by a parabolic flow profile. For separation of particles, a field is applied in a direction perpendicular to the flow axis. In symmetrical flow field-flow fractionation (SFFFF or SF^4) or asymmetrical flow field-flow fractionation (AF^4), the separation is achieved by a vertical flow field. A detailed theoretical background of flow field-flow fractionation is given by Schimpf et al. (2000).

Due to the pre-concentration of the analyte in the channel during the injection and focusing step and the separation of (dissolved) sample matrix (filtration through a membrane), coupling of AF^4 and highly sensitive detectors such as ICP-MS and MALLS has been shown to be a powerful tool for the sensitive detection of nanoparticles and nanomolecules (e.g., humic substances) (Baalousha et al., 2005, 2006a, b; Beckett et al., 1987; Hassellöv et al., 2006; Magnuson et al., 2001; Prestel et al., 2005; von der Kammer et al., 2005). In particular, the stability of ENP and their interactions with heavy metals (for example Cu, Pb, and Zn) can be investigated by means of coupling AF^4/MALLS/ICP-MS.

Laponite (Laponite RD, Laporte Industries Ltd.) may be used as a model ENP in order to investigate the interactions between inorganic nanoparticles and heavy metals in the liquid phase. Laponite is a synthetic three-layer clay mineral with a distinct content of Mg and Si, a diameter of 25 nm and a thickness of 0.92 nm. Aqueous suspensions of laponite (concentration: 200 mg/L) were equilibrated for 2 weeks with solutions containing heavy metals (Cu, Pb, and Zn, respectively). The total metal concentration in the resulting suspensions was 10 μmol/L, each. As reference, laponite suspensions without addition of heavy metals were used. In the suspensions, pH values of 5 and 7 were adjusted. The suspensions were injected into a AF^4/UV/ICP-MS system. As eluent, a solution of NaCl (1 mmol/L) was used. The separation conditions for the AF^4 experiments are given in Table 9.1.

For the calibration of the AF^4 channel, monodisperse particle standards with diameters of 21 ± 1.5, 60 ± 2.5, 102 ± 3, 220 ± 6, and 486 ± 5 nm (polystyrene, Nanosphere, Duke Scientific Corporation) were used.

Laponite particles were detected in the channel outlet by means of UV light scattering detection (LCD 500, GAT Gamma Analysen Technik GmbH) at $\lambda = 254$ nm. The subsequent ICP-MS instrument enabled the detection of laponite (via the Mg and Si signals) and the detection of metals bound onto the surface of laponite particles. For determination of recovery rates, the suspensions were diluted (1:10) and directly detected (bypass injection) without a separation step.

9.3.2.2 Interactions Between Nanoparticles and Heavy Metals

The results of the coupling experiments for laponite suspensions without addition of heavy metals show a good agreement of the signals obtained by the different detectors (particle diameter) (Fig. 9.6). The hydrodynamic diameter of the particle suspensions derived using the signals of UV, Mg, and Si are similar. Thus, UV, Mg, and Si can be used as quantitative indicator for laponite in these experiments.

Table 9.1 Experimental conditions of AF^4 experiments

Parameter	Value
Flush time	300 s
Injection time	200 s
Injection flow	0.1 mL/min
Injection volume	100 μL
Laminar flow during injection	1.8 mL/min
Laminar flow during focusing	1.8 mL/min
Focusing time	60 s
Laminar flow at the channel outlet	0.3 mL/min
Channel thickness	350 μm
Membrane molecular weight cutoff	300 Da
Membrane material	Polyethersulfone

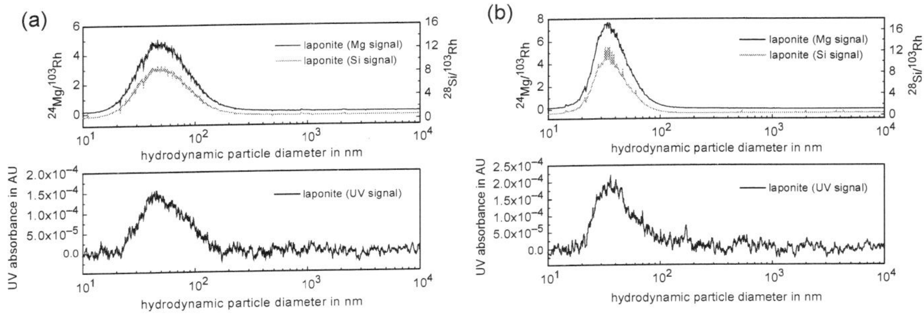

Fig. 9.6 AF^4 fractograms of laponite suspensions without addition of heavy metals at pH = 5 (**a**) and pH = 7 (**b**)

Laponite particles show a slight agglomeration at pH = 5 where a mean particle diameter of 50 nm could be observed (34 nm at pH = 7).

Recovery rates of laponite strongly depend on the pH value of the solution: decreasing pH leads to a lower recovery rate (Table 9.2). At high pH values, the surface of the membrane used in the AF^4 channel is more negatively charged, which causes repulsion forces between the membrane surface and the negatively charged laponite particles. At lower pH values, the charge of the membrane surface is less negative and the laponite particles are more likely to adsorb onto the membrane surface and to be retained in the channel.

The recovery rates obtained by the UV signals differ from those obtained using Mg and Si (Table 9.2). This deviation can be explained by the light scattering properties of particles, which is strongly dependent on the particle diameter. Scattering intensity decreases with decreasing particle diameter to the sixth power. Thus, UV detection is less sensitive for the detection of small laponite particles than the detection of their chemical components by ICP-MS.

Table 9.2 Recovery rates of laponite and heavy metals calculated from AF^4 and bypass measurements

	Recovery in %					
Sample	UV	Mg	Si	Cu	Pb	Zn
Laponite without metals (pH = 5)	31	61	64	–	–	–
Laponite without metals (pH = 7)	44	79	76	–	–	–
Laponite with Cu (pH = 5)	55	50	46	51	–	–
Laponite with Cu (pH = 7)	41	78	82	93	–	–
Laponite with Pb (pH = 5)	44	55	53	–	58	–
Laponite with Pb (pH = 7)	48	79	72	–	96	–
Laponite with Zn (pH = 5)	80	53	54	–	–	45
Laponite with Zn (pH = 7)	33	76	75	–	–	86

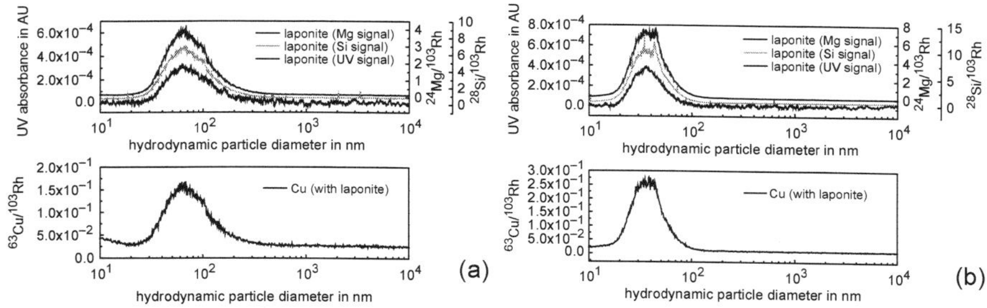

Fig. 9.7 AF^4 fractograms of laponite in presence of Cu at pH = 5 (**a**) and pH = 7 (**b**)

The fractograms of the AF^4/ICP-MS experiments for Cu-doted suspensions of laponite (Fig. 9.7) show a fraction of Cu occurring at the same elution time (or hydrodynamic particle diameter) as the peak signals indicating the laponite particles (at pH = 5 and pH = 7). It is attractive to assume that Cu is bound onto the laponite surface. By the AF^4 fractionation, the dissolved Cu species being smaller than the membrane cutoff (300 Da) are filtrated during the initial injection and focusing step. The fractograms indicate that the Cu is regularly bound to the laponite particles.

Higher recovery rates for Cu were found at pH = 7 than at pH = 5 (Table 9.2). This is due to the enhanced deprotonation of hydroxyl groups at higher pH values which leads to a more negatively charged laponite surface. This is reflected by increased absolute values of the zeta potential of the laponite particles (data not shown). Thus, the attraction of positively charged Cu cations is enhanced in the same way. For Pb and Zn, the results obtained are similar. At pH = 7, Pb and Zn are more strongly bound to laponite than at pH = 5 (Table 9.2). The agglomeration of laponite at pH = 5 could also be observed in the presence of heavy metals and confirm the results obtained with the pure laponite suspensions.

9.3.3 Transport Columns with UV and ICP-MS Detection

9.3.3.1 Background and Experimental Details

Transport columns are widely used to experimentally simulate the transport behavior of particles and to assess transport of heavy metals bound to particles in porous media. For this purpose, particle suspensions are injected as "pulses" into a column filled with a porous medium, e.g., quartz sand or natural soil material.

As mentioned in the previous sections, particles in the column outlet can be identified by means of different detectors either *on-line* or *off-line*. There are numerous examples for transport experiments with *off-line* detection of particles and inorganic compounds (Grolimund et al., 1996; Jordan et al., 1997; Lenhart and Saiers, 2003; Leon Morales et al., 2007; Schmitt et al., 2003).

The *off-line* detection, especially of heavy metals and particles, shows some significant disadvantages:

1. the collection and preparation of the collected samples (including labeling, stabilization, and dilution) is time consuming,
2. the time resolution of the breakthrough curves is low and strongly dependent on the frequency of sample collection,
3. the quantification of heavy metals and particles is difficult as samples have to be diluted in order to get a sufficient volume of sample solution for the different analyses,
4. during sample preparation, the samples might be contaminated or they might alter (agglomeration, precipitation of calcite due to the absorption of carbon dioxide into alkaline solutions).

On-line coupling of a transport column and ICP-MS can be advantageously used to investigate the transport behavior of laponite particles through porous media and to assess the co-transport of adsorbed heavy metals (Metreveli et al., 2005).

As transport column, a glass column filled with quartz sand (length: 171 mm, inner diameter: 20 mm) was used (flow rate: 1 mL/min, injection volume: 500 μL). In order to characterize the hydraulic properties of the column (porosity, residence time), a solution of $NaNO_3$ (concentration: 2 mmol/L) was used as tracer. For determination of recovery rates, samples were diluted (1:20) and directly injected in the bypass flow. The suspensions and detection systems were the same as in the AF^4 fractionation experiments (Sect. 9.3.2). Furthermore, solutions of Cu, Pb, and Zn without laponite were injected in the sand column to study the effect of laponite on metal mobility in the sand column (concentrations of Cu, Pb, and Zn: 10 μmol/L, each). As eluent, demineralized water was used. pH values in the samples and in the eluent were adjusted to 5 and 7.

9.3.3.2 Influence of Nanoparticles on the Transport of Heavy Metals

The breakthrough behavior of Pb through the column in presence and absence of laponite is shown in Fig. 9.8. As in Sect. 9.3.2, laponite was detected by UV absorption as well as via the Mg signal of the ICP-MS. At pH = 5, three fractions can be obtained for laponite exiting the column. The first fraction eluted earlier (0.85 pore volumes) than the tracer ions (NO_3^-, 1 pore volume). This effect is similar to the size exclusion effect in SEC. Similar as in the AF^4 experiments, laponite agglomerates at low pH values. During transport through the column at pH = 5, agglomerated laponite particles are mainly moving via the large pores, whereas the small dissolved ions (water molecules, tracer ions) will diffuse into smaller pores and therefore be retained. This effect has already been described in literature (Baumann et al., 2002; Harvey et al., 1989; Huber et al., 2000; Kretzschmar and Sticher, 1997; Kretzschmar et al., 1997; Metreveli et al., 2002, 2005).

The second fraction of laponite eluted after 0.98 pore volumes (no retardation), the third fraction after 1.1 pore volumes. Lowering of the pH value leads to an enhanced positive charge of the edges of the laponite particles. It can be assumed that electrostatic interactions between these positively charged edges and negatively charged quartz sand particles cause partial retardation of laponite particles. The

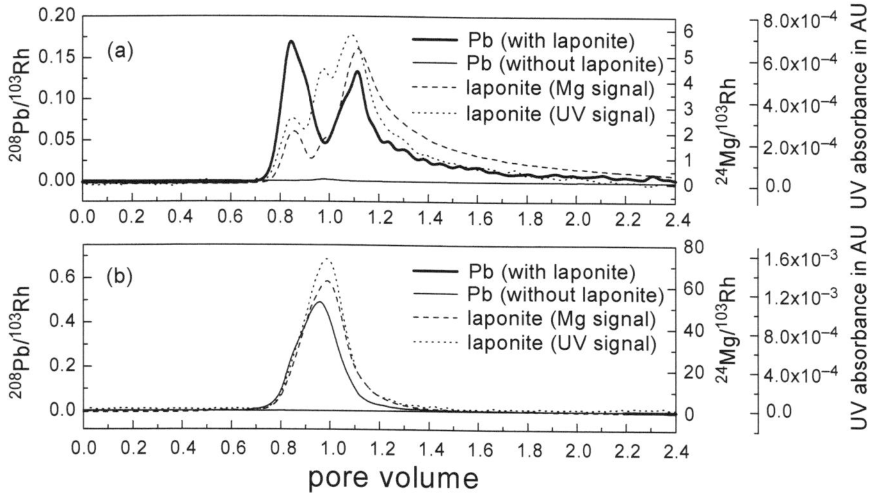

Fig. 9.8 Breakthrough of Pb in presence and in absence of laponite at pH = 5 (**a**) and pH = 7 (**b**)

noticeable tailing of the UV and the Mg signal indicates a reversible sorption of the laponite nanoparticles onto the stationary phase. In contrast to the results obtained at pH = 5, only one fraction of laponite could be detected at pH = 7 after 0.99 pore volumes (no retardation).

In order to quantify the interactions and the deposition of laponite in the column, UV and Mg signals were used to calculate the recovery rates for laponite. The recovery at pH = 5 was lower than at pH = 7. Higher pH values induce more negatively charged surface of the laponite particles. Therefore, they are more mobile due to the repulsion by the negatively charged surface of the quartz sand matrix (Table 9.3).

No significant transport of Pb through the column could be observed after injection of solutions of Pb without laponite (at pH = 5 and pH = 7). Pb^{2+} ions are most likely almost entirely adsorbed onto the quartz sand surface. In the presence

Table 9.3 Recoveries of laponite and heavy metals in the transport experiments. Recoveries of all signals were calculated for 2.4 pore volumes

	Recovery (%)				
Sample	UV	Mg	Pb	Zn	Cu
Laponite with Pb (pH = 5)	32	28	16	–	–
Laponite with Pb (pH = 7)	54	57	3.3	–	–
Laponite with Zn (pH = 5)	27	8	–	36	–
Laponite with Zn (pH = 7)	69	63	–	28	–
Laponite with Cu (pH = 5)	28	11	–	–	17
Laponite with Cu (pH = 7)	31	21	–	–	2.9

of laponite, Pb could be detected in the column outlet. At pH = 5, two peaks for Pb could be observed in the breakthrough curves after 0.84 and 1.1 pore volumes, corresponding to the first and third fraction obtained for laponite. At pH = 7, only one peak for the breakthrough of Pb was obtained occurring at the same elution volume as laponite. The results indicate a transport of Pb through the column which is facilitated by laponite. Although the mobility of laponite at pH = 7 is higher than at pH = 5, recovery of Pb was lower at pH = 7 than at pH = 5 (Table 9.3). Obviously, competing adsorption processes between laponite and the quartz sand matrix play an important role for the transport of Pb. In the experiments with Cu and Zn, a similar transport behavior of laponite and the heavy metals was observed. Laponite enhanced the transport of Cu and Zn through the column.

The results indicate that nanoparticles act as carriers for inorganic pollutants and therefore lead to their enhanced mobility in the aquatic environment.

9.3.4 Coupling of a Column Leaching Unit with a Nanoparticle Analyzer Based on Laser-Induced Breakdown Detection (NPA-LIBD)

9.3.4.1 Theoretical Background of NPA-LIBD

Many analysis techniques for characterization of nanoparticles in liquid samples are invasive and cause severe artifacts. Their application has an influence on the properties of the usually complex original system. A promising method to minimize these problems is the nanoparticle analyzer based on laser-induced breakdown detection (NPA-LIBD). The method allows the characterization of particles in liquid media down to around 10 nm in size, and of concentrations down to a few ng/L.

The fundamental idea behind the method is to focus a pulsed laser beam into a sample cell (Fig. 9.9). For igniting a plasma (dielectric breakdown of matter) on solid particles (here nanoparticles and colloids), a significantly lower energy density is required than for plasma generation in liquids without particles such as "pure" water (Radziemski and Cremers, 1989). The laser energy can be adjusted so that every time the pulsed laser beam hits a particle in the sample, the particle will be evaporated, accompanied by a plasma flash and an acoustic shock wave. The more particles there are in the sample and the bigger those particles are, the more often such plasma events can be detected and the higher the breakdown probability is (number of hits per total number of laser pulses) (Bundschuh, 1999; Kitamori et al., 1988).

Mathematically, the breakdown probability can be given as

$$W_{\mathrm{Bd}} = 1 - (1 - c_{\mathrm{P}} V_{\mathrm{P}})^{\frac{V_{\mathrm{F,eff}}(P)}{V_{\mathrm{P}}}} \tag{9.1}$$

W_{Bd}: breakdown probability
c_{P}: particle concentration in $1/\mathrm{m}^3$

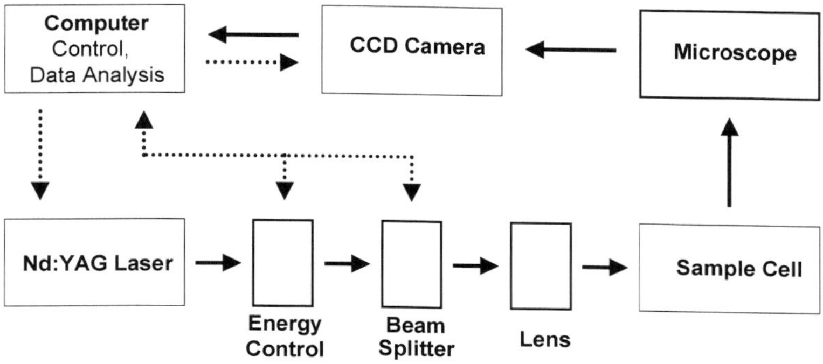

Fig. 9.9 Basic setup of the laser-induced breakdown detection (LIBD) for characterizing colloids regarding size, number density, and mass concentration

V_P: particle volume in m^3
$V_{F,eff}(P)$: effective focal volume of particle P in m^3

A detailed theoretical description is given in the literature (Bundschuh et al., 2005; Scherbaum et al., 1996). For characterization of an unknown sample, the mean particle size has to be determined first, and subsequently the particle concentration and number density can be calculated from the breakdown probability (Scherbaum et al., 1996).

Compared to laser light scattering and obscuration, the NPA-LIBD shows a higher sensitivity by several orders of magnitude, especially for particles <100 nm (Bundschuh et al., 2001; Wagner, 2005). For this reason, the application of NPA-LIBD is attractive to characterize liquid samples containing nanoparticles at concentrations typical of natural systems. Due to the sensitivity of the instrument, the NPA-LIBD usually does not require sample preparation (e.g., pre-concentration), except for dilution if the breakdown probability is too high. As the number of colloids evaporated due to plasma ignition in the course of a measurement is negligible compared to the total number of particles in the sample, the method is quasi non-invasive. In particular, NPA-LIBD can be used as highly sensitive detector for nanoparticles in analytical coupling experiments (Thang et al., 2000).

For further details concerning the instrumentation of a NPA-LIBD, we refer to the literature (Wagner, 2005).

9.3.4.2 Experimental Details

NPA-LIBD was used in order to detect the release of colloids from waste materials (Köster et al., 2007). This is of particular interest to assess the stability of these materials which were also used as secondary raw materials in road construction. In column leaching tests, a municipal waste incinerator product (MWIP) was eluted with water (Köster et al., 2007). The material is a reference material from the German Federal Institute for Materials Research and Testing. Further details about the material are given by Berger et al. (2004).

The packed columns (length: 237 mm, inner diameter: 40 mm) were eluted at a constant flow rate of 1.0 mL/min with ultrapure water (initial pH = 5) over a period of several hours, then the elution was stopped several times for a certain time (2 h up to 120 h). After the flow interruption, the leaching was continued by switching on the pump again. The interruption of the elution process experimentally simulates stagnation phases of flow.

Electrical conductivity and pH value were determined *on-line*. The highly sensitive particle analysis method NPA-LIBD has been used to monitor the release of particles *on-line* using a flow-through cell coupled directly to the column leaching unit and to assign absolute values (polystyrene equivalents) to particle size and concentration (*off-line*). For *off-line* measurements, eluates were collected in 10 mL fractions. The samples were analyzed by means of ICP-OES in order to determine the concentration of inorganic compounds (without differentiation between particles and liquid).

9.3.4.3 Particle Release Pattern

In the initial phase of the elution, high breakdown probabilities are recorded indicating the presence of high amounts of particles and colloids being washed out (Fig. 9.10). In this elution stage, the electrical conductivity is at its maximum compared to the further course of the experiment. The high electrical conductivity is caused by soluble ions (solution and desorption processes) such as sodium, potassium, and chloride (see Lager et al., 2006). pH value is highly alkaline due to dissolution of alkaline minerals.

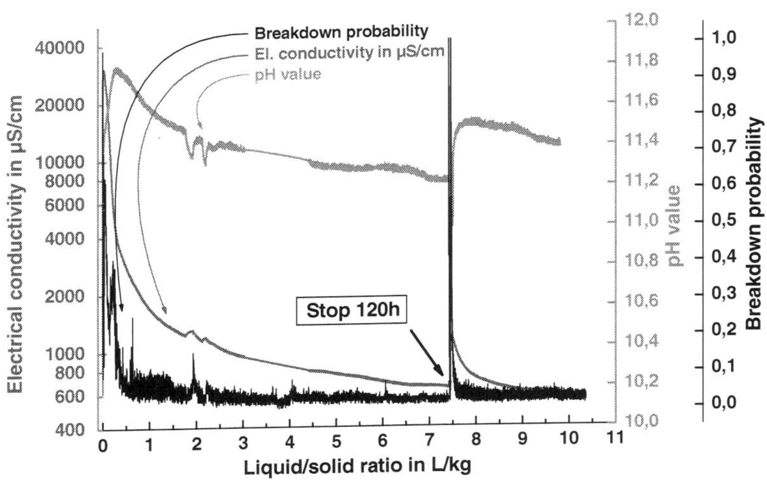

Fig. 9.10 Development of breakdown probability (particle signal), electrical conductivity, and pH value during the elution of MWIP. The flow was interrupted once for 120 h at a liquid/solid ratio of 7.5 L/kg (modified from Köster et al., 2007)

The breakdown probability as well as electrical conductivity and pH value decrease with increasing time suggesting that the initial saturation of the column with the eluent might not lead to a thorough wetting of the MWIP. This is reflected by subsequent fluctuations of the measurement signals in the initial stage of the elution (mainly seen in the particle signal). Additionally, the decreasing ionic strength causes the widening of the electrical double layer of the colloids which opposes their agglomeration (Stumm and Morgan, 1996). This leads to secondary peaks in the breakdown signal caused by smaller particles (stabilization).

Stopping the elution for a certain time (here for 120 h, column remains saturated with eluent) and subsequent continuation of the experiment shows again a sharp increase of the breakdown probability. The duration and the intensity of the temporary increase in particle mobilization strongly depend on the time of the flow interruption. The longer the stagnancy period, the bigger and longer the observed mobilization effect is (Fig. 9.11).

For interruption times shorter than a certain threshold time (with this experimental setup: about 2 h) no significant increase in the particle signal can be observed. Thus, the increasing release of particles after flow interruptions is no artifact caused by switching on the pump leading to a short and sudden increase in pressure. It rather seems likely that during the elution break silicate phases dissolve (incongruent dissolution of calcium silicate hydrates) and subsequently release colloidal fragments. The release of particles can also be observed when the electrolyte concentration is decreased (Ferstl, 2002).

The *on-line* monitoring of the particle release is highly attractive for it could be observed that the collected samples (*off-line* measurements) were not stable over

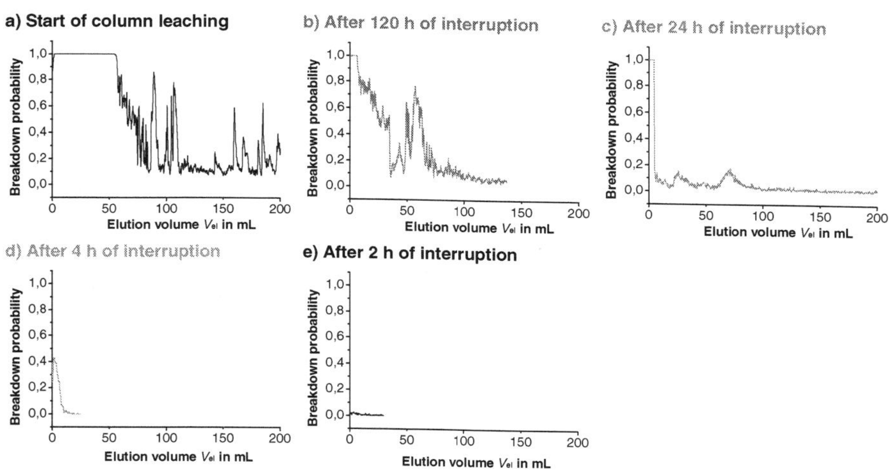

Fig. 9.11 Influence of the flow interruption time during column leaching experiments on the breakdown probability (particle signal) after re-starting the eluent flow (**b–e**). Figure (**a**) ("start of column leaching") corresponds to the beginning of the experiment and not to a flow interruption test

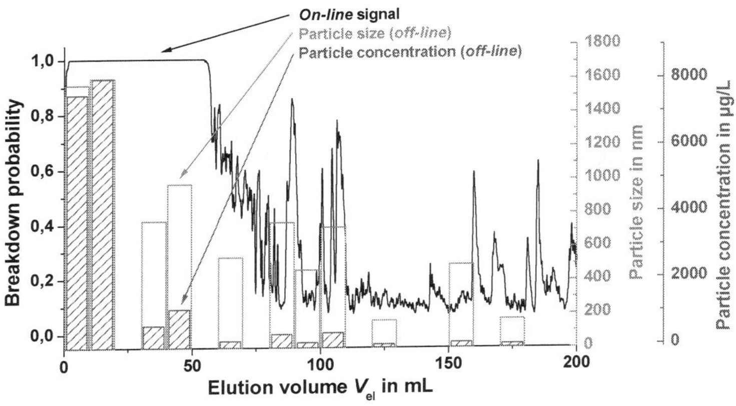

Fig. 9.12 Particle concentrations and particle sizes (*off-line*) in the effluent samples of the leaching column after first start of the experiment. For comparison, the *on-line* signal of the breakdown probability is given (modified from Köster et al., 2007)

time regarding the colloid population (Delay et al., 2006; Köster et al., 2007). For this reason, selected fractions were analyzed by NPA-LIBD in order to obtain information on particle size and concentration immediately after completion of the column leaching experiments. Figure 9.12 gives an example of the results obtained for the *off-line* measurements in comparison to the *on-line* detection.

The results for the collected sample fractions are mean values with regard to the achievable resolution defined by the number of sample fractions per time interval. In the figure, this is symbolized by columns of a certain width. Time resolution in the *on-line* mode is higher than in the *off-line* mode. Correspondingly, the particle diameters and the particle concentrations in the fractions are mean values not reflecting the maximum values possible in the column outlet.

After the start of the column leaching experiments, a high release of silicon, calcium, chromium, and copper was observed (Fig. 9.13). The concentrations decreased with increasing elution time. The concentrations determined by ICP-OES include both dissolved and colloidal metals and metal compounds. After a flow interruption of 120 h, the element release follows the same pattern again although the initial concentrations are lower compared to those obtained after the first start. The possible explanation for the increase in concentrations during flow interruptions is the above-mentioned dissolution of solid phases such as silicates (Ferstl, 2002). Since the breakdown probability is increased at the same time, part of the released heavy metals may be bound to colloids. To finally examine the composition of the particles released, further experiments are necessary.

Coupling a column leaching unit with NPA-LIBD allows the *on-line* characterization of solid materials regarding their stability (corrosion and weathering) and their release of (nano)particles. This is of special interest for the examination of nanocomposites such as polymer matrices containing nanoparticles.

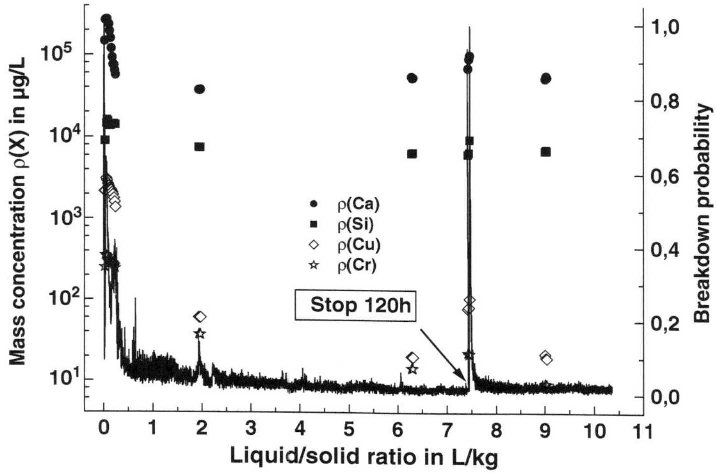

Fig. 9.13 Heavy metal mass concentrations versus liquid/solid ratio. Both concentration and breakdown probability (particle signal) fall off simultaneously, thus the mobilized heavy metals may be partly colloid bound (modified from Köster et al., 2007)

9.4 Future Challenges

The examples of analytical coupling techniques for the quantification of nanoparticles and to characterize their interactions with water constituents reflect the state of the art in this field.

One of the key challenges will be the detection of ENP (concentration, elemental composition) at levels typical of environmental samples, including disturbances by natural nanoparticles. For this purpose, the further development and improvement of adequate sample pretreatment methods (separation, pre-concentration) and highly sensitive detection systems is necessary. Clear windows for influential parameters on the surface interactive forces have to be defined and included in interpretation of particle distribution and transport.

In this context, detailed investigations of the influence of NOM on the agglomeration and transport behavior of ENP in environmental aqueous systems are highly important, as NOM is omnipresent in aquatic environmental systems.

Finally, the methods and coupling techniques for the characterization of ENP have to be standardized (e.g., DIN) in order to have a common base for data collection and data evaluation.

References

Baalousha M, von der Kammer F, Motelica-Heino M, Le Coustumer P (2005) Natural sample fractionation by FlFFF-MALLS-TEM: sample stabilization, preparation, pre-concentration and fractionation. J Chromatogr A 1093: 156–166.

Baalousha M, von der Kammer F, Motelica-Heino M, Hilal HS, Le Coustumer P (2006a) Size fractionation and characterization of natural colloids by flow-field flow fractionation coupled to multi-angle laser light scattering. J Chromatogr A 1104: 272–281.

Baalousha M, von der Kammer F, Motelica-Heino M, Baborowski M, Hofmeister C, Le Coustumer P (2006b) Size-based speciation of natural colloidal particles by flow field flow fractionation, inductively coupled plasma-mass spectroscopy, and transmission electron microscopy/X-ray energy dispersive spectroscopy: colloids-trace element interaction. Environ Sci Technol 40: 2156–2162.

Baumann T, Müller S, Niessner R (2002) Migration of dissolved heavy metal compounds and PCP in the presence of colloids through a heterogeneous calcareous gravel and a homogeneous quartz sand - pilot scale experiments. Water Res 36: 1213–1223.

Beckett R, Jue Z, Giddings JC (1987) Determination of molecular weight distributions of fulvic and humic acids using flow field-flow fractionation. Environ Sci Technol 21: 289–295.

Benn TM, Westerhoff P (2008) Nanoparticle silver released into water from commercially available sock fabrics. Environ Sci Technol 42: 4133–4139.

Berger W, Kalbe U, Eckardt J, Fischer H, Jansky HJ (2004) Aufbereitung von Referenzmaterialien zur Untersuchung der Eluierbarkeit von Schadstoffen. (Processing of reference materials for examination of contaminant leaching behaviour). Aufbereitungstechnik 45: 37–43.

Betorelle F, Wilhelm C, Roger J, Gazeau F, Ménager C, Cabuil V (2006) Fluorescence-modified superparamagnetic nanoparticles: intracellular uptake and use in cellular imaging. Langmuir 22: 5385–5391.

Bissen M, Frimmel FH (2000) Speciation of As(III), As(V), MMA and DMA in contaminated soil extracts by HPLC-ICP/MS. Fres J Anal Chem 367: 51–55.

Biswas P, Wu CY (2005) Nanoparticles and the environment. J Air Waste Manage Assoc 55: 708–746.

Braun AM, Maurette MT, Oliveros E (1991) Photochemical Technology. Wiley, Chichester.

Brunner T, Wick P, Manser P, Spohn P, Grass R, Limbach L, Bruinink A, Stark W (2006) In vitro cytotoxicity of oxide nanoparticles: comparison to asbestos, silica, and the effect of particle solubility. Environ Sci Technol 40: 4374–4381.

Bundschuh T (1999) Entwicklung und Anwendung der Laser-induzierten Breakdown-Detektion zur Quantifizierung aquatischer Kolloide und Actinidenkolloide. Thesis, TU München.

Bundschuh T, Hauser W, Kim JI, Knopp R, Scherbaum FJ (2001) Determination of colloid size by 2-D optical detection of laser induced plasma. Coll Surf A 180: 285–293.

Bundschuh T, Wagner T, Köster R (2005) Laser-induced breakdown detection (LIBD) for the highly sensitive quantification of aquatic colloids. Part I: Principle of LIBD and mathematical model. Particle and Particle Sys Char 22: 172–180.

Chithrani D, Ghazani A, Chan W (2006) Determining the size and shape dependence of gold nanoparticle uptake into mammalian cells. Nano Lett 6: 662–668.

Cumbal L, Sengupta AK (2005) Arsenic removal using polymer-supported hydrated iron(III) oxide nanoparticles: role of Donnan membrane effect. Environ Sci Technol 39: 6508–6515.

Delay M, Wagner TU, Köster R, Frimmel FH (2006) Kopplung einer Säulenelutions-Einheit mit Laser-induzierter Breakdown-Detektion (LIBD) zur empfindlichen Detektion von Kolloiden. GIT 50: 552–555.

Diegoli S, Manciulea AL, Begum S, Jones IP, Lead JR, Preece JA (2008) Interaction between manufactured gold nanoparticles and naturally occurring organic macromolecules. Sci Tot Envir 402: 51–61.

Doll T, Frimmel FH (2005) Cross-flow microfiltration with periodical back-washing for photocatalytic degradation of pharmaceutical and diagnostic residues – Evaluation of the long-term stability of the photocatalytic activity of TiO_2. Water Res 39: 847–854.

Dunemann L, Begerow J (1995) Kopplungstechniken zur Elementspeziesanalytik. VCH Verlagsgesellschaft mbH, Weinheim.

Dunphy Guzmán K, Taylor M, Banfield J (2006) Environmental risks of nanotechnology: national nanotechnology initiative funding 2000–2004. Environ Sci Technol 40: 1401–1407.

Fahrner W (ed.) (2005) Nanotechnology and nanoelectronics. Materials, Devices, Measurement Techniques. Springer, Berlin.

Ferstl W (2002) Physikalisch-chemische Charakterisierung von Kolloiden in Wasser/Reststoff-Systemen: Kolloidgetragene Schwermetallmobilisierung in Schlacken. FZKA-Bericht 6736. Forschungszentrum Karlsruhe, Karlsruhe.

Fleischer T (2002) Technikfolgenabschätzung zur Nanotechnologie – Inhaltliche und konzeptionelle Überlegungen. Technikfolgenabschätzung – Theorie und Praxis 11: 112–124.

Frimmel FH (1998) Characterization of natural organic matter as major constituents in aquatic systems. J Contam Hydrol 35: 201–216.

Giddings JC (1966) A new separation concept based on a coupling of concentration and flow nonuniformities. Sep Sci 1: 123–125.

Grolimund D, Borkovec M, Barmettler K, Sticher H (1996) Colloid-facilitated transport of strongly sorbing contaminants in natural porous media: a laboratory column study. Environ Sci Technol 30: 3118–3123.

Grüne M, Kernchen R, Kohlhoff J, Kretschmer T, Luther W, Neupert U, Notthoff C, Reschke R, Wessel H, Zach HG (2005) Nanotechnologie. Grundlagen und Anwendungen. Fraunhofer IRB Verlag, Stuttgart.

Harvey RW, George LH, Smith RL, LeBlanc DR (1989) Transport of microspheres and indigenous bacteria through a sandy aquifer: results of natural- and forced-gradient tracer experiments. Environ Sci Technol 23: 51–56.

Hasselöv M, Lyvén B, Haraldsson C, Sirinawin W (1999a) Determination of continuous size and trace element distribution of colloidal material in natural water by on-line coupling of flow field-flow fractionation with ICPMS. Anal Chem 71: 3497–3502.

Hasselöv M, Lyvén B, Beckett R (1999b) Sedimentation field-flow fractionation coupled online to inductively coupled plasma mass spectrometry – New possibilities for studies of trace metal adsorption onto natural colloids. Environ Sci Technol 33: 4528–4531.

Hassellöv M, von der Kammer F, Beckett R (2006) Characterization of aquatic colloids and macromolecules by field-flow fractionation. In: Wilkinson KJ, Lead JR (eds.) Environmental Colloids and Particles Behaviour, Separation and Characterisation. IUPAC Series on Analytical and Physical Chemistry of Environmental Systems. John Wiley and Sons, Chichester, pp. 223–276.

Huber SA, Balz A, Frimmel FH (1994) Identification of diffuse and point sources of dissolved organic carbon (DOC) in a small stream (Alb, Southwest Germany), using gel filtration chromatography with high-sensitivity DOC detection. Fresen J Anal Chem 350: 496–503.

Huber N, Baumann T, Niessner R (2000) Assessment of colloid filtration in natural porous media by filtration theory. Environ Sci Technol 34: 3774–3779.

Hyung H, Kim JH (2008) Natural organic matter (NOM) adsorption to multi-walled carbon nanotubes: effect of NOM characteristics and water quality parameters. Environ Sci Technol 42: 4416–4421.

Hyung H, Fortner JD, Hughes JB, Kim JH (2007) Natural organic matter stabilizes carbon nanotubes in the aqueous phase. Environ Sci Technol 41: 179–184.

Joo SH, Cheng IF (2006) Nanotechnology for Environmental Remediation. Springer, Berlin.

Jordan RN, Yonge DR, Hathhorn WE (1997) Enhanced mobility of Pb in the presence of dissolved natural organic matter. J Contam Hydrol 29: 59–80.

Kitamori T, Yokose K, Suzuki K, Sawada T, Goshi Y (1988) Laser breakdown acoustic effect of ultrafine particle in liquids and its application to particle counting. Jpn J Appl Phys 27: L983–L985.

Köster R, Wagner T, Delay M, Frimmel FH (2007) Release of contaminants from bottom ashes – colloid facilitated transport and colloid trace analysis by means of laser-induced breakdown detection (LIBD). In: Frimmel FH, von der Kammer F, Flemming H-K (eds) Colloidal Transport in Porous Media. Springer-Verlag, Berlin.

Kretzschmar R, Borkovec M, Grolimund D, Elimelech M (1999) Mobile subsurface colloids and their role in contaminant transport. Advan Agron 66: 121–193.

Kretzschmar R, Barmettler K, Grolimund D, Yan Y, Borkovec M, Sticher H (1997) Experimental determination of colloid deposition rates and collision efficiencies in natural porous media. Water Resour Res 33: 1129–1137.

Kretzschmar R, Sticher H (1997) Transport of humic-coated iron oxide colloids in a sandy soil: influence of Ca^{2+} and trace metals. Environ Sci Technol 31: 3497–3504.

Krug H (2005) Auswirkungen nanotechnologischer Entwicklungen auf die Umwelt. Ecomed Verlag, Landsberg.

Krug HF, Kern K, Diabaté S (2004) Toxikologische Aspekte der Nanotechnologie. Versuch einer Abwägung. Technikfolgenabschätzung – Theorie und Praxis 13:58–64.

Kutsuna S, Toma M, Takeuchi K, Ibusuki T (1999) Photocatalytic degradation of some methyl perfluoroalkyl ethers on TiO_2 particles in air: the dependence on the dark-adsorption, the products, and the implication for a possible tropospheric sink. Environ Sci Technol 33: 1071–1076.

Lager T, Delay M, Karius V, Hamer K, Frimmel FH, Schulz HD (2006) Determination and quantification of the release of inorganic contaminants from municipal waste incineration ash. Acta Hydrochim Hydrobiol 34: 73–85.

Lankes U, Müller MB, Weber M, Frimmel FH (2009) Reconsidering the quantitative analysis of organic carbon concentrations in size exclusion chromatography. Water Res, doi:10.1016/j.watres.2008.11.025.

Lead JR, Wilkinson KJ (2007) Environmental colloids and particles: current knowledge and future developments. In: Wilkinson KJ, Lead JR (eds.) Environmental Colloids and Particles: Behaviour, Separation and Characterisation. John Wiley & Sons, New York.

Lecoanet H, Wiesner M (2004a) Velocity effects on fullerene and oxide nanoparticle deposition in porous media. Environ Sic Technol 38: 4377–4382.

Lecoanet H, Bottero J, Wiesner M (2004b) Laboratory assessment of the mobility of nanomaterials in porous media. Environ Sci Technol 38: 5164–5169.

Lenhart JJ, Saiers JE (2003) Colloid mobilization in water-saturated porous media under transient chemical conditions. Environ Sci Technol 37: 2780–2787.

Leon Morales CF, Strathmann M, Flemming H-C (2007) Influence of biofilms on the movement of colloids in porous media. Implications for colloid facilitated transport in subsurface environments. Water Res 41: 2059–2068.

Limbach LK, Li Y, Grass R, Brunner T, Hintermann M, Muller M, Gunther D, Stark W (2005) Oxide nanoparticle uptake in human lung fibroblasts: effects of particle size, agglomeration, and diffusion at low concentrations. Environ Sci Technol 39: 9370–9376.

Limbach LK, Wick P, Manser P, Grass RN, Bruinink A, Stark WJ (2007) Exposure of engineered nanoparticles to human lung epithelial cells: influence of chemical composition and catalytic activity on oxidative stress. Environ Sci Technol 41: 4158–4163.

Liu Y, Majetich SA, Tilton RD, Sholl D, Lowry GV (2005) TCE Dechlorination rates, pathways, and efficiency of nanoscale iron particles with different properties. Environ Sci Technol 39: 1338–1345.

Löchtefeld S, Kühr AK, Claus F (2005) Synthetische Nanopartikel. Ergebnisse der Stakeholderbefragung, im Auftrag des Umweltbundesamtes. iku GmbH, Dortmund, im Auftrag des Umweltbundesamtes.

Long T, Saleh N, Tilton R, Lowry G, Veronesi B (2006) Titanium dioxide (P25) produces reactive oxygen species in immortalized brain microglia (BV2): implications for nanoparticle neurotoxicity. Environ Sci Technol 40: 4346–4352.

Luther W (ed.) (2004) Industrial Application of Nanomaterials – Chances and Risks. Technology Analysis. VDI Technologiezentrum, Düsseldorf.

Lyon D, Adams L, Falkner J, Alvarez P (2006) Antibacterial activity of fullerene water suspensions: effects of preparation method and particle size. Environ Sci Technol 40: 4360–4366.

Magnuson ML, Lytle DA, Frietch CM, Kelty CA (2001) Characterization of submicrometer aqueous iron(III) colloids formed in the presence of phosphate by sedimentation field flow fractionation with multiangle laser light scattering detection. Anal Chem 73: 4815–4820.

Masciangioli T, Zhang W-X (2003) Environmental technologies at the nanoscale. Environ Sci Technol 37: 102A–108A.

Mattigod SV, Fryxell GE, Alford K, Gilmore T, Parker K, Senre J, Engelhard M (2005) Functionalized TiO_2 nanoparticles for use for in situ anion immobilization. Environ Sci Technol 39: 7306–7310.

Metreveli G, Kaulisch EM, Frimmel FH (2005) Coupling of a column system with ICP-MS for the characterisation of colloid-mediated metal(loid) transport in porous media. Acta Hydrochim Hydrobiol 33: 337–345.

Metreveli G, Specht C, Kaulisch E-M, Delay M, Frimmel FH (2002) Kolloidaler Transport von Metallionen an Laponit (Modellsilikat). Mitteilungen der Deutschen Bodenkundlichen Gesellschaft 99:199–200.

Mueller NC, Nowack B (2008) Exposure modeling of engineered nanoparticles in the environment. Environ Sci Technol 42: 4447–4453.

Nel A, Xia T, Li N (2006) Toxic potential of materials at the nanolevel. Science 311: 622–627.

Nowack B, Bucheli TD (2007) Occurrence, behavior and effects of nanoparticles in the environment. Environ Pollut 150: 5–22.

Obare SO, Meyer GJ (2005) Nanostructured materials for environmental remediation of organic contaminants in water. J Environ Sci Health A 39: 2549–2582.

Oberdörster G, Oberdörster E, Oberdörster J (2005) Nanotoxicology: an emerging discipline evolving from studies of ultrafine particles. Environ Health Perspect 113: 823–839.

Oppenländer T (2003) Photochemical Purification of Water and Air. Wiley VCH, Weinheim.

Pal S, Tak YK, Song JM (2007) Does the antibacterial activity of silver nanoparticles depend on the shape of the nanoparticle? A study of the gram-negative bacterium *Escherichia coli*. Appl Environ Microbiol 73: 1712–1720.

Paschen H, Coenen C, Fleischer T, Grünwald R, Oertel D, Revermann C (2004) Nanotechnologie. Forschung, Entwicklung, Anwendung. Springer, Berlin.

Pelley AJ, Tufenkji N (2008) Effect of particle size and natural organic matter on the migration of nano- and microscale latex particles in saturated porous media. J Colloid Interface Sci 321(1): 74–83.

Prestel H, Schott L, Niessner R, Panne U (2005) Characterization of sewage plant hydrocolloids using asymmetrical flow field-flow fractionation and ICP-mass spectrometry. Water Res 39: 3541–3552.

Radziemski LJ, Cremers DA (1989) Laser-Induced Plasmas and Applications. Marcel Dekker Inc, New York.

Roco MC (2005) Environmentally responsible development of nanotechnology. Environ Sci Technol 39: 106A–112A.

Rothen-Rutishauser BM, Schurch S, Haenni B, Kapp N, Gehr P (2006) Interaction of fine particles and nanoparticles with red blood cells visualized with advanced microscopic techniques. Environ Sci Technol 40: 4353–4359.

Scherbaum FJ, Knopp R, Kim JI (1996) Counting of particles in aqueous solutions by laser induced breakdown photoacoustic detection. Appl Phys B 63: 299–306.

Schimpf ME, Caldwell K, Giddings JC (2000) Field-Flow Fractionation Handbook. Wiley-Interscience, New York.

Schmitt D, Müller MB, Frimmel FH (2000) Metal distribution in different size fractions of natural organic matter. Acta Hydrochim Hydrobiol 28: 400–410.

Schmitt D, Saravia F, Frimmel FH, Schuessler W (2003) NOM-facilitated transport of metal ions in aquifers: importance of complex-dissociation kinetics and colloid formation. Water Res 37: 3541–3550.

Schmitt D, Taylor HE, Aiken GR, Roth DA, Frimmel FH (2002) Influence of natural organic matter on the adsorption of metal ions onto clay minerals. Environ Sci Technol 36: 2932–2938.

Specht C, Frimmel FH (2000) Specific interactions of organic substances in size-exclusion chromatography. Environ Sci Technol 34: 2361–2366.

Stumm W, Morgan JJ (1996) Aquatic Chemistry: Chemical Equilibria and Rates in Natural Waters, 3rd ed. John Wiley & Sons, New York.

Tercero Espinoza LA, ter Haseborg E, Weber M, Frimmel FH (2008) Investigation of the photocatalytic degradation of brown water natural organic matter by size exclusion chromatography. Appl Catal B, doi:10.1016/j.apcatb.2008.08.013.

Thang NM, Knopp R, Geckeis H, Kim JI, Beck HP (2000) Detection of nanocolloids with flow-field flow fractionation and laser-induced breakdown detection. Anal Chem 72: 1–5.

Theodore L, Kunz RG (2005) Nanotechnology: Environmental Implications and Solutions. John Wiley and Sons, Chichester.

von der Kammer F, Baborowski M, Friese K (2005) Field-flow fractionation coupled to multi-angle laser light scattering detectors: applicability and analytical benefits for the analysis of environmental colloids. Analytica Chimica Acta 552: 166–174.

Wagner T (2005) Kolloidchemie in aquatischen Systemen – Technische und methodische Weiterentwicklung der Laser-induzierten Breakdown-Detektion (LIBD) von Nano-Teilchen. Thesis, Regensburg.

Wagner V, Wechsler D (2004) Nanobiotechnologie II: Anwendungen in der Medizin und Pharmazie. Technologienanalyse. VDI Technologienzentrum, Düsseldorf.

Wang CB, Zhang WX (1997) Synthesizing nanoscale iron particles for rapid and complete dechlorination of TCE and PCBs. Environ Sci Technol 31: 2154–2156.

Wiesner M, Lowry G, Alvarez P, Dionysiou D, Biswas P (2006) Assessing the risks of manufactured nanomaterials. Environ Sci Technol 40: 4336–4345.

Wilkinson KJ, Lead JR (eds.) (2006) Environmental Colloids and Particles Behaviour, Separation and Characterisation. IUPAC Series on Analytical and Physical Chemistry of Environmental Systems. John Wiley and Sons, Chichester.

Xie B, Xu Z, Guo W, Li Q (2008) Impact of natural organic matter on the physicochemical properties of aqueous C_{60} nanoparticles. Environ Sci Technol 42: 2853–2859.

Zhou JL, Liu R, Wilding A, Hibberd A (2007) Sorption of selected endocrine disrupting chemicals to different aquatic colloids. Environ Sci Technol 41: 206–213.

Chapter 10
Nanoparticles: Interaction with Microorganisms

Heiko Schwegmann and Fritz H. Frimmel

10.1 Introduction

Nanotechnology is concerned with materials and systems whose structures and components exhibit novel physical, chemical and biological properties and processes due to their nanoscale size (1–100 nm) (U.S. National Nanotechnology Initiative). The promising new technology supplies the basis for many new products and processes offering the ability to reduce pollution and minimise the use of resources. These immense improvements will impact society, especially in the health and manufacturing technologies. It is expected to become a US $1 trillion market by 2015 (Nel et al., 2006). In this context, nanotechnology is already discussed as the new key technology of the 21st century (Woyke, 2007).

Reason enough also to care about the backstage of the scene, i.e. used nanoparticle (NP) treatment and impact on the environment, to avoid or at least minimise adverse effects and to guarantee a sustainable NP application.

10.2 Biological Impact of Nano-products

Different products with advertised nanoparticles are already in the market (Table 10.1). They find use in a variety of different products such as electronics, cosmetics, pharmaceuticals and fields of biomedical, energy, environmental, catalytic and material applications (Nowack and Bucheli, 2007).

Single-walled carbon nanotubes (SWCNTs) have a very broad commercial application potential due to their superior mechanical, electrical and magnetic properties (Dreher, 2004). By 2011 the annual worldwide production of SWCNT is estimated to exceed 1000 t (Lekas, 2005). They are used in water purification, as catalysts and in the automobile industry. The increased production can lead to an enhanced

F.H. Frimmel (✉)
Engler Bunte Institut, Chair of Water Chemistry, Karlsruhe Institute of Technology (KIT), Engler-Bunte-Ring 1, D-76131 Karlsruhe, Germany
e-mail: fritz.frimmel@kit.edu

F.H. Frimmel, R. Niessner (eds.), *Nanoparticles in the Water Cycle*,
DOI 10.1007/978-3-642-10318-6_10,

Table 10.1 Nanoparticles in industrial products and possible exposure pathways into the water cycle

Nanoparticle	Products	Potential release into the water cycle
Ag	Coatings of textiles	Abrasion during washing
	Cosmetics, bandages	Application and removal through washing
	Paints, sprays, cleaning agents	Runoff
TiO_2	UV-protector in sunscreens	Application and removal through washing
	Paints	Runoff
CeO_2	Additive in fuels	Diesel exhausts
SiO_2	Additive in polymers, dental fillings	
Al_2O_3	Energy production	Disposal
C_{60}	Lubrication grease	Abrasion
CNT	Plastics, sporting equipment	Disposal
	Electronics	By-product emission into wastewater
Fe^0	Soil remediation	

risk of releasing the NP in the environment by discharges or spillages (Klaine et al., 2008).

The thermodynamically stable form of most metals are their oxides. Metal oxides have been produced as bulk material for many years by the industry. Nowadays nanoparticulate versions of these materials find their way into products and are widely used in a number of chemical and biological applications and in food industry. Several of the economically most important nanoparticulate metal oxides are TiO_2, ZnO, Fe_2O_3, Fe_3O_4, CeO_2 and SiO_2 (Nam and Lead, 2008).

TiO_2 and ZnO show powerful photocatalytic properties and due to their ultraviolet-blocking ability they are part of sunscreens, paints and coatings. The production of these metal oxides is estimated to reach 1000 t/year in 2010 (Pitkethly, 2004).

Iron oxide NPs have been widely used in biological applications and manufacturing pigments (Cornell and Schwertmann, 1997), but due to the superparamagnetic properties there are also arising markets in biomedical applications, e.g. as magnetic contrast agents (Thierry et al., 2007).

SiO_2 is mainly used as an additive for polymers and the bulk material is regarded to be safe. It plays an important role in the ecotoxicological assessment due to the broad usage. Most companies which use NP, work in their production steps with SiO_2-NP (Schmied and Riediker, 2007).

Zero valent metal nanoparticles are of major concern because of their ability to function in a particle-specific way. They are typically made by reduction of solutions of metal salts. The most prominent agents of these NP are Ag, Fe and Au.

Ag-nanometal has been used in many consumer applications, mostly because of its well-demonstrated use as an antimicrobial agent. Although they have been widely found in a variety of products, a concrete assessment of Ag-NP regarding environmental implications is still missing (Carlson et al., 2008). They are perhaps the most worrying NP because of their bactericidal capability and most likely access into the environment through the consumer products (Eckelman and Graedel, 2007). The most publications referring to the toxicity of NP report the effects of silver to microorganisms.

In recent years, zero valent Fe-NPs have increasingly been utilized in groundwater remediation and hazardous waste treatment (Sun et al., 2007). The power of Fe-NP is their ability to degrade chlorinated organic solvents and organic dyes (Zhang, 2003).

Besides the technical potential of NP and their promising properties for daily life applications there is a severe lack of reliable procedures and standardised methods (Mueller and Nowack, 2008) to follow their fate in the environment and their influence on complex technical systems like waste water treatment plants.

10.3 Nanoparticle Characteristics and Entry into the Environment

As of now, the risk posed by NP for the environment has not been well examined. However, it is beyond doubt, that an increased entry of these particles into soil, water and air at some stage between the production and the disposal has to be expected due to the increasing use of NP (Krug, 2005). Due to the lack of regulations and because the bulk material was regarded as safe, the majority of NP-producing companies have not done any risk assessment for NP so far (Helland et al., 2008). However, changes in size at the nanometre scale imply changes in the surface properties of the particles, which in turn can lead to a quite different behaviour in comparison to the one of the bulk material. Therefore, it is crucial to investigate the fate and behaviour of NP in the environment. Some of the main questions are Do the NP maintain their size and structure or do they agglomerate under environmental conditions? Do they undergo chemical or microbial transformations due to oxidation and reduction reactions? The distinct properties of NP, e.g. large surface to volume ratio and functional surfaces, which make them attractive in technical and economic applications, can also present a potential hazard after application.

Another important aspect is the gradual change of the NP and their properties in the environment. In case of Al_2O_3, it was shown that natural organic matter (NOM) as humic acids can build an outer layer around the NP, increasing the colloidal stability at neutral pH values (Ghosh et al., 2008). The adsorption of NOM onto the

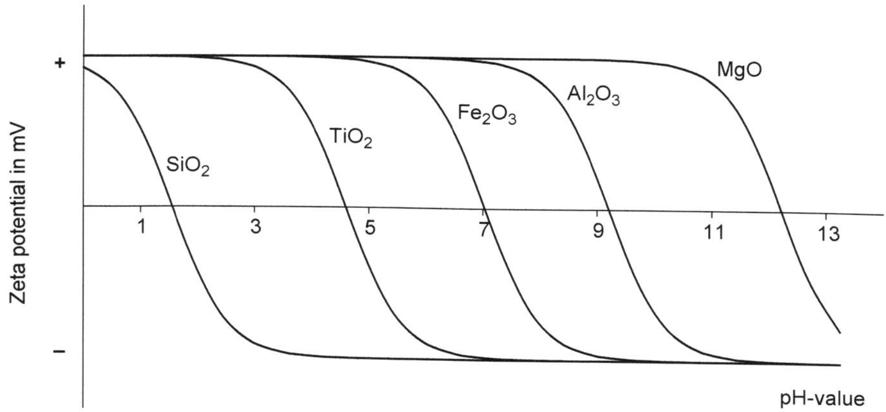

Fig. 10.1 Idealized zeta potential characteristics of various metal oxides at different pH values

surface of the NP also leads to changes in the surface charge. For example, Fe_2O_3 has a zeta potential of +19 mV in a 10 mM KCl solution at pH 8.0, but the addition of 0.5 mg/L NOM changes the potential at the same pH to –37 mV (Westerhoff et al., 2008). Absolute values of zeta potentials above 30 mV are indicators for a stable dispersion, because the electrostatic rejection inhibits the particles to agglomerate. Water bodies with a high ionic strength reduce the zeta potential of the NP resulting in larger aggregates (Müller, 1996). Figure 10.1 depicts typical zeta potential characteristics of some metal oxides.

10.4 Importance of Microorganisms

The effect of NP on different cell lines of mammalian cells has been investigated in various studies in recent years. An internalisation of NP was demonstrated resulting in miscellaneous effects (Bennat and Müller-Goymann, 2000; de la Fuente et al., 2006; Limbach et al., 2005; Limbach et al., 2007). However, the impact of NP on microorganisms is sparsely researched. Nevertheless, the effect of nanoparticles on bacteria is important because bacteria constitute the lowest level of life and hence the entrance to the food chain in many ecosystems. Furthermore, microorganisms are major contributors in the carbon-, nitrogen- and phosphorus cycle for recycling biomass in nutrients (Lyon et al., 2007). For example, nitrification and denitrification are important steps in the nitrogen cycle performed by microorganisms and therefore play a major role in wastewater treatment processes (ter Haseborg and Frimmel, 2007).

The major difference between mammalian cells (epithelial cells, fibroblast cells) and microorganisms (*E. coli*, *B. subtilis*) in connection with the impact of NP is the cell wall. While, microorganisms possess a cell wall consisting of a rigid structure surrounding the cell membrane, mammalian cells lack this envelope. The rigid structure protects the cell from environmental influences and provides their shape. Bacteria are divided into two classes according to the structure of the cell wall: Gram-positive bacteria (*B. subtilis*) exhibit a cell wall consisting of a thick peptidoglycan layer (20–80 nm); Gram-negative bacteria (*E. coli*) have only a small layer of peptidoglycan (5–10 nm), but possess an outer membrane with porins.

A further remarkable difference between bacterial and mammalian cells is the ability of endocytosis by mammalian cells. Endocytosis encompasses several diverse mechanisms by which cells internalise macromolecules and particles into transport vesicles derived from the plasma membrane. It controls entry into the cell and enables cells to take up particles up to 120 nm (Conner and Schmid, 2003). Bacteria only have porins in their membrane with a diameter of up to 5 nm.

Most eukaryotes, gram-positive and gram-negative bacteria have at neutral pH value a negative surface charge and several of these bacteria do not exhibit an iso-electric point, they are negatively charged at all pH values (Kleijn and van Leeuwen, 2000). The electrostatic attraction and repulsion between the NP and microorganisms play an important role in the adhesion of the NP to the microorganisms, and hence the observed toxicity. Goodman et al. (2004) demonstrated this with Au-NP functionalised with cationic and anionic side chains. They used vesicles composed of phosphatidylcholine (SOPC) and phosphatidylserine (SOPS) (net negative charge), and a second preparation of SOPC only (no net charge) as a reference model for cell membranes. The positively charged Au-NP lysed the (SOPC/SOPS) vesicles more efficiently. Dillen et al. (2008) also showed a greater adhesion of positively charged NP on *Pseudomonas aeruginosa* and *Staphylococcus aureus* resulting in an increase of size of the microorganisms.

The bacteriostatic and bactericidal effects of various NPs on microorganisms are presented in Tables 10.2 and 10.3. Many NPs do not exhibit a bacteriostatic effect at the applied concentrations during growth measurements. In complex growth media, high contents of proteins, polysaccharides and salts can influence the stability of the NP leading to agglomeration. Additionally, the intracellular substances of the first killed bacteria can interact with the NP causing aggregation and consequently a growth delay (Sondi and Salopek-Sondi, 2004). It seems that zero valent metals have the highest impact on microorganisms during their growth phase. While the methods of growth determination vary as depicted in Table 10.2, measurements for the determination of the lethal concentration (LC) are dominated by the principle of colony forming units (cfu) (Table 10.3). However, there are also differences in the test design for cfu. In some cases the NP are dispersed in the agar in prior to the transfer into the petri dishes. Otherwise, the NP and the microorganism are mixed in different solutions for various contact times. Afterwards the suspensions are diluted and inoculated on agar. Due to the varying cell concentrations and diverse NP dispersions with diverse ionic strengths, a direct comparison of the toxicity of the individual NP is hardly possible.

Table 10.2 Overview of EC_{50} values of various NPs on different microorganisms dependent on the particle size measurement method

Microorganism	[a]Growth media	NP	Size in nm	[b]Size determination	[e]EC_{50} ($\mu g\,mL^{-1}$)	Method	References
Chlamydomonas reinhardtii	MOPS 10 mM	Ag	44	DLS[c]	3.6	Photosynthetic yield	Navarro et al. (2008)
Escherichia coli	Luria Bertani	SiO_2	300	DLS[c]	>33,000	Optical density	Williams et al. (2006)
		SiO_2/Fe_3O_4	500		>2,200		
		Au	45		110		
Nitrifying culture	NH_4NO_3 8.3 mM	Ag	9	TEM[d]	0.5	Oxygen uptake rate	Choi and Hu (2008)
Pseudokirchneriella subcapitata	According to OECD	SiO_2	12.5	DLS[c]	>20	Cell density	Van Hoecke et al. (2008)
			27		>28.8		
Escherichia coli	BBL	Ag	14	STEM	0.45	Membrane damage by SYTO9/PI	Choi et al. (2008)
Nitrifying culture	NH_4NO_3 8.3 mM				0.3	Oxygen uptake rate	
Desmodesmus subspicatus	According to OECD	TiO_2 (irradiated in sun light simulator)	25	TEM[d]	44	Fluorescence	Hund-Rinke and Simon (2006)
			100		>50		
Escherichia coli	Luria Bertani	ZnO	250	DLS[c]	100–250	Optical density	Zhang et al. (2007)

Table 10.2 (continued)

Microorganism	[a]Growth media	NP	Size in nm	[b]Size determination	[e]EC_{50} ($\mu g\,mL^{-1}$)	Method	References
Escherichia coli	M9	Ag	9	TEM[d]	1–2	Optical density	Lok et al. (2006)
Bacillus subtilis	Nutrient broth	C_{60}/PVP (irradiated)	–	–	<13	Optical density	Kai et al. (2003)
Vibrio fischeri		C_{60}	–	–	1	Luminescence	Lyon et al. (2005)

[a] Growth media: mainly the abbreviations of the used growth media are mentioned. The exact composition of the media can be found in each article.
[b] Size determination: type of measurement to determine the size of the nanoparticles.
[c] DLS: dynamic light scattering determines the hydrodynamic diameter of the nanoparticles in the suspension.
[d] TEM: transmission electron microscopy determines the size of the primary particles in a dried state. Size distribution can be calculated by image processing.
[e] EC_{50}: refers to the concentration at which 50% of the population shows a defined effect other than lethality.

Table 10.3 Examples for the measured toxicity of various NPs on different microorganisms dependent on the particle size, the cell concentration and the type of dispersant

Microorganism	Cells (mL^{-1})	Solution	NP	Size (nm)	[a]Size determination	[b]LC_{90} ($\mu g\ mL^{-1}$)	[c]$LC_{99.9}$ ($\mu g\ mL^{-1}$)	Method	References
E. coli	10^5	Luria Bertani Agar[g]	Ag		TEM[e]	15	60	Cfu[f]	Sondi and Salopek-Sondi (2004)
E. coli	10^7	Nutrient broth	Ag (rod)	150*16	TEM[e]	8	35	Cfu[f]	Pal et al. (2007)
			Ag (spherical)	39		5	11		
Ag (triangular)	40	4	6						
E. coli	1.5×10^9	KNO_3 100 mM	CeO_2	7	DLS[d]	100–230	<500	Cfu[f]	Thill et al. (2006)
E. coli	5×10^7	Water (ultrapure)	Fe_2O_3	380	DLS[d]	>700	>700	Cfu[f]	Auffan et al. (2008)
			Fe_3O_4	50	DLS[d]	700	>700		
Fe^0	320	DLS[d]	175	>700					
E. coli	10^6	Carbonate buffer	Fe^0 (air saturated)	35	TEM[e]	70	80	Cfu[f]	Lee et al. (2008)
			Fe^0 (deaerated)			1.2	4		
S. aureus	10^3	Phosphate buffer	MgO	14	TEM[e]	10,000	10,000	Cfu[f]	Huang et al. (2005)
E. coli	6.4×10^7	NaCl 150 mM	MgO	11	TEM[e]	<1,000	<1,000	Cfu[f]	Makhluf et al. (2005)
				18		<1,000	<1,000		
				23		<1,000	>1,000		
S. aureus	9×10^7			11		<1,000	<1,000		
				18		<1,000	<1,000		
				23		<1,000	>1,000		

Table 10.3 (continued)

Microorganism	Cells (mL^{-1})	Solution	NP	Size (nm)	[a]Size determination	[b]LC_{90} ($\mu g\ mL^{-1}$)	[c]$LC_{99.9}$ ($\mu g\ mL^{-1}$)	Method	References
E. coli	5×10^7	NaCl 150 mM	Single-walled CNT	2,000*0.9	TEM[e]	5			Kang et al. (2008)
B. subtilis	10^3	MD Agar[g]	SiO_2	205	DLS[d]	2,000–5,000	5,000	Cfu[f]	Adams et al. (2006)
E. coli						>5,000	>5,000		
B. subtilis			TiO_2	330		1,000–2,000	2,000		
E. coli						>5,000	>5,000		
B. subtilis			ZnO	480		10	>500		
E. coli			>1,000	>1,000					
E. coli	2×10^2	Luria Bertani Agar[g]	ZnO	11	TEM[e]	210	>210	Cfu[f]	Brayner et al. (2006)
E. coli	3×10^5	MD Medium	C_{60}	50	DLS[d]		1.5 – 3	Cfu[f]	Lyon et al. (2005)
B. subtilis							2 – 4		

[a] Size determination: type of measurement to determine the size of the nanoparticles.
[b] LC_{90}: the concentration of nanoparticles at which 90% of the microorganisms are killed.
[c] $LC_{99.9}$: the concentration of nanoparticles at which 99.9% of the microorganisms are killed.
[d] DLS: dynamic light scattering determines the hydrodynamic diameter of the nanoparticles in the suspension.
[e] TEM: transmission electron microscopy determines the size of the primary particles in a dried state. Size distribution can be calculated by image processing.
[f] cfu: colony forming units.
[g] The nanoparticles were suspended in the agar resulting in a reduced diffusion of the NP in comparison to liquid media.

10.5 Test Systems for Toxicity Measurements

Most toxicological experiments with NP and microorganisms run the same way: various concentrations of different nanoparticles are mixed in a test system of pure water, model salt solutions, real environmental samples or culture medium and are exposed to the target microorganism or culture. After a distinct time, ranging from 10 min to a few days, the chosen endpoint or biomarker is measured. Depending on the biochemical hypothesis being tested, different endpoints can be used to investigate and measure the impact of nanoparticles on microorganisms in the experimental set-up. In most cases toxicity is determined by counting grown colonies on agar plates, but for clear evidence of the impairing influence of NP the tests have to be verified by further toxicity measurements. The principles of some of these methods are shortly introduced here.

Colony forming units (cfu): The plate count of cfu is the most frequently used method for the measurement of viable cells in bacterial populations. This method takes 24–48 h to form visible colonies. The count of cfu is based on the assumptions that under suitable culture conditions each bacterium grows and proliferates to produce a single colony. Therefore, it is assumed that the inoculum is homogeneous and that no agglomerates of cells are present (Tortora et al., 1989). In the case of measurements in the presence of NP one has to insure that the adsorbed NPs on the cells do not lead to an aggregation of the microorganisms. This would lead to excessively high toxicity values and positively false results (Fig. 10.2).

Optical density (OD): The optical density (OD) is the measurement of the turbidity of a bacterial suspension in a clear liquid medium and provides a quick estimation of the number of bacteria present in solution. The increase of the turbidity

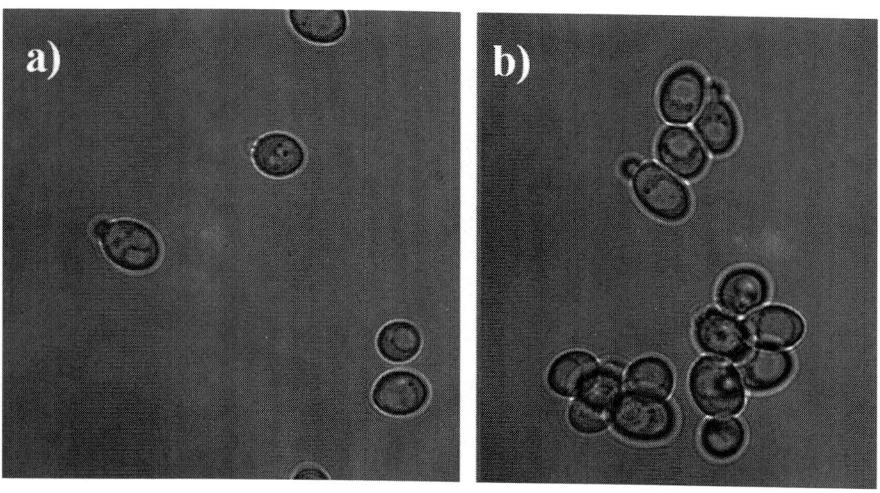

Fig. 10.2 *Saccharomyces cerevisiae* in 0.75 mM NaCl solution (**a**) without NP (**b**) with 60 mg L^{-1} Fe_2O_3

over time at growth measurements correlates with the number of living microorganisms; therefore, OD values obtained from spectrophotometric readings are directly related to the concentration of cells in the media under the Lambert–Beer law. Consequently a NP-dependent growth inhibition can be calculated (Weir et al., 2008).

Membrane damage via fluorescence: The assessment of the membrane integrity of microorganisms is often determined by using two different fluorescence dyes. One dye, mainly green fluorophores (DAPI, SybrGreen, SYTO®), is able to penetrate into all cells and binds to the nucleic acid. When used alone, it labels all microorganisms and the cell number can be determined. The second dye, in most cases the red fluorophore propidium iodide, can penetrate only into cells with a damaged membrane and has an excitation maximum close to the emission maximum of the green fluorescent dye. Therefore damaged cells reduce the green fluorescence intensity and enhance the red fluorescence intensity. The percentage of dead cells can be estimated based on the ratio of the two intensities (Singh, 2006).

Oxygen uptake rate: In aerobic cultures the oxygen uptake rate correlates with cell concentration. The oxygen uptake can be measured in sealed flasks. The produced CO_2 is absorbed by soda lime, and hence from the drop of pressure the consumed quantity of oxygen can be calculated.

Luminescence: Bioluminescent bacteria can be used to assess toxicity of inhibitory substances. Bioluminescence is produced by the enzyme luciferase, which catalyses the oxidation of the reduced luciferin by oxygen, thereby being elevated to its excited state. In non-toxic matrices the enzyme is regenerated under light emission (Fent, 1998). Toxic matter reduces the bioluminescence underlying various mechanisms like interactions with cell surface receptors, disruption of cell membrane functions, chemical reactions with cellular components or inhibition of enzyme systems (Arain, 2006). Tests with the bioluminescent bacteria *V. fischeri* require a high salt concentration which can interfere with the stability of NP.

Measurement of reactive oxygen species (ROS): 2′,7′-Dichlorodihydrofluorescein diacetate (DCDFA) is the most popular probe for measuring ROS. DCDFA can enter the cell and accumulates in the cytosol. There it will be deacetylated by esterases to 2, 7-dichlorohydrofluorescein (DCFH). This nonfluorescent product is converted by ROS to 2, 7-dichlorofluorescein DCF, which can be visualized quantitatively by fluorescence measurements (Halliwell and Whiteman, 2004).

Bacteria with specific marker genes: The use of specified marker genes can help as an additional tool to understand the mechanisms that are the reason for toxic effects. A well-working method is based on the application of recombinant bioluminescence bacteria with different promoters fused with *lux* genes. In such a set-up the defence mechanism of the cell by expression of the specific enzyme can be measured by luminescence because of the fused *lux* genes (Hwang et al., 2008). Recombinant bacteria lacking in genes for the expression of defence enzymes (e.g. the *sod*A gene expresses the enzyme superoxide dismutase which catalyses superoxide radicals in oxygen and hydrogen peroxide) can be used as marker for the damaging effects of ROS (Ruiz-Laguna et al., 2000).

10.5.1 Mode of Antibacterial Action

Nanosilver (Ag): Ag is one of the metals most commonly used in nanoparticulate form because of its strong antimicrobial activity. The mechanism by which Ag nanoparticles kill microorganisms is a controversial issue. Some researchers claim that the dissoluted Ag^+ ions from the nanoparticle are the more active species, whereas in some studies the nanoparticle itself was assumed to lead to a higher toxicity.

Most of the studies were focused on *E. coli* as an indicator microorganism. Lok et al. (2006) deduced that during growth of *E. coli* the effective antimicrobial concentration was 1,000-fold higher for Ag^+ than for Ag nanoparticles coated with BSA. The nano-Ag treated cells showed an accumulation of envelope protein precursors identified by proteomic analyses. Therefore nano-Ag may target the cell membrane, leading to a dissipation of the proton motive force. However, in growth studies with a nitrifying culture exposed to nano-Ag, there was no evidence for changes in the membrane integrity at 1 mg/L Ag measured with the fluorescence dyes SYTO 9™ and PI (Choi et al., 2008), and it was demonstrated that the inhibition of a nitrifying culture correlated well with the fraction of nano-Ag less than 5 nm from different Ag stock solutions (Choi and Hu, 2008). In addition, Morones et al. (2005) found that mainly nano-Ag in the range of 1–10 nm dispersed into different growing cultures of gram-negative bacteria, attach to the cell surface and penetrate inside the cell. Pal et al. (2007) also concluded that a reduction in the size of the nano-Ag will increase the contact area with the microorganisms. Furthermore, Smetana et al., (2008) observed that highly dispersed nanosilver coated with mercapto-1,2-propanediol could enter the cell without any toxic effects.

It is interesting to note that Ag^+ generated less intracellular ROS compared to nano-Ag. Choi and Hu, (2008) concluded that nanoparticles smaller than 5 nm can enter more easily into the cells through porins than Ag^+ because of the hydrophobic properties of the NP without damaging the cell membrane. In contrast, Sondi and Salopek-Sondi (2004) detected significant changes and damages in the cell membrane of nanosilver-treated bacteria. Obviously the formation of pits on the surfaces caused cell death by the incapability of regulating transports through the plasma membrane. Whereas Choi and Hu (2008) claim that cell death occurred through intracellular ROS generation due to nanosilver incorporation. Hwang et al. (2008) deduced from their experiments with nano-Ag and *E. coli* that Ag^+, produced by the particles outside the cell, is the main source of toxicity, because these silver ions can move into the cell and lead to a production of ROS. They further suggest that the cell dies due to the denaturation of DNA and sulphur containing proteins. The particles themselves stay outside and damage to a less extent the cell membrane, which leads to a disruption of the ion efflux system. Therefore the cell cannot extrude the harmful Ag^+ ions from the cytoplasm.

Smetana et al. (2008) reasoned that the reduced silver ions inside the cell agglomerate and reform silver metal nanoparticles. It was also concluded that very small, irregular surfaces are necessary for the high biocidal activity as Pal et al. (2007) investigated a shape-dependent interaction.

Biofilm formation on stainless steel could be reduced by depositing silver NP embedded in an organosilicon matrix. The modified surface showed anti-adhesive properties against fungal biofilms in spite of the identical hydrophobic surface properties (Guillemot et al., 2008). However, the disinfection of biofilms in tap water was ineffective (Silvestry-Rodriguez et al., 2008), and the ineffectiveness was reasoned by complexing the silver cations with the anionic extracellular polymeric substances of the biofilm.

Carbon nano tubes (CNT) and fullerenes (C_{60}): In case of C_{60} and CNT the function of the generated ROS has been discussed as the main source of damage. Kai et al. (2003) speculated that the toxic effect is based on the light-induced generation of reactive oxygen species ($O_2^{-\bullet}$ and $^{\bullet}OH$) affecting the cell membrane. However, Lyon et al. (2006) demonstrated that in the absence of light and under anaerobic conditions toxic effects also occur. Photocatalyzed ROS production is probably not the sole antibacterial mechanism associated with C_{60}. In the case of CNT it was demonstrated by Kang et al. (2008) that damage was caused by direct cell contact and physicochemical/mechanical interaction with the outer cell membrane of *E. coli*. The higher toxicity of single-walled CNT compared to multiwalled CNT can be attributed to the smaller diameter.

Iron (Fe^0) and iron oxides (Fe_2O_3, Fe_3O_4): Iron and its oxides can act as a source of ROS through oxidation. Therefore, the highest cytotoxicity towards *E. coli* was observed by Fe^0 and further decreased for Fe_3O_4 to Fe_2O_3. Mutants of *E. coli,* where the enzyme production of superoxide dismutase, an important antioxidant enzyme, is disabled, showed increased cytotoxicity. These results support the hypothesis that the oxidation of the reduced iron (Fe^0, Fe^{II}) resulting in generation of ROS is one of the main reasons for toxic effects. But no internalised NP could be observed by TEM analysis (Auffan et al., 2008). In contrast, Lee et al. (2008) demonstrated an internalisation of Fe-NP into *E. coli* by TEM analysis suggesting that Fe^{II} ions released from metallic iron can pass into the cell and form via oxidation Fe^{III} oxide particles inside. Furthermore, it appears that nano-Fe^0 penetrates the cells through the vulnerable membranes after the chemical disruption, causing more serious physical damages.

Magnesium oxide (MgO): MgO-NPs are slowly converted to hydroxides in water. Upon dissociation and the loss of hydroxide anions the NPs become positively charged (Stoimenov and Klabunde, 2005). In combination with microorganisms, this causes a coagulation into large aggregates (Stoimenov et al., 2002). Oxygen, dissolved in the solution, can generate superoxide anions $O_2^{\;-}$ which are stable and can exist in high concentrations at the surface of the NP covered with a layer of $Mg(OH)_2$ resulting in basic conditions (Huang et al., 2005). Therefore, the microorganisms suffer considerable cell wall damage upon contact (Stoimenov and Klabunde, 2005). Hence, it was demonstrated that MgO loaded with Cl_2 is very effective in disinfecting *B. cereus* spores (Koper et al., 2002).

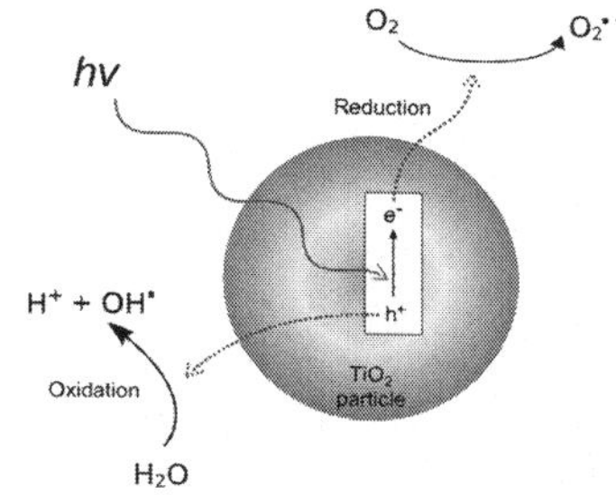

Fig. 10.3 Generation of ROS in water at irradiated surface of TiO_2

Titanium dioxide (TiO_2) and zinc oxide (ZnO): Both NPs are photosensitive and can produce ROS in the presence of light. Irradiated TiO_2 in water generates electrons in the conduction band and positive holes through irradiation with UV light. The electron excitation can produce ROS in water (Fig. 10.3). The oxidation process with the generated radicals by light irradiation is called photocatalysis.

In irradiation experiments it was demonstrated that the photocatalytic disinfection of *E. coli* is correlating with the amount of OH radicals formed (Cho et al., 2004). Additionally the toxic effect is dependent on the growth rate of the microorganism. It was shown that *E. coli* with higher specific growth rates were more susceptible than slow growing *E. coli* cells (Berney et al., 2006). Referring to Rincon and Pulgarin (2004), $HPO_4{}^{2-}$ and $HCO_3{}^{-}$ ions can retard the photocatalytic effect towards *E. coli* whereas the presence of Fe^{3+} ions in water does promote the toxic effect (Rincon and Pulgarin, 2007). In the case of ZnO, Jones et al. (2008) found a higher toxicity under normal ambient lab light and for smaller ZnO-NP (8 nm) than in darkness and with 50 nm particles. According to Brayner et al. (2006) an internalisation of ZnO-NP with a primary particle size of 11 nm into *E. coli* cells was observed by TEM analysis. In contrast, 200 nm ZnO aggregates determined by DLS measurements were not internalised into *E. coli* and led to damages at the membrane only (Zhang et al., 2007). In comparison ZnO-NP were more toxic against *B. subtilis* than TiO_2-NP. In comparison the toxicity towards *E. coli* was lower for TiO_2 and ZnO-NP. In all cases the antibacterial activity was higher under solar irradiation than in the dark (Adams et al., 2006).

10.6 Conclusion

The influence of various NPs on different microorganisms in the growth phase is evident and toxic effects in model solutions could be observed. Through the growing markets for NPs their appearance in the aquatic environment is inevitable. Lab results showed negative effects to environmentally relevant microorganisms. However, the application of the results from the lab onto the real ecosystem needs precaution. So far it is obvious that many factors will affect the toxicity of the NPs such as ionic strength of the water body, presence of NOM, surface charge and size distribution of the particles. The comparison of the results of various publications suffers from the different experiment designs applied. The problematic situation asks for the development of sound standardized protocols and methods. With respect to the controversial findings in the mode of bactericidal action or in the internalisation of the particles into living cells more systematic research is needed to clarify the dominating particle transfer mechanisms and the NP-specific potential of eco-toxicity. So far as threat for the biologically working waste water treatment and the aquatic environment cannot be excluded. On the other hand, the bactericidal potential and applicability of NPs in water treatment has not been elucidated.

Acknowledgement We thank Luis Tercero for the revision of the manuscript.

References

Adams LK, Lyon DY, Alvares PJJ (2006) Comparative eco-toxicity of nanoscale TiO_2, SiO_2, and ZnO water suspensions. Water Res 20: 3527–3532.

Arain S (2006) Microrespirometry with Sensor Equipped Microtiter plates. Dissertation, Regensburg.

Auffan M, Achouak W, Rose J, Roncato MA, Chaneac C, Waite DT, Masion A, Woicik JC, Wiesner MR, Bottero JY (2008) Relation between the redox state of iron-based nanoparticles and their cytotoxicity toward *Escherichia coli*. Environ Sci Technol 42: 6730–6735.

Bennat C, Müller-Goymann CC (2000) Skin penetration and stabilization of formulations containing microfine titanium dioxide as physical UV filter. J Cosmetic Sci 22: 271–283.

Berney M, Weilenmann HU, Ihssen J, Bassin C, Egli T (2006) Specific growth rate determines the sensitivity of *Escherichia coli* to thermal, UVA, and solar disinfection. Appl Environ Microbiol 72: 2586–2593.

Brayner R, Ferrai-Iliou R, Brivois N, Djediat S, Benedetti MF, Fievet F (2006) Toxicological impact studies based on *Escherichia coli* bacteria in ultrafine ZnO nanoparticles colloidal medium. Nano Lett 6: 866–870.

Carlson C, Hussain SM, Schrand AM, Braydich-Stolle LK, Hess KL, Jones RL, Schlager JJ (2008) Unique cellular interaction os silver nanoparticles: size dependent generation of reactive oxygen species. J Phys Chem B 112: 13608–13619.

Cho M, Chung H, Choi W, Yoon J (2004) Linear correlation between inactivation of *E. Coli* and OH radical concentration in TiO_2 photocatalytic disinfection. Water Res 38: 1069–1077.

Choi O, Hu Z (2008) Size dependent and reactive oxygen species related nanosilver toxicity to nitrifying bacteria. Environ Sci Technol 42: 4583–4588.

Choi O, Deng KK, Kim NJ, Ross L Jr., Surampalli RY, Hu Z (2008) The inhibitory effects of silver nanoparticles, silver ions, and silver chloride colloids on microbial growth. Water Res 42: 3066–3074.

Conner SD, Schmid SL (2003) Regulated portals of entry into the cell. Nature 422: 37–44.

Cornell RM, Schwertmann U (1997) The Iron Oxides. VCH Publishers, Weinheim.

De la Fuente J, Berry CC, Riehle MO, Curtis ASG (2006) Nanoparticle targeting at cells. Langmuir 22: 3286–3293.

Dillen K, Bridts C, van der Veken P, Cos P, Vandervoort J, Augustyns K, Stevens W, Ludwig A (2008) Adhesion of PLGA or Eudragit/PLGA nanoparticles to *Staphylococcus* and *Pseudomonas*. Int J Pharm 349: 234–240.

Dreher KL (2004) Health and environmental impact of nanotechnology: toxicological assessment of manufactured nanoparticles. Toxicol Sci 77: 3–5.

Eckelman MJ, Graedel TE (2007) Silver emissions and their environmental impacts: a multilevel assessment. Environ Sci Technol 41: 6283–6289.

Fent K (1998) Ökotoxikologie, Thieme, Stuttgart

Ghosh S, Mashayekhi H, Pan B, Bhowmik P, Xing B (2008) Colloidal behavior of aluminium oxide nanoparticles as affected by pH and natural organic matter. Langmuir 24: 12385–12391.
Goodman CM, McCusker CD, Yilmaz T, Rotello VM (2004) Toxicity of gold nanoparticles with cationic and anionic side chains. Bioconjugate Chem 15: 897–900.
Guillemot G, Despax B, Raynaud P, Zanna S, Marcus P, Schmitz P, Mercier-Bonin M (2008) Plasma deposition of silver nanoparticles onto stainless steel for the prevention of fungal biofilms: a case study on *Saccharomyces cerevisiae*. Plasma Process Polym 5: 228–238.
Halliwell B, Whiteman M (2004) Measuring reactive species and oxidative damage in vivo and in cell culture: how should you do it and what do the results mean? Brit J Pharmacol 142: 231–255.
Helland A, Scheringer M, Siegrist M, Kastenholz HG, Wiek A, Scholz RW (2008) Risk assessment of engineered nanomaterials: a survey of industrial approaches. Environ Sci Technol 42: 640–646.
Huang L, Li DQ, Lin YL, Wei M, Evans DG, Duan X (2005) Controllable preparation of Nano-MgO and investigation of its bactericidal properties. J Inorg Biochem 99: 986–993.
Hund-Rinke K, Simon M (2006) Ecotoxic effect of photocatalytic active nanoparticles on algae and daphnia. Environ Sci Pollut Res 13: 225–232.
Hwang ET, Lee JH, Chae Yj, Kim YS, Kim BC, Sang BI, Gu MB (2008) Analysis of the toxic mode of action of silver nanoparticles using stress-specific bioluminescent bacteria. Small 4: 746–750.
Jones N, Ray B, Ranjit KT, Manna AC (2008) Antibacterial activity of ZnO nanoparticle suspensions on a broad spectrum of microorganisms. FEMS Microbiol Lett 279: 71–76.
Kai Y, Komazawa Y, Miyajima A, Miyata N, Yamakoshi Y (2003) Fullerene as a novel photoinduced antibiotic. Fullerenes, Nanotubes and Carbon Nanostructures 11: 79–87.
Kang S, Herberg M, Rodrigues DF, Elimelech M (2008) Antibacterial effects of carbon nanotubes: size does matter. Langmuir 24: 6409–6413.
Klaine SJ, Alvarez PJJ, Batley GE, Fernandes TF, Handy RD, Lyon DY, Mahendra S, McLaughlin MJ, Lead JR (2008) Nanomaterials in the environment: behavior, fate, bioavailability and effects. Environ Toxicol Chem 27: 1825–1851.
Kleijn JM, van Leeuwen HP (2000) Chapter 2: Electrostatic and electrodynamic properties of biological interfaces. In: Bazkin A, Norde W (eds.) Physical Chemistry of Biological Interfaces. Marcel Dekker, New York.
Koper OP, Klabunde JS, Marchin GL, Klabunde KJ, Stoimenov P, Bohra L (2002) Nanoscale powders and formulations with biocidal activity toward spores vegetative cells of Bacillus species, viruses and toxins. Curr Microbiol 44: 49–55.
Krug HF (2005) Impact of nanotechnological developments on the environment. Z. Umweltchem. Ökotox.
Lee C, Kim JY, Lee WI, Nelson KL, Yoon J, Sedlak DL (2008) Bactericidal effect of zero valent iron nanoparticles on *Escherichia coli*. Envron Sci Technol 42: 4927–4933.
Lekas D (2005) Analysis of nanotechnology from an industrial ecology perspective. Part II: substance flow analysis of carbon nanotubes. Project on emerging nanotechnologies report.
Limbach LK, Li Y, Grass RK, Brunner TJ, Hintermann MA, Muller M, Gunther D, Stark WJ (2005) Oxide nanoparticle uptake in human lung fibroblasts: effects of particle size, agglomeration, and diffusion at low concentrations. Environ Sci Technol 39: 9370–9376.
Limbach LK, Wick P, Manser P, Grass RN, Bruinink A, Stark WJ (2007) Exposure of engineered nanoparticles to human lung epithelial cells: influence of chemical composition and catalytiic activity on oxidative stress. Envron Sci Technol 41: 4158–4163.
Lok CN, Ho CM, Chen R, He QH, Yu WY, Sun H, Tam KH, Chiu JF, Che CM (2006) Protemic analysis of the mode of antibacterial action of silver nanoparticles. J Proteome Res 5: 916–924.
Lyon DY, Adams LK, Falkner JC, Alvarez PJJ (2006) Antibacterial activity of Fullerene water suspensions: effects of preparation method and particle size. Environ Sci Technol 40: 4360–4366.

Lyon DY, Fortner JD, Sayes CM, Colvin VL, Hughes JB (2005) Bacterial cell association and antimicrobial activity of a C_{60} water suspension. Environ Tox Chem 24: 2757–2762.
Lyon DY, Thill A, Rose J, Alvarez PJJ (2007) Ecotoxicological impacts of nanomaterials. In: Wiesner MR, Bottero JY (eds.) Environmental Nanotechnology: Applications and Impacts of Nanomaterials. McGraw Hill, New York.
Makhluf S, Dror R, Nitzan Y, Abramovich Y, Jelinek R, Gedanken A (2005) Microwave-assisted synthesis of nanocrystalline MgO and its use as a bactericide. Adv Func Mater 15: 1708–1715.
Morones JR, Elechiguerra JL, Camacho A, Holt K, Kouri JB, Ramirez JT, Yacaman MJ (2005) The bacteriocidal effect of silver nanoparticles. Nanotechnology 16: 2346–2353.
Mueller NC, Nowack B (2008) Exposure modeling of engineered nanoparticles in the environment. Environ Sci Technol 42: 4447–4453.
Müller RH (1996) Zetapotential und Partikelladung in der Laborpraxis. Wissenschaftliche Verlagsgesellschaft mbH, Stuttgart.
Nam YJ, Lead JR (2008) Manufactured nanoparticles: an overview of their chemistry, interactions and potential environmental implications. Sci Total Environ 400: 396–414.
Navarro E, Piccapietra F, Wagner B, Marconi F, Kaegi R, Odzak N, Sigg L, Behra R (2008) Toxicity of silver nanoparticles to *Chlamydomonas reinhardtii*. Environ Sci Technol 42: 8959–8964.
Nel A, Xia T, Mädler L, Li N (2006) Toxic potentials of materials at the nanolevel. Science 311: 622–627.
Nowack B, Bucheli TD (2007) Occurrence, behavior and effects of nanoparticles in the environment. Environ Pollut 150: 5–22.
Pal S, Tak YK, Song JM (2007) Does the antibacterial activity of silver nanoparticles depend on the shape of the nanoparticle. A study of the Gram-negative bacterium *Escherichia coli*. Appl Environ Microbiol 73: 1712–1720.
Pitkethly MJ (2004) Nanomaterials – the driving force. Mat Today 7 S1: 20–29.
Rincon AG, Pulgarin C (2004) Effect of pH, inorganic ions, organic matter and H_2O_2 on *E. coli* photocatalytic inactivation by TiO_2 Implications in solar water disinfection. Appl Catal B: Environ 51: 283–302.
Rincon AG, Pulgarin C (2007) Absence of *E. coli* regrowth after Fe^{3+} and TiO_2 solar photoassisted disinfection of water in CPC solar photoreactor. Catal Today 124: 204–214.
Ruiz-Laguna J, Prieto-Alamo MJ, Pueyo C (2000) Oxidative mutagenesis in *Escherichia coli* strains lacking ROS-scavenging enzymes and/or 8-Oxoguanine defenses. Environ Mol Mutagen 35: 22–30.
Schmied K, Riediker M (2008) Use of nanoparticles in swiss industry: a targeted survey. Environ Sci Technol 42: 2253–2260.
Silvestry-Rodriguez N, Bright KR, Slack DC, Uhlmann DR, Gerba CF (2008) Silver as a residual disinfectant to prevent biofilm formation in water distribution systems. Appl Environ Microbiol 74: 1639–1641.
Singh MP (2006) Rapid test for distinguishing membrane active antibacterial agents. J Microbiol Meth 67: 125–130.
Smetana AB, Klabunde KJ, Marchin GR, Sorensen CM (2008) Biocidal activity of nanocrystalline silver powders and particles. Langmuir 24: 7457–7464.
Sondi I, Salopek-Sondi B (2004) Silver nanoparticles as antimicrobial agent a case study on *E. coli* as a model for Gram-negative bacteria. J Colloid Interface Sci 275: 177–182.
Stoimenov PK, Klabunde KJ (2005) Nanotechnology in biological agent decontamination.. In: Kumar CSSR, Hormes J, Leuschner C (eds.) Nanofabrication Towards Biomedical Applications. Wiley-VCH, Weinheim.
Stoimenov PK, Klinger RL, Marchin GL, Klabunde KJ (2002) Metal oxide nanoparticles as bactericidal agents. Langmuir 18: 6679–6686.
Sun YP, Li XQ, Zhang WX, Wang HP (2007) A method for the preparation of stable dispersion of zero-valent iron nanoparticles. Colloid Surf A: Physicochem Eng Asp 308: 60–66.

Ter Haseborg E, Frimmel FH (2007) Impact of selected pollutants in synthetic industrial wastewater on nitrifying biofilms in fixed bed biofilmreactors – visualized with fluorescence in situ hybridization. Anal Lett 40: 1473–1486.
Thierry B, Majewski P, Ngothai Y, Shi Y (2007) Preparation of monodisperse functionalised superparamagnetic nanoparticles. Int J Nanotechnol 4: 523–530.
Thill A, Zeyons O, Spalla O, Chauvat F, Rose J, Auffan M, Flank AM (2006) Cytotoxicity of CeO_2 Nanoparticles for *Escherichia coli*. Physico-chemical insight of the cytotoxicity mechanism. Environ Sci Technol 40: 6151–6156.
Tortora GJ, Funke BR, Case CL (Eds.) (1989) Microbiology: An Introduction. Benjamin/ Cummings Publishing Company, Redwood City, Chapter 6.
Van Hoecke K, Schamphelaere KAC, van der Meeren P, Lucas S, Janssen CR (2008) Ecotoxicity of silica nanoparticles to the green alga Pseudokirchneriella subcapitata – importance of surface area. Environ Toxicol Chem 27: 1948–1957.
Weir E, Lawlor A, Whelan A, Regan F (2008) The use of nanoparticles in anti-microbial materials and their characterization. Analyst 133: 835–845.
Westerhoff P, Zhang Y, Crittenden J, Chen Y (2008) In: Grassian VH (eds.) Nanoscience and nanotechnology. Wiley, New Jersey, Chapter 4.
Williams DN, Ehrman SH, Pulliam Holoman TR (2006) Evaluation of the microbial growth response to inorganic nanoparticles. J Nanobiotechnol 4: 3.
Woyke A (2007) "Nanotechnology" as a new key technology? – an attempt of a historical and systematical comparison with other technologies. J Gen Philos Sci 38: 329–345.
Zhang L, Jiang Y, Ding Y, Povey M, York D (2007) Investigation into the antibacterial behaviour of suspensions of ZnO nanoparticles. J Nanopart Res 9: 479–489.
Zhang WX (2003) Nanoscale iron particles for environmental remediation: an overview. J Nanopart Res 5: 323–332.

Chapter 11
Ecotoxicology of Engineered Nanoparticles

Karl Fent

11.1 Introduction

Nanotechnology opens new applications in many fields including medicine, material science, manufacturing and various technologies. A increasing variety of products on the market contain engineered nanoparticles (ENP) (Schmid and Riediker, 2008). Metal nanoparticles are among the most widely used, but Buckminster fullerenes, carbon nanotubes and also carbon black have widespread use. ENP are in use in many industrial sectors and products, including cosmetics, paints, fillers, catalysts, semiconductors, microelectronics and drug carriers in medicine to name a few. Considerable quantities of silver, aluminium oxide, iron oxide, silica oxide, titanium dioxide (TiO_2) and zinc oxide (ZnO) ENP are applied. The specific design and synthesis of ENP that create particular physico-chemical properties has influence on their biological and environmental behaviour.

Nanomaterials enter the environment in the production process and via applications and weathering of materials, but presently, the amount of ENP in nanotechnology-based materials and their concentrations are unknown. ENP may have particular properties that may be associated with higher health and environmental risks than natural NP formed by geochemical and combustion processes and found in the atmosphere and aquatic systems. Human health risks may arise from inhalation of ENP with associated inflammation, dispersion in the human body and exposure of vulnerable organs (e.g. heart, brain) and tissues with associated toxicity. Adverse effects on the immune system, damage of cells as well as cancerogenicity may represent health risks. In the environment, ENP are dispersed, and organisms may be exposed to these nanoparticles in the aquatic environment or they may get inhaled via air. Exposure of organisms in the environment may result in adverse

---

K. Fent (✉)
School of Life Sciences, University of Applied Sciences Northwestern Switzerland, CH-4132 Muttenz, Switzerland; Department of Environmental Sciences, Swiss Federal Institute of Technology (ETH Zürich), CH-8092 Zürich, Switzerland
e-mail: karl.fent@bluewin.ch

F.H. Frimmel, R. Niessner (eds.), *Nanoparticles in the Water Cycle*,
DOI 10.1007/978-3-642-10318-6_11,

effects in cells and organs, similarly as in the human body. In addition ENP may be accumulated by organisms and they may even be accumulated in the food chain.

Due to their unique physical and chemical properties nanomaterials are extensively studied. Natural nanoparticles (NP) were investigated for their physico-chemical behaviour in the environment (Buffle and Leppard, 1995; Buffle, 2006; Novak and Bucheli, 2007), but less for human health and environmental consequences with the exception of small-sized dust particles. ENP may pose environmental risks not associated with NP due to their specific properties. Whereas nanotechnology is gathering increasing attention, some concerns have arisen with respect to potential health effects on humans and the environment, particularly in the public opinion and media. It is important to consider them as early warnings of nanotechnology for its acceptance (Hansen et al., 2008). At present there is a need to evaluate the health and environmental risks associated with ENP as an exponentially increasing number of nanoparticle-based products are developed and brought to the market.

To date, little is known about the occurrence, fate and potential effects of ENP in the environment. The potential for effects on ecosystems will depend not only on the amount of nanoparticles emitted but also on their physico-chemical characteristics (size, surface/volume ratio, shape, chemical composition), which can be influenced by environmental parameters. Exposure of organisms in the environment to ENP will occur not only via wastewater from manufacturers, but also via municipal wastewater and direct inputs from weathering and ageing of products containing nanomaterials. The exposure of organisms will largely depend not only on the concentration, size and surface characteristics of dissolved NP but also on adsorption and aggregation phenomena. Knowledge on the chemistry of natural NP and colloids in the aquatic system (Buffle and Leppard, 1995) may in part be transferred to ENP. Adsorption to environmental media such as sediments, biofilms and micro-surface layers of open waters may influence the bioavailability and thus exposure. To date, studies devoted to evaluate the bioavailability in a systematic fashion are lacking. As extrapolated from natural colloids, the ionic strength is of influence for the aggregation tendency; the high ionic strength in seawater will cause aggregation of ENP (Novak and Bucheli, 2007).

11.2 Bioavailability, Uptake and Distribution in Organisms

Methods of nanoparticle preparation for experiments and the behaviour of ENP during exposure influence the bioavailability and therefore the activity and toxicity. Surface chemistry, shape and size of ENP are important factors for the uptake, toxicokinetics and toxicity. ENP exist as emulsions or suspensions in the aqueous environment, they are not in true dissolution as are chemicals. They may form precipitate and agglomerates that differ in toxicity.

Aquatic organisms come in contact with ENP mainly through their surfaces. Uptake routes include direct ingestion and entry across epithelial surfaces such as gills or, to some extent, through body surfaces. Uptake via epithelia in gills and gut is most prominent (as with chemicals), although ENP also cover surface epithelia (e.g. cornea, skin) and the mucus layer of fish, which may result in uptake to some extent. Carbon nanotubes (CNT), for instance, were found to readily associate with gill mucus in trout (Smith et al., 2007). The same kind of nanoparticles gets taken up by various cell types through diverse pathways (Nel et al., 2006).

As aquatic organisms are exposed to ENP both via water and ingestion, gills (oxygen uptake) and gut (ingestion) are the primary and prominent tissues and organs exposed to ENP, where toxicity may occur. As in case of environmental pollutants, the bioavailability of NP is an important factor for uptake and toxicity. Absorption to surfaces, aggregation behaviour of the particles and their environmental chemistry play a dominant role. The bioavailability will be influenced by the presence of organic matter, changes of pH and ions in environmental media. Carbon black acts as an effective sorbent for organic contaminants. It was shown to reduce toxicity of pesticide to algae (Knauer et al., 2007). Based on colloid chemistry, it seems that increases in salinity influence aggregation. The high ionic strength of not only seawater but also Ca^{2+} will cause ENP to aggregate (Handy et al., 2008b). Although probably to a lesser extent, aggregates will also be bioavailable, as found with TiO_2 and C_{60} in *Daphnia* (Baun et al., 2008a). Uptake and depuration of radioactive-labelled carbon nanotubes have been studied in an oligochaete, showing that CNT are associated with sediments remaining in the guts of organisms and not adsorbed in cellular tissues, which is in contrast to the behaviour of chemical compounds. This indicates that CNT are not readily absorbed into tissues of oligochaetes (Petersen et al., 2008).

The uptake of ENP is concentration and size dependent. Passive diffusion seems most important, although in the gut, vesicular uptake (endocytosis) may be important (Moore, 2006). Transport of colloidal particles across the cell wall is not occurring in bacteria. This is possible in eukaryotes having cellular uptake processes such as endocytosis or phagocytosis. Active uptake via ATP-binding cassette (ABC)-transporters is not probable. ENP may cause inflammation of the gill and injury to the gut at milligram concentrations as found for TiO_2 (Federici et al., 2007) or CNT (Smith et al., 2007). This may facilitate the uptake, because severe inflammation or erosion harms epithelial layers. Larger moieties such as aggregates are less transferred than smaller particles. As with chemicals, uptake through the skin mucus in fish, or carapax in crustacean, and insects is not or less probable.

After crossing the cell membrane, ENP may not only be stored in vesicles, mitochondria and additional organelles within epithelial cells but also induce adverse effects. Indeed, epithelia are targeted by ENP. The small size and large surface area and the ability to generate reactive oxygen species play a major role in the toxicity of ENP; they may generate oxidative stress and cytotoxicity (Oberdörster, 2004). In human epidermal keratinocytes, quantum dots (QD) were shown to be internalized into early endosomes and then transferred to late endosomes or lysosomes (Zhang and Monteiro-Riviere, 2009). QD endocytic pathways were primarily regulated by the G protein-coupled receptor-associated pathway and low-density lipoprotein receptor/scavenger receptor. The surface coating, size and charge of QD nanoparticles are important parameters in determining how nanoparticle uptake occurs in mammalian cells. Silica ENP led to formation of protein aggregates in the cell nucleus, thus impairing nuclear function (Chen et al., 2004). Uptake of tungsten carbide nanoparticles has been studied in fish gill cell line RTgill-W1 (Kühnel et al., 2009). In these cell cultures, the localization seemed to be restricted to the cytoplasm. The current literature suggests that the uptake of ENP depends on the characteristics of the ENP (size, surface characteristics) and the specific cell type.

After absorption ENP are incorporated and they cross the epithelial cells and are further transported to the blood or haemolymph (in invertebrates) or, alternatively, through tight junctions into the blood stream, where they get transported and distributed to tissues and organs of the body. ENP are probably transported via sorption to blood proteins to other organs and tissues. They may also form aggregates in the blood. Distribution may also occur directly via cellular transport; ENP may be transported directly into the brain via the olfactory nerve through axonal transport (Oberdörster et al., 2006).

Uptake of ENP has experimentally been investigated in several organisms. Nanosized ZnO was internalized by bacteria (Brayner et al., 2006). It was also shown that multi-walled carbon nanotubes were taken up by protozoans and localized in mitochondria (Zhu et al., 2006a). Uptake of NP into zooplankton and fish has also been studied. CNT, nC_{60}, fullerene (C_{60}), single-walled carbon nanotubes (SWCNT) and TiO_2 are taken up by *Daphnia* and incorporated (Oberdörster, 2004; Oberdörster et al., 2006). Uptake of fluorescent carboxylated ENP by *Daphnia* was found to be followed by translocation from the gut to fat droplets (Fernandes et al., 2007). C_{60} nanoparticles in the digestive tract were also excreted. Translocation of polystyrene nanoparticles (20 nm) from the digestive tract to other parts of the body was observed (Baun et al., 2008b). After 48 h exposure of *Daphnia magna* to a 0.4 mg/L solution of dispersed nanotubes, the CNT comprised of 6% of the residual organism dry mass. *Daphnia* were unable to excrete the nanotubes unless they feed on algae. Residual accumulated CNT remained in the gut and were not absorbed into cellular tissues (Petersen et al., 2009). SWCNT were shown to be internalized by the ciliata *Tetrahymena thermophila*, which also causes these protozoa to aggregate, which in turn impedes their ability to ingest and digest their prey bacteria (Ghafari et al., 2008).

Water-suspended dispersed non-ionized fluorescent latex ENP (polystyrene microspheres) were used to investigate uptake, distribution and effects of nanoparticles in the eggs and bodies of medaka (Kashiwada, 2006). Different sizes of NP were adsorbed to the egg shell (chorion) of medaka eggs and accumulated in oil droplets. Interestingly, 474 nm particles and not smaller ENP (39 nm particles) showed highest uptake into eggs. These nanoparticles accumulated in the yolk and gall bladder during embryonic development. However, other ENP such as SWCNT (Cheng et al., 2007) and fluorescent silica ENP did not penetrate the chorion of fish eggs (Fig. 11.1). In contrast, in vivo imaging of transport of silver nanoparticles (5–46 nm) shows that they are transported into and out of zebrafish embryos through chorion pore canals and exhibit Brownian diffusion (Lee et al., 2007). Individual Ag nanoparticles were observed inside embryos at each developmental stage. Dose-dependent abnormalities (deformities) and mortality was a consequence.

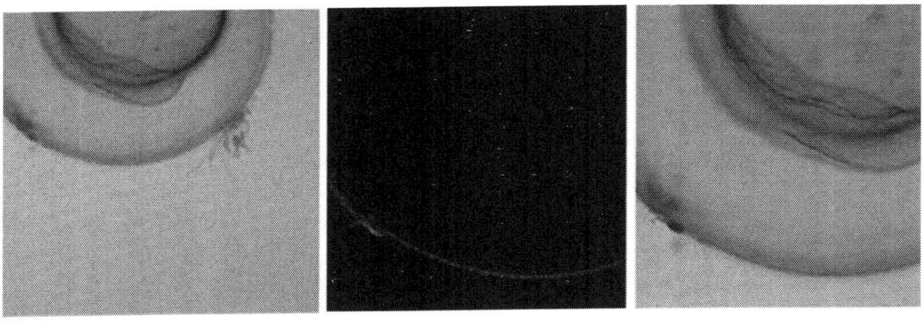

Fig. 11.1 Engineered fluorescent 200-nm silica nanoparticles (SNP) do not penetrate the egg chorion of fish and adhere to the surface. *Left*, SNP on the chorion surface, *middle* and *right*, fluorescent SNP at the surface shown in the confocal and transmission microscope (K. Fent, C. Weisbrod, A. Heller-Wirth, unpublished)

Adult fish accumulated the 39-nm sized ENP mainly in the gills and intestine when exposed to 10 mg/L ENP. They were also distributed into blood and detected in testis, liver and brain. Obviously, these ENP may penetrate the blood–brain barrier. In summary, the data from several organisms indicate that ENP may be bioavailable and taken up by organisms mainly via gill and intestine. They get distributed without metabolism into the body similar to chemical compounds. Therefore, the concept used for studying kinetics and adverse effects of chemicals applies also to ENP.

11.3 Determining Exposure and Effect Concentration

Thus far, toxicity experiments with ENP were performed using nominal concentrations; effective concentrations have not or very rarely been determined. This is not satisfactory, as similar to ecotoxicological experiments with chemicals, adsorption and aggregation processes may lower the actual exposure concentrations. Aggregation occurs when nanomaterials are introduced into media in exposure tanks and containers; hence, exposure concentrations may differ from nominal concentrations. It seems not completely clear, whether effects should be based on the total mass of ENP, its number or surface area in environmental media. Cell culture experiments indicate that the effects are dependent on the total mass, number of particles and their surface area, as well as on their concentration (Lison et al., 2008).

More emphasis is placed in recent studies to determine actual exposure concentrations (and forms) of ENP. Electron microscopy and various spectrophotometric methods applied such as MS, LC-MS and ICP-MS were used to determine organic and inorganic constituents of nanoparticles, but the latter destroy the particles. In exposure vessels ENP settle and concentrate on the bottom. Therefore, exposure concentrations fall relatively fast in the exposure containers. For instance, fullerenes form dispersions of water-soluble aggregates when agitated in fresh or deionized water for prolonged periods of time. Increasing the ionic strength of an aqueous solution (such as NaCl) containing nC_{60} results in the formation of large-sized aggregates (Brant et al., 2005b; Brant et al., 2006).

Characterization of experimental test material is important, since without characterization, it is difficult or impossible to compare studies and recognize parameters that influence toxicity. An important challenge in experimental investigations is the dissolution of NP in exposure vessels, and its characterization. Solubilizing ENP is critical, as aggregates have a different bioavailability, uptake and toxicity than single ENP. Sonication methods may be the best for solublizing SWCNT, but much easier are fullerenes (C_{60}) to test. Solubilization can be achieved by either using an organic solvent such as tetrahydrofuran (THF) or stirring for a long period of time (often 2 months), which is environmentally more accurate and relevant (Oberdörster et al., 2006). During solubilization ENP form aggregates or clusters. The solubilization method has influence on the toxicity. Tetrahydrofuran-nC_{60} were more toxic than stirred nC_{60} in Daphnia and fish (Oberdörster et al., 2006), likely due to presence of solvent residues (Brant et al., 2005a). Hence, it remains unclear whether effects are related to this solvent rather than to the particles itself (Oberdörster et al., 2006). The same is true for cell culture experiments, where formation of aggregates in cell culture media may occur. In conclusion, solubilization is a critical issue that must carefully be controlled in ecotoxicological experiments (Klaine et al., 2008).

In case of metallic ENP, toxicity may be based rather on metal residues than the nanoparticle feature itself. It was demonstrated that the toxicity of bulk metals such as ZnO is similar than that of ZnO nanoparticles. Both nano-ZnO and ZnO aqueous suspensions delayed zebrafish embryo and larva development, decreased survival and hatching and caused tissue damage at similar concentrations (Zhu et al., 2007).

11.4 Ecotoxicity

Nanomaterials have been studied in toxicology, mainly for their acute effects in rodents. The LD50 values are in the high mg/kg to g/kg range, which means that they are moderately acutely toxic. Target organs are members of the reticuloendothelial system, including liver and spleen (Stern and McNeil, 2008). Also the kidneys are targeted as they are the primary clearance route for many nanoparticles. Of course the major target of ultrafine air particles is the lung. In contrast to acute effects, the long-term effects are only little known, and the risks of ENP associated with chronic toxicity are unknown.

In ecotoxicology, ENP have been studied for uptake and toxicity only in a small number of organisms belonging to bacteria, algae, protozoa, zooplankton (*Daphnia*), oligochaetes, mussels, amphibians (*Xenopus laevis*) and fish (for review Baun et al., 2008a; Klaine et al., 2008; Navarro et al., 2008a). They belong to the classical species used in ecotoxicological testing, but whether they are the most important ecological species remains open. In general, species differences in sensitivity will also occur in the sensitivity against ENP. Furthermore, exaggerated concentrations of ENP are studied, but the potential effects at environmental realistic concentrations remain elusive.

In general, toxicity may be dependent on particle concentration, particle size and shape and surface properties (surface chemistry). A toxicological review of quantum dots (QD) suggests that the toxicity depends on physico-chemical and environmental factors (Hardman, 2006). QD size charge, concentration, outer coating bioactivity have been implicated as determining factors in toxicity. Most data on the ecotoxicology of NP were generated in classical ecotoxicity tests using established test systems similar to the regulatory framework of testing of chemicals. Thus far, mainly acute toxicity was studied in a few species, particularly *D. magna*, but chronic toxicity data are largely lacking. The acute toxicity data indicate that the acute systemic toxicity of many ENP in aquatic and terrestrial organisms appears to be rather low. This is in contrast to the pulmonary toxicity of certain ENP such as CNT that show significant toxicity in mammals and humans. However, it should be noted that the risks of ENP are related to chronic and long-term effects that are not yet adequately studied. It should also be noted that the behaviour of ENP and associated toxicity may be different at environmentally relevant concentrations as compared to the high levels used in acute toxicity studies (Gao et al., 2009). Furthermore, partial oxidation ("ageing") and surface modification in the environment may alter the toxicity of EHP (Phenrat et al., 2009).

11.4.1 Toxicity Mechanisms and Target Organs

The mechanistic basis of NP toxicity is not well understood, both in toxicology and in ecotoxicology. Toxicity studies in mammals demonstrated various types of cellular effects, mainly on the respiratory tract (Oberdörster, 2004; Oberdörster et al., 2006). Ultrafine particles and nanomaterials have experimentally quite well been studied in mammals due to their occurrence in air dust. Exposure analysis of humans via air has mainly been performed in epidemiological studies. At present air pollution is by far the most important factor for human exposure occurring via natural and traffic-derived NP. Pulmonary effects of ultrafine particles are known for long, yet the effects of ENP have been investigated only in the last few years (Oberdörster et al., 2006). Hence, respiratory toxicity and inflammation reactions to NP are known. Epidemiological studies clearly support an association between particulate air pollutants and pulmonary, cardiovascular and CNS disease (Stern and McNeil, 2008).

In short, the effects of nanoparticles in mammals (and humans) can be summarized as follows: inflammation, cytokine production, cytoskeletal changes, altered vesicular trafficking, oxidative stress, apoptosis, as well as changes in gene expression and cell signalling in response to different types of ultrafine particles. This was found in a number of mammalian cells in vitro and rodents in vivo (Oberdörster, 2004; Oberdörster et al., 2006). The targets within cells include the lysosomes. Change in lysosomal permeability and the subsequent release of lysosomal enzymes is one of the mechanisms involved in the induction of alveolar macrophages by silica microparticles (Thibodeau et al., 2004). Recently, it was shown in mice that inhaled multi-walled CNT led to suppression of systemic immune function (Mitchell et al., 2007), and this is based on the activation of cyclooxygenase enzymes in the spleen in response to a signal from the lungs (Mitchell et al., 2009).

ENP induce cytotoxicity in vitro in cell lines at high concentrations (Herzog et al., 2007). As an example, single-walled carbon nanotubes (SWCNT) were found to have a very low acute toxicity (Davoren et al., 2007). Silver ENP (15 and 30 nm) lead to more than 10-fold increase of reactive oxygen species (ROS) levels in cells exposed to 50 mg/L, suggesting that the cytotoxicity is likely to be mediated through oxidative stress. In addition, a significant inflammatory response was observed by the release of TNF-alpha, MIP-2 and IL-1beta to cell culture media (Carlson et al., 2008). In vitro cytotoxicity of many ENP has well been documented (Brunner et al., 2006). Different ENP were investigated for cytotoxicity and ability to cause DNA damage and oxidative stress in human lung epithelial cell line A549 (Karlsson et al., 2008). A high variation among different ENP was observed. CuO nanoparticles were most potent regarding cytotoxicity and DNA damage. The effects were not due to soluble metal impurities.

Mechanisms of actions known in mammals may similarly apply for other organisms. Silver ENP are well-known antibacterial agents causing toxicity via multiple mechanisms (Morones et al., 2005). They may alter the membrane properties, affecting the permeability and respiration of bacterial cells. They can also penetrate into bacteria and cause DNA damage and release toxic Ag^+ ions. In addition to direct effects, indirect effects by ENP are important. ENP may lead to physical blockage of the gills of aquatic organisms and/or digestive tract. For instance masses of black CNT were observed in the intestine of amphibian larvae, which lead to acute toxicity at high concentrations of 10–500 mg/L in addition to blockage of gill respiration (Mouchet et al., 2008).

Prime targets of ENP in aquatic organisms including fish (Handy et al., 2008a) are uptake organs such as gills and intestine. Their epithelia are exposed directly to the particles via water and food, respectively. Adsorption to and interaction with epithelial cells may harm cell membranes and cellular structures. In fact, it has been demonstrated that fish gills are sensitive to ENP such as TiO_2 (Federici et al., 2007) or single-walled carbon nanotubes (SWCNT) (Smith et al., 2007). Adsorption and attachment to the cell surface compromise its integrity and function. ENP may also indirectly cause membrane damage through the generation of ROS, leading to lipid peroxidation and damage of proteins and nucleic acids. Therefore, cell membrane damage is an important mechanism by which ENP act on the cellular level. The cytotoxic response of cells is dependent on the degree of functionalization in case of SWNT (Sayes et al., 2006). Exposure of cells in vitro increased ROS levels and reduced glutathione levels (Lin et al., 2006). The dose-dependent cytotoxicity was closely correlated to increased oxidative stress.

Additional modes of actions of ENP have been determined in mammals. Oxidative stress is induced by iron nanoparticles in human bronchial epithelial cells, leading to cytotoxicity and cell damage (Keenan et al., 2009). The response is equivalent to the response observed when cells are exposed to the same concentration of dissolved Fe(II), however. Fullerenes (C_{60}) may act as antioxidants; however, a strong oxidative potential through photoactivation has also been shown. ENP may interact with proteins directly or by producing ROS or other damaging radicals. Some ENP (e.g. fullerenes) are photoreactive. Induction of oxidative stress seems to be an important mechanism also in aquatic and terrestrial species. Fullerenes induced potentially harmful lipid peroxidation in the brain of fish (largemouth bass) (Oberdörster, 2004). In contrast, no such effects have been found in *Fundulus heteroclitus*, however (Blickley and McClellan-Green, 2008).

ENP also induce pulmonary inflammation, fibrosis, granulomatosis and emphysema in mammals. CNT have been compared to asbestos, due to their needle-like shape, raising concerns that they may lead to mesothelioma and lung cancer in mammals (Stern and McNeil, 2008). Indeed multi-walled CNT resulted in asbestos-like pathology in mice such as inflammation and formation of lesions known as granulomas (Poland et al., 2008). The asbestos-like hazard of such CNT is a cause of concern important for the production and application of such ENP, in particular at the working place. This is because CNT may have unusual toxicity properties, such as a special ability to stimulate mesenchymal cell growth and to cause granuloma formation and fibrogenesis (Donaldson et al., 2006). CNT may exhibit some of their effects through oxidative stress and inflammation.

Less is known so far in aquatic and terrestrial toxicology. Similar to mammals, possible toxicity mechanisms of ENP may include among others the following:

1. Disruption of membranes or membrane potential
2. Formation of reactive oxygen species and oxidative stress
3. Induction of apoptosis and necrosis; induction of stress-related genes
4. Oxidation and denaturation of proteins and other biomolecules

Recently it was demonstrated that fullerenes C_{60} elicit oxidative stress response in embryonic zebrafish (Usenko et al., 2008). Gene expression analysis with zebrafish using microarrays on the effects of C_{60} demonstrated alterations in several key stress response genes including glutathione-*S*-transferase, glutamate cysteine ligase, ferritin, alpha-tocopherol transport protein and heat-shock protein 70 (Usenko et al., 2008). These alterations in gene expression patterns clearly indicate induction of oxidative stress in zebrafish. Furthermore, there was an upregulation of development, cell cycle (including induction of apoptosis) and signal transduction genes.

Genotoxicity has been investigated with TiO_2 in fish cells in vitro (Vevers and Jha, 2008). Increased levels of DNA strand breaks were observed only in combination with UVA light. Obviously, genotoxicity was photo-induced. Fullerenes were found to be nonmutagenic in the Ames test and non-genotoxic in a Chinese hamster lung cell chromosomal aberration assay up to 5 mg/mL (Mori et al., 2006).

Silver ENP led to cellular and DNA damage as well as to carcinogenic and axidative stresses in medaka, as suggested by analysis of gene expression alterations (Chae et al., 2009). Silver resulted in a lower overall transcriptional stress response in the liver than silver ENP.

Recently, molecular and histological effects following exposure to 0.1 mg/L nanocopper and 1 mg/L nanosilver and nano-TiO_2 (particle sizes between 45 and 688 nm) have been studied in gills after exposure of zebrafish for 24 and 48 h (Griffitt et al., 2009). Nanocopper (100 μg/L) increased mean gill filament width. Gene expression analysis demonstrated that the exposure to each nanometal or soluble metal produced a distinct gene expression profile, suggesting that each exposure is producing a biological response by a different mechanism. The effect of each nanoparticle does not appear to be based solely on the release of soluble metal ions into the water, but by the particulate feature as well. Although nano-TiO_2 did not show toxicity, different genes are altered, demonstrating a transcriptomal effect. Whereas 1 mg/L nano-TiO_2 altered a number of genes involved in ribosome structure and activity, nanocopper upregulated over 100 genes involved in the cellular processes such as apoptosis, mitogenesis and cell proliferation, as well as in cancer progression. Interestingly, no genes involved in oxidative stress were altered. Therefore, the extent to which oxidative stress can generally be assumed as a key molecular process by which ENP act on the cellular level remains elusive.

11.4.2 Effects on Aquatic Organisms: Acute Toxicity

There are a few major classes of nanomaterials: metal and metal oxide nanoparticles, carbon-based nanoparticles and nanotubes, fullerenes and macromolecules. At present knowledge about the toxicity of these ENP is clearly not sufficient. Acute toxicity of ENP on a variety of different organisms has been studied. Most studies were performed with filter-feeding *Daphnia*, where general toxicity signs such as immobilization were usually assessed. Here, some of the most relevant findings are summarized in the following and in Table 11.1, other literature reviews are given elsewhere (Baun et al., 2008a; Klaine et al., 2008; Novak and Bucheli, 2007). So far, almost all studies deal with the acute toxicity of very high concentrations of ENP. In general, they found acute toxicities of various ENP in the range of around 1 to several mg/L (Table 11.1), and therefore it is not very high. Only a few studies show toxicity below 0.1–0.5 mg/L. Furthermore, toxicity seems to be dependent on the physico-chemical characteristics of the nanoparticles, therefore a range of non-toxic to toxic nanomaterials can be found.

High concentrations of ENP inhibit the growth of algae with EC_{50} in tens of mg/L (Hund-Rinke and Simon, 2006). Aggregates may be formed at the surface including cell walls leading to indirect effects on growth and photosynthesis. Mortality and toxicity in *Daphnia* was induced by single-walled nanotubes (SWNT) at 10 mg/L (Templeton et al., 2006), whereas the lowest observed effect concentration of filtrated fullerenes was 0.26 mg/L. Acute toxicity was also observed with inorganic ENP with decreasing toxicity for ZnO, SiO_2, TiO_2 (Adams et al., 2006).

Exposure of aquatic crustaceans (*D. magna, Artemia salina* and Gammarids) to several ENP (TiO_2, fluorescent polystyrene particles, carbon black) indicates that particles are readily ingested and accumulated in the gastrointestinal tract and distributed into body lipid droplets (Fernandes et al., 2007; Roberts et al., 2007) and

Table 11.1 Acute and chronic toxicity of engineered nanoparticles to aquatic organisms (selected examples)

Engineered nanoparticle	Size (nm)	Species	Effect	EC50/LC50/LOEC (Exposure time)	References
Ag	10–200 (mostly 25)	Algae *Chlamydomonas reinhardtii*	Inhibition of photosynthesis	829 nM (5 h)	Navarro et al. (2008b)
		Daphnia pulex	Mortality	40 μg/L (48 h)	Griffitt et al. (2008)
	90–2,000	*Ceriodaphnia dubia*	Mortality	2-46 μg/L (LC50, 48 h)	Gao et al. (2009)
	45–216	Zebrafish (*Danio rerio*)	Alteration of gene expression	1 mg/L (24 h, 48 h)	Griffitt et al. (2009)
TiO_2	25, 100	Algae *D. subspicatus*	Growth inhibition	14 mg/L	Hund-Rinke and Simon (2006) Lovern and Klaper (2006)
	10–20	*Daphnia magna*	Mortality	5.5 mg/L (48 h)	Hund-Rinke and Simon (2006)
	25	*D. magna*	Mortality	40% at 20 mg/L	Lovern et al. (2007)
	30	*D. magna*	Behavioural and physiological alterations	12.5 mg/L (48 h)	Federici et al. (2007)
		Rainbow trout (*O. mykiss*)	Alterations in gills, liver, intestine	0.1–1 mg/L (LOEL) (14 d)	
	221–688	Zebrafish (*D. rerio*)	Alteration of gene expression	1 mg/L	Griffitt et al. (2009)
ZnO	73–408 50–70	Algae *Pseudokirchneriella subcapitata*	Growth inhibition	60 μg Zn/L (72 h)	Franklin et al. (2007)
		D. magna	Mortality	3.2 mg/L (24 h)	Heinlaan et al. (2008)
		Zebrafish embryos	Development, hatching, mortality	1.79 mg/L (LC50, 96 h)	Zhu et al. (2008a)

Table 11.1 (continued)

Engineered nanoparticle	Size (nm)	Species	Effect	EC50/LC50/LOEC (Exposure time)	References
CuO		*D. pulex*	Mortality	60 µg/L (48 h)	Griffitt et al. (2008)
	30	*D. magna*	Mortality	3.2 mg/L (24 h)	Heinlaan et al. (2008)
	100–8,000	*Ceriodaphnia dubia*	Mortality	7–48 µg/L (LC50, 48 h)	Gao et al. (2009)
	95–447	Zebrafish (*D. rerio*)	Alteration of gene expression	100 µg/L (24 h, 48 h)	Griffitt et al. (2009)
CeO_2	14, 20, 29 (aggregates 275–552)	Algae *Pseudokirchneriella subcapitata* *D. magna*	Growth inhibition Mortality Reproduction	5.6 mg/L (LOEC) (72 h) 56–100 mg/L (LC50); 18–32 mg/L (21 d)	Van Hoecke et al. (2009)
SiO2	14, 60, 930	*D. magna*	Mortality	70% at 10 mg/L (24 h??)	Adams et al. (2006)
Fullerene (C_{60})	720	*D. magna*	Mortality	0.46 mg/L (48 h) (THF), 7.9 mg/L (sonicated)	Lovern and Klaper (2006)
	10–20	*D. magna*	Heart rate, behavioural changes	0.26 mg/L (1 h)	Lovern et al. (2007)
	10–200 (aggregates)	*D. magna*	Delay in moulting	2.5 and 5 mg/L (21 d)	Oberdörster et al. (2006)
	10–200 (aggregates)	Fathead minnows	Reproduction effect	1 mg/L (THF-nC_{60}), no	Oberdörster et al. (2006)
	100	Largemouth bass	Mortality	mortality in	
	not reported	Zebrafish embryos	Lipid peroxidation in brain	water-stirred	
			Mortality, delayed	1.5 mg/L	Oberdörster (2004)
		Zebrafish embryos (dechorionated)	development, edema,	0.5, 1 mg/L	Zhu et al. (2007)
			heartbeat, hatching	1.5 mg/L (96)	Usenko et al. (2008)
			Mortality, edema	0.2 mg/L (5 d)	
			Gene expression (oxidative stress)		

Table 11.1 (continued)

Engineered nanoparticle	Size (nm)	Species	Effect	EC50/LC50/LOEC (Exposure time)	References
Single-walled carbon nanotubes (SWCNT)	1.2	*D. magna*	Mortality	100% at 20 mg/L (48 h) 36% at 10 mg/L	Roberts et al. (2007)
	1	*Amphiascus tenuiramis*	Life cycle mortality Ventilation rate	0.1 mg/L (3 d)	Templeton et al. (2006) Smith et al. (2007)
	1.1	Rainbow trout (*O. mykiss*)	Gill alterations Behavioural change	0.25 mg/L (10 d)	
Multi-walled carbon nanotubes (MWCNT)	1–3	*Xenopus laevis* larvae	Mortality	85% at 500 mg/L (12 d)	Mouchet et al. (2008)
	11	Zebrafish embryos	Delayed hatching	120 mg/L	Cheng et al. (2007)
Quantum dots (QD)		*Pseudokirchneriella subcapitata*	Growth inhibition	37.1 μg/L (EC_{50}, 96 h)	Bouldin et al. (2008)
		Ceriodaphnia dubia	Mortality Lipid peroxidation	1.6 mg/L	Bouldin et al. (2008)
	CdTe QD	Mussel *Elliptio complanata*	Immune system	> 0.11 mg/L (48 h)	Gagne et al. (2008)
	CdSecore/ZnSshell-PLL, 382	Zebrafish embryo	Mortality	7 μM (Cd equiv.)	King-Heiden et al. (2009)

other parts of the body (Baun et al., 2008a). The ENP also adhered to the surfaces of organisms and may induce effects such as increased moulting of the carapace by neonates (Fernandes et al., 2007). *Daphnia* can also biomodify single-walled CNT by stripping the lipid coating, which is used as a nutrient (Roberts et al., 2007).

Thus far, only a few experiments have been conducted with saltwater species. The toxicity of fullerenes has been investigated in adult and larval fish *F. heteroclitus* (Blickley and McClellan-Green, 2008). Very little effects were noted and no median lethal concentrations could be calculated up to 10 mg/L. Aggregates of fullerenes adhered to the chorion of eggs and did not affect embryonic development or hatching. Exposed larvae exhibited a dose-dependent increase in total glutathione, which was accompanied by a decreasing trend in lipid peroxidation. Increases in glutathione were also noted in liver but not gills of adult fish.

Current knowledge of different categories of ENP on various organisms indicates the following:

Metallic ENP. High concentrations of various metallic and inorganic NP have been shown to inhibit the growth of bacteria (Morones et al., 2005) and algae (Hund-Rinke and Simon, 2006), and cause mortality to invertebrates and fish. Due to their use as bactericides, nano-sized silver particles are quite well studied in bacteria. The cells of bacteria are damaged and this is dependent on the concentration, shape and size of the particles (Morones et al., 2005; Pal et al., 2007). Nanosilver appears to be more toxic than silver anions (Lok et al., 2006). Silver nanoparticles can be toxic because of their size and shape, but on the other hand, they could be toxic because they release silver ions, which are well known for their antibacterial activity and toxicity to algae. Exposure of algae to silver nanoparticles showed that in the first hour silver ions inhibit the photosynthesis about 18 times more than nanosilver did (Navarro et al., 2008b). But thereafter, the nanoparticles continued to be even more toxic than the ions alone. Obviously, the nanoparticles contributed to the toxicity as a source of silver, which is formed in the presence of algae.

Other inorganic ENP such as TiO_2, SiO_2 and ZnO show toxicity to bacteria and other organisms. Nano-sized filtered TiO_2 induced acute toxicity to *D. magna* at 2 mg/L and higher (Lovern and Klaper, 2006). The effect was greater for these and C_{60} particles when prepared in THF than with sonication, and C_{60} was more toxic than TiO_2. Toxicity of nano-sized and bulk ZnO, CuO and TiO_2 was studied in bacteria (Microtox assay), *D. magna* and the crustacean *Thamnocephalus platyurus*. Metal-specific recombinant biosensors were used to differentiate the effects induced by the metal oxides or nanoparticles (Heinlaan et al., 2008). Suspensions of nano and bulk TiO_2 were not toxic even at 20 g/L, whereas ZnO and CuO of bulk and nano-compounds showed acute toxicity. It was shown that the toxicity was largely due to soluble Zn and Cu ions, respectively.

A comparative toxicity study of bulk and nano-sized ZnO, TiO_2 and aluminium oxide (Al_2O_3) to zebrafish early life stages was undertaken to understand potential effects of such nanoparticles (Zhu et al., 2008a). Both bulk and nano-ZnO aqueous suspensions delayed zebrafish embryo and larval development, decreased hatching rate and survival and caused tissue damage. The 96-h LC_{50} of nano-ZnO and ZnO was 1.79 and 1.55 mg/L, respectively. No effects were found for the other two ENP at the concentrations tested. This study showing no different toxicities between bulk and nano-ZnO indicates that the toxicity is mainly based on the metal oxide concentration rather than the nature of nanoparticles.

Similarly, a series of metallic ENP was tested in an algae, daphnids and zebrafish and compared to metal ions (Griffitt et al., 2008). Nanosilver and nanocopper caused toxicity in all organisms with a 48-h median lethal concentration as low as 40 and 60 μg/L, respectively, in *Daphnia pulex*, whereas TiO_2 did not cause toxicity. Susceptibility differed among species, with filter-feeding invertebrates being more susceptible compared with zebrafish. This study also found that soluble metals were more toxic than the nano-particulate forms.

Effects of a range of metallic ENP were assessed on rainbow trout (Federici et al., 2007) and zebrafish (Griffitt et al., 2007; Griffitt et al., 2008). Exposure of rainbow trout to dispersed nanoTiO_2 (0.1–1 mg/L) for 14 d in a static-renewal experiment caused some effects on gills such as oedema and thickening of lamellae (Federici et al., 2007). Decrease in Na^+K^+-ATPase activity in gills and intestine and a trend of decreased activity in the brain were also observed. An increase in total glutathione levels the gills is a sign of anti-oxidative defence and depletion in the liver. Liver cells showed minor fatty change and lipidosis, and some liver cells showed condensed nuclear bodies. Fish probably ingested TiO_2 during exposure leading to erosion on the intestinal epithelia. Obviously, ENP are not major iono-regulatory toxicants, although they lead to oxidative stress, organ pathologies and induction of anti-oxidative defences in form of glutathione level increase.

Silica nanoparticles show low acute toxicity. No cytotoxicity occurred in cell cultures up to 0.1 mg/mL silica nanoparticles (Jin et al., 2007). The growth of algae is only little affected by 20 and 28.8 mg/L (Van Hoecke et al., 2008). The 12.5 and 27 nm-sized ENP were found to adhere to the outer cell surface of algae. No signs of uptake were found. This is also the case with 50 nm and 200 nm fluorescent nanoparticles in fish eggs (Fig. 11.1).

Fullerenes (C_{60}) have been studied to a larger extent than other non-inorganic ENP. Increasing the salinity or ionic strength of aqueous media causes aggregations of fullerenes to be formed (Fortner et al., 2005). However, an increase in humic acids stabilizes the suspension allowing them to remain in solution. Hence, organisms may be exposed to C_{60} dissolved in water (Chen and Elimelech, 2007). The dispersion of fullerenes and their toxicity in natural waters vary significantly with water chemistry and the reactivity, as shown in two toxicity tests with *Ceriodaphnia* and a bioassay; C_{60} did not exhibit toxicity even at concentrations of >3 mg/L (Gao et al., 2009).

Acute toxicity to bacteria (Lyon et al., 2008), *D. magna* (Lovern and Klaper, 2006; Oberdörster, 2004) and fish (Oberdörster, 2004; Oberdörster et al., 2006; Zhu et al., 2006b) has been studied. The exposure of juvenile largemouth bass to 0.5 and 1 mg/L nC_{60} resulted in signs of lipid peroxidation in the brain (Oberdörster, 2004). These were assumed to reach the brain via the olfactory nerve. Cytochrome P450 enzymes were not affected by these ENP. THF-pre-treated C_{60} induced mortality in fathead minnows, whereas stirring of ENP in water did not affect survival. However, lipid peroxidation was observed in the gill (Zhu et al., 2006b). This shows that the method of ENP preparation is important for their biological activity. The formation of aggregates usually reduces the bioavailability and toxicity. The use of solvents may also alter the toxicity as it may contribute to the effect. In case of metal ENP, the dissolution of metal ions (in addition to the nanoparticles) may contribute to the toxicity. Buckminster fullerene aggregates suspended in water (nC_{60}) affected zebrafish embryo survival, hatching rate, heartbeat and caused pericardial oedema at 1.5 mg/L and higher (Zhu et al., 2007). Moreover, embryo and larval development was delayed. Subsequent treatment with antioxidant glutathione mitigated toxicity, suggesting that oxidative stress was the cause for observed toxicity. Fullerol (a hydroxylated C_{60} derivative, $C_{60}(OH)_{16-18}$ had no such effect even at 50 mg/L (Zhu et al., 2007). An oxidative stress response was also observed in embryonic zebrafish (Usenko et al., 2008). Exposure of fish *F. heteroclitus* embryos larvae and adults to nC_{60} resulted in very little mortality up to 7 mg/L (Blickley and McClellan-Green, 2008). Fullerenes adsorbed to the chorion of eggs but did not affect development of the embryos or their hatching success. In larvae and adults, increase in total glutathione and a decreasing trend in lipid peroxidation were the only signs of fullerene exposure.

Baun et al. (2008b) studied the effect of combinations of ENP and chemical contaminants as a mixture. In *D. magna* and an algal species it was observed that C_{60} may serve as a contaminant carrier (Baun et al., 2008b). Suspensions of C_{60} were made by a 2-month stirring in medium and subsequently mixed with different contaminants. Depending on the contaminant, different toxicities occurred in *D. magna* and the algae *Pseudokirchneriella subcapitata*. The possibility of ENP serving as contaminant carriers has environmental relevance and should further be explored.

CNT induced dose-dependent growth inhibition in a protozoan (Zhu et al., 2006a). When CNT were coated with lipids they were readily taken up by *Daphnia*. They modified the solubility of the CNT through digestion of the lipids coating (Roberts et al., 2007). In fish, single-walled CNT (SWCNT) exposure caused a dose-dependent rise in ventilation rate, alteration in gills (oedema, altered mucocytes, hyperplasia), which points to a respiratory toxicant (Smith et al., 2007). In addition, increases in glutathione levels in gills and liver and condensed nuclear bodies in the liver were observed. SWCNT were ingested by fish resulting in precipitates in the gut and intestine. The brain showed swelling in parts of the cerebellum. Aggressive behaviour and gill irritation/damage led to the conclusion that SWCNT are respiratory toxicants in trout.

Effects of CNT have also been investigated in developing zebrafish (Cheng et al., 2007). Delay in hatching was observed at high concentrations for both single- and multi-walled CNT, but the toxicity may have been in part due to possible contamination of CNT. At very high concentrations of 120 mg/L hatching was delayed, whereas carbon black had no effect (Cheng et al., 2007). SWCNT did not influence embryonic development and survival. It was found that the chorion of zebrafish embryos acts as a protective barrier to prevent the passage of large nanotube aggregates through the pores. Delayed hatching was regarded as an indirect effect of hypoxia, because the nanotube aggregates hindered oxygen uptake. In contrast, 40 nm fluorescent latex nanoparticles are taken up by embryos and larvae of medaka and accumulate in the yolk and gallbladder (Kashiwada, 2006). Double-walled CNT induced acute toxicity at 10–200 mg/L in larvae of the frog *X. laevis* due to physical blockage of the gills and/or digestive tract. This material has been taken up into the intestine (Mouchet et al., 2008).

Quantum dots (QD) are semiconductor nanocrystals which contain a cadmium crystalline core (Cd/Se and CD/Te) coated with a shell of ZnS and an outer organic polymer coating. QD are incorporated by primary rat hepatocytes, which may result in toxicity (Derfus, 2004). Aqueous toxicity and food chain transfer of QD were studied in freshwater algae and *Ceriodaphnia dubia* (Bouldin et al., 2008). No lethality was observed at 48-h exposure to *C. dubia* up to 110 μg/L; however, the alga *P. subcapitata* was affected with a 96 h EC_{50} of 37.1 μg/L. It was further shown that QD were transferred from dosed algae to *C. dubia*, indicating that these nanoparticles are transferred via food chain. It has to be further investigated to what extent bioaccumulation or even biomagnification in the aquatic food chain will occur (Holbrook et al., 2008). Effects of QD (cadmium telluride) were also studied in freshwater mussel *Elliptio complanata*, a filter feeder (Gagne et al., 2008). Exposures led to oxidative stress in gills and to DNA damage. Increased oxidative stress has also been indicated by increased carbon black concentrations in *D. magna* (Zhu et al., 2006b). Toxicity was influenced in zebrafish embryos by the QD coating, which also contributed to the QD suspension stability (King-Heiden et al., 2009). At sublethal concentrations, many QD produced not only signs of cadmium toxicity but also distinctly different toxicity that could not be explained by cadmium release.

11.4.3 Effects on Aquatic Organisms: Chronic Toxicity

In constrast to studies on the acute mortality in various organisms including *Daphnia* (Lovern and Klaper, 2006; Lovern et al., 2007; Oberdörster, 2004; Oberdörster et al., 2006; Zhu et al., 2006b), an estuarine meiobenthic copepod (Templeton et al., 2006) and fish (Kashiwada, 2006; Oberdörster, 2004), the long-term toxicity (sublethal alterations or subtle chronic effects) of nanomaterial has been studied only marginally. Potential impacts of environmentally realistic exposures (low concentrations) or chronic toxicity of ENP are not or inadequately known. Currently, there is a lack of knowledge about the long-term risks and potential mechanisms of toxicity of ENP, and therefore, there is a clear need to address these questions instead of focusing on acute toxicity only.

Similar to the situation with chemical compounds such as pharmaceuticals (Fent et al. 2006), chronic toxicity data of ENP are very sparse or lacking. There are only very few studies so far. Some data are given in Table 11.1. *D. magna* were exposed for 21 d to 2.5 and 5 mg/L fullerenes, resulting in a delay of moulting and reduced offspring production, a clear sign of chronic toxicity (Oberdörster et al., 2006). The full life cycle effects of SWNT on an estuarine copepod (*A. tenuiremis*) were studied over 28 d exposure. No significant mortality, development and reproduction effects were observed with purified SWNT, but effects occurred at the highest concentration of 10 mg/L of more complex SWNT mixtures (Templeton et al., 2006).

Growth reduction and signs of oxidative stress, as indicated by a decrease in glutathione and an increase in antioxidant enzymes superoxide dismutase and catalase in gills and liver were observed in juvenile carp that were exposed to 0.04–1 mg/L fullerene aggregates for 32 days (Zhu et al., 2008b). In addition, lipid peroxidation was altered in liver and gills. Single-walled carbon nanotubes induced effects on gills (increase of Na^+K^+-ATPase activity) in rainbow trout when exposed up to 10 days. Increase of glutathione was observed in gills, liver and intestinal tract. Furthermore, aggressive behaviour was observed (Smith et al., 2007).

11.4.4 Terrestrial Plants and Soil Organisms

Less attention has been paid so far on implication of ENP to the terrestrial environment. Agglomerates of C_{60} applied to soil (up to 50 mg/kg dry soil) had little effects (Johansen et al., 2008). Respiration and microbial biomass were unaffected, whereas the number of fast-growing bacteria was decreased after ENP application. Also protozoans reacted not very sensitive to fullerenes.

Phytotoxicity was studied in ryegrass. In the presence of ZnO ENP, ryegrass biomass significantly reduced, root tips shrank and root epidermal and cortical cells were destroyed (Lin and Xing 2008). ZnO ENP greatly adhered on to the root surface. Individual ZnO ENP were observed present in apoplast and protoplast of the root endodermis and stele, but little ZnO ENP could translocate up in the ryegrass.

Reproduction of earthworm *Eisenia veneta* was affected by double-walled nanotubes (DWNT) administered through food at concentrations above 37 mg DWNT/kg food. The most sensitive toxicological parameter was reproduction (cocoon production), with no effect on hatchability, survival or mortality up to 495 mg DWNT/kg and 1000 mg C_{60}/kg. The toxicity of ENP and bulk ZnO, Al_2O_3 and TiO_2 has been evaluated in the nematode *Caenorhabditis elegans*. Both ENP and their bulk counterparts inhibited growth and the reproductive capability at mg/L concentrations (Wang et al., 2009). The toxicity of Al_2O_3 and TiO_2 nanoparticles was higher than that of their bulk counterparts. Silver nanoparticles decreased reproduction potential in *C. elegans*. Increased expression of the superoxide dismutase 3 and other genes occurred with 0.1 and 0.5 mg/L exposures concurrently with decreases in reproduction (Roh et al., 2009). Oxidative stress was indicated to be an important mechanism of silver ENP toxicity.

11.5 Conclusions

Recent studies suggest a risk of nanotubes that may act similar as asbestos fibres with a potential cancerogenic activity (Poland et al., 2008). Moreover adverse effects on the immune system have been indicated (Mitchell et al., 2009). This suggests a toxic potential for human exposure, previously known with respect to inhalation of nanoparticles. Much less is known, however, on the potential risks to the environment. Thus far, the acute toxicity of a variety of ENP with different physico-chemical properties and sizes has been studied in various organisms at high concentrations. However, the effects at low and environmentally relevant concentrations remain largely unknown. Furthermore, it is almost unknown whether or not, and which chronic effects occur at realistic environmental concentrations. In addition, basic questions such as the bioavailability from environmental media, uptake and distribution in the body of organisms remain to be studied in more detail. The current knowledge is yet too restricted to draw conclusions about potential environmental hazards and risks of ENP. Further investigations are needed on the mode of actions and the toxic mechanisms of different ENP, particularly at low and environmentally realistic concentrations. Emphasis should be placed particularly on long-term effects, chronic toxicity and risks for bioaccumulation and transfer in the food chain. All measures should be taken to avoid or at least to reduce the introduction of ENP into the environment. Precaution is needed, because environmental consequences are largely unknown, in particular with respect to long-term effects.

References

Adams LK, Lyon DY, McIntosh A, Alvarez PJ (2006) Comparative toxicity of nano-scale TiO_2, SiO_2 and ZnO water suspensions. Water Sci Technol 54(11–12): 327–334.

Baun A, Hartmann NB, Grieger K, Kusk KO (2008a) Ecotoxicity of engineered nanoparticles to aquatic invertebrates: a brief review and recommendations for future toxicity testing. Ecotoxicology 17: 387–395.

Baun A, Sorensen SN, Rasmussen RF, Hartmann NB, Koch CB (2008b) Toxicity and bioaccumulation of xenobiotic organic compounds in the presence of aqueous suspensions of aggregates of nano-C_{60}. Aquat Toxicol 86: 379–387.

Blickley TM, McClellan-Green P (2008) Toxicity of aqueous fullerene in adult and larval *Fundulus heteroclitus*. Environ Toxicol Chem 27: 1964–1971.

Bouldin JL, Ingle TM, Sengupta A, Alexander R, Hannigan RE, Buchanan RA (2008) Aqueous toxicity and food chain transfer of quantum dots in freshwater algae and *Ceriodaphnia dubia*. Environ Toxicol Chem 27: 1958–1963.

Brant JA, Lecoanet H, Hotze M, Wiesner MR (2005a) Comparison of electrokinetic properties of colloidal fullerenes (n-C_{60}) formed using two procedures. Environ Sci Technol 39: 6343–6351.

Brant JA, Lecoanet H, Wiesner MR (2005b) Aggregation and deposition characteristics of fullerene nanoparticles in aqueous systems. J Nano Res 7: 545–553.

Brant JA, Labille J, Bottero JY, Wiesner MR (2006) Characterizing the impact of preparation method on fullerene cluster structure and chemistry. Langmuir 22: 3878–3885.

Brayner R, Ferrari-Illiou R, Brivois N, Djediat S, Benedetti MF, Fievet F (2006) Toxicological impact studies based on *Escherichia coli* bacteria in ultrafine ZnO nanoparticles colloidal medium. Nano Lett 6: 866–870.

Brunner TJ, Wick P, Manser P, Spohn P, Grass RN, Limpach LK, Bruinink A, Stark WJ (2006) In vitro cytotoxicity of oxide nanoparticles: comparison to asbestos, silica, and the effect of particle solubility. Environ Sci Technol 40: 4374–4381.

Buffle J, Leppard GG (1995) Characterisation of aquatic colloids and macromolecules. 1. Structure and behavior of colloidal material. Environ Sci Technol 29: 2169–2175.

Buffle J (2006) The key role of environmental colloids/nanoparticles for the sustainability of life. Environ Chem 3: 155–158.

Carlson C, Hussain SM, Schrand AM, Braydich-Stolle L, Hess KL,, Jones RL, Schlager JJ (2008) Unique cellular interaction of silver nanoparticles: size-dependent generation of reactive oxygen species. J Phys Chem B 112: 13608–13619.

Chae YJ, Pham CH, Lee J, Bae E, Yi J, Gu MB (2009) Evaluation of the toxic impact of silver nanoparticles on *Japanese* medaka *(Oryzias latipes)*. Aquat Toxicol 94: 320–327.

Chen KL, Elimelech M (2007) Influence of humic acid on the aggregation kinetics of fullerene (C_{60}) nanoparticles. Langmuir 22: 10994–11001.

Chen Y, Chen J, Dong J, Jin Y (2004) Comparing study of the effect of nanosized silicon dioxide and microsized silicon dioxide on fibrogenesis in rats. Toxicol Ind Health 20: 21–27.

Cheng J, Flahaut E, Cheng SH (2007) Effect of carbon nanotubes on development zebrafish (*Danio rerio*) embryos. Environ Toxicol Chem 26: 708–716.

Davoren M, Herzog E, Casey A, Cottineau B, Chambers G, Byrne HJ, Lyng FM (2007) In vitro toxicity evaluation of single walled carbon nanotubes on human A549 lung cells. Toxicol in Vitro 21: 438–448.

Derfus A (2004) Probing the cytotoxicity of semiconductor quantum dots. Nano Lett 4: 11–18.

Donaldson K, Aiken R, Tran L, Stone V, Duffin R, Forrest G, Alexander A (2006) Carbon nanotubes: A review of their properties in relation to pulmonary toxicology and workplace safety. Toxicol Sci 92: 5–22.

Federici G, Shaw BJ, Handy RH (2007) Toxicity of titanium dioxide to rainbow trout (*Oncorhynchus mykiss*): Gill injury, oxidative stress, and other physiological effects. Aquat Toxicol 84: 415–430.

Fent K, Weston AA, Caminada D (2006) Ecotoxicology of human pharmaceuticals. Aquat Toxicol 76: 122–159.

Fernandes TF, Christofi N, Stone V (2007) The environmental implications of nanomaterials. In: Monteiro-Riviere N, Lang Tran C (eds.) Nanotoxicology: Characterization, Dosing and Health Effects. CRC Press, Boca Raton, FL, pp. 405–420.

Fortner JD, Lyon DY, Sayes CM et al. (2005) C_{60} in water: nanocrystal formation and microbial responses. Environ Sci Technol 39: 4307–4316.

Franklin NM, Rogers NJ, Apte SC, Batley GE, Gadd GE, Casey PS (2007) Comparative toxicity of nanoparticulate ZnO, bulk ZnO, and $ZnCl_2$ to a freshwater microalga *(Pseudokirchneriella subcapitata)*: The importance of particle solubility. Environ Sci Technol 41: 8484–8490.

Gagne F, Auclair J, Turcotte P, Fournier M, Gagnon C, Sauve S, Blaise C (2008) Ecotoxicity of CdTe quantum dots to freshwater mussels: impacts on immune system, oxidative stress and genotoxicity. Aquat Toxicol 86: 333–340.

Gao J, Youn S, Hovsepyan A, Llaneza VL, Wang Y, Bitton G, Bonongo JCJ (2009) Dispersion and toxicity of selected manufactured nanomaterials in natural river water samples: effects of water chemical composition. Environ Sci Technol 43: 3322–3328.

Ghafari P, St-Denis CH, Power ME, Xu J, Tsou V, Mandal HS, Bols NC, Xiaowu T (2008) Impact of carbon nanotubes on the ingestion and digestion of bacteria by ciliated protozoa. Nature Nanotech 3: 347–351.

Griffitt RJ, Hyndman K, Denslow ND, Barber DS (2009) Comparison of molecular and histological changes in zebrafish gills exposed to metallic nanoparticles. Toxicol Sci 107: 404–415.

Griffitt RJ, Weil R, Hyndman KA, Denslow ND, Powers K, Taylor D, Barber DS (2007) Exposure to copper nanoparticles causes gill injury and acute lethality in zebrafish (*Danio rerio*). Environ Sci Technol 41: 8178–8186.

Griffitt RJ, Luo J, Gao J, Bonzongo JC, Barber DS (2008) Effects of particle composition and species on toxicity of metallic nanomaterials in aquatic organisms. Environ Toxicol Chem 27: 1972–1978.

Handy RD, Henry TB, Scown TM, Johnston BD, Tyler CR (2008a) Manufactured nanoparticles: their uptake and effects on fish—a mechanistic analysis. Ecotoxicology 17: 369–409.

Handy RH, Owen R, Valsami-Jones E (2008b) The ecotoxicology of nanoparticles and nanomaterials: current status, knowledge gaps, challenges, and future needs. Ecotoxicology 17: 315–325.

Hansen SF, Maynard A, Baun A, Tickner JA (2008) Late lessons from early warnings for nanotechnology. Nat Nanotechnol 3: 444–447.

Hardman R (2006) A toxicologic review of quantum dots: toxicity depends on physicochemical and environmental factors. Environ Health Perspect 114: 165–172.

Heinlaan M, Ivask A, Blinova I, Dubourguier HC, Kahru A (2008) Toxicity of nanosized and bulk ZnO, CuO and TiO_2 to bacteria *Vibrio fischeri* and crustaceans *Daphnia magna* and *Thamnocephalus platyurus*. Chemosphere 71: 1308–1316.

Herzog E, Casey A, Lyng FM, Chambers G, Byrne HJ, Davoren M (2007) A new approach to the toxicity testing of carbon-based nanomaterials-The clonogenic assay. Toxicol Lett 174: 49–60.

Holbrook RD, Morphy KE, Morrow JB, Cole KD (2008) Trophic transfer of nanoparticles in a simplified invertebrate food web. Nat Nanotechnol 3: 352–355.

Hund-Rinke K, Simon M (2006) Ecotoxic effect of photocatalytic active nanoparticles (TiO_2) on algae and daphnids. Environ Sci Poll Res 13: 225–232.

Jin Y, Kannan S, Wu M, Zhao JX (2007) Toxicity of luminescent silica nanoparticles to living cells. Chem Res Toxicol 20: 1126–1133.

Johansen A, Pedersen AL, Jensen KA, Karlson U, Hansen BM, Scott-Fordsmand JJ, Winding A (2008) Effects of C_{60} fullerene nanoparticles on soil bacteria and protozoans. Environ Toxicol Chem 27: 1895–1903.

Karlsson HL, Cronholm P, Gustafsson J, Müller L (2008) Copper oxide nanoparticles are highly toxic: a comparison between metal oxide nanoparticles and carbon nanotubes. Chem Res Toxicol 21: 1726–1732.

Kashiwada S (2006) Distribution of nanoparticles in see-trough medaka (*Oryzias latipes*). Environ Health Perspect 114: 1697–1702.

Keenan CR, Goldstein R, Lucas D, Sedlak DL (2009) Oxidative stress induced by zero-valent iron nanoparticles and Fe(II) in human bronchial epithelial cells. Environ Sci Technol 43: 4555–4560.

King-Heiden TC, Wiecinski PN, Mangham AN, Metz KM, Nesbit D, Pedersen JA, Hamers RJ, Heideman W, Peterson RE (2009) Quantum dot nanotoxicity assessment using the zebrafish embryo. Environ Sci Technol 43: 1605–1611.

Klaine SJ, Alvarez PJJ, Batley GE, Fernandes TF, Handy RH, Lyon DY, Mahendra S, McLaughlin MJ, Lead JR (2008) Nanomaterials in the environment: Behavior, fate, bioavailability and effects. Environ Toxicol Chem 27: 1825–1851.

Knauer K, Sobek A, Bucheli TD (2007) Reduced toxicity of diuron to the freshwater green alga *Pseudokirchneriella subcapitata* in the presence of black carbon. Aquat Toxicol 83: 143–148.

Kühnel D, Buscha W, Meissner T, Springer A, Potthoff A, Richter V, Gelinskyc M, Scholz S, Schirmer K (2009) Agglomeration of tungsten carbide nanoparticles in exposure medium does not prevent uptake and toxicity toward a rainbow trout gill cell line. Aquat Toxicol 93: 91–99.

Lee KL, Nallathamby PD, Browning LM, Osgood CJ, Xu XHN (2007) In vivo imaging of transport and biocompatibility of single silver nanoparticles in early development of zebrafish embryos. ACS Nano 1: 133–143.

Lin D, Xing B (2008) Root uptake and phytotoxicity of ZnO nanoparticles. Environ Sci Technol 42: 5580–5585.

Lin W, Huang YW, Zhou XD, Ma Y (2006) In vitro toxicity of silica nanoparticles in human lung cancer cells. Toxicol Appl Pharmacol 217: 252–259.

Lison D, Thomassen LC, Rabolli V et al. (2008) Nominal and effective dosimetry of silica nanoparticles in cytotoxicity assays. Toxicol Sci 104: 155–162.

Lok CN, Ho CM, Chen R, He QY, Yu WY, Sun HZ, Tam PKH, Chiu JF, Che CM (2006) Proteomic analysis of the mode of antibacterial action of silver nanoparticles. J Proteome Res 5: 916–924.

Lovern SB, Klaper R (2006) *Daphnia magna* mortality when exposed to titanium dioxide and fullerene (C-60) nanoparticles. Environ Toxicol Chem 25: 1132–1137.

Lovern SB, Strickler JR, Klaper R (2007) Behavioral and physiological changes in *Daphnia magna* when exposed to nanoparticle suspensions (titanium dioxide, nano-C_{60}, and $C_{60}HxC_{70}Hx$). Environ Sci Technol 41: 4465–4470.

Lyon DY, Brunet L, Hinkal GW, Wiesner MR, Alvarez PJJ (2008) Antibacterial activity of fullerene water suspensions (nC_{60}) is not due to ROS-mediated damage. Nano Lett 6: 1261–1268.

Mitchell LA, Gao J, Vander Wal R, Gigliotti A, Burchiel SW, McDonald JD (2007) Pulmonary and systemic immune response to inhaled multiwalled carbon nanotubes. Toxicol Sci 100: 203–214.

Mitchell LA, Lauer FT, Burchiel SW, McDonald JD (2009) Mechanisms for how inhaled multiwalled carbon nanotubes suppress systemic immune function in mice. Nat Nanotechnol 4: 409–410.

Moore MN (2006) Do nanoparticles present ecotoxicological risks for the health of the aquatic environment? Environ Int 32: 967–976.

Mori T, Takada H, Ito S, Matsubayashi K, Miwa N, Sawaguchi T (2006) Preclinical studies on safety of fullerene upon acute oral administration and evaluation for no mutagenesis. Toxicology 225: 48–54.

Morones JR, Elechiguerra JL, Camacho A, Holt K, Kouri JB, Ramirez JT, Yacaman MJ (2005) The bactericidal effect of silver nanoparticles. Nanotechnology 16: 2346–2353.

Mouchet F, Landois P, Sarremejean E, Bernard G, Puech P, Pinelli E, Flahaut E, Gauthier L (2008) Characterisation and in vivo ecotoxicity evaluation of double-wall carbon nanotubes in larvae of the amphibian *Xenopus laevis*. Aquat Toxicol 87: 127–137.

Navarro E, Baun A, Behra R, Hartmann NB, Filser J, Miao AJ, Quigg A, Santschi PH (2008a) Environmental behavior and ecotoxicity of engineered nanoparticles to algae, plants, and fungi. Ecotoxicology 17: 372–386.

Navarro E, Piccapietra F, Wagner B, Marconi F, Kaegi R, Odzak N, Sigg L, Behra R (2008b) Toxicity of silver nanoparticles to *Chlamydomonas reinhardtii*. Environ Sci Technol 42: 8959–8964.

Nel A, Xia T, Madler L, Li N (2006) Toxic potential of materials at the nanolevel. Science 311: 622–627.

Novak B, Bucheli T (2007) Occurrence, behavior and effects of nanoparticles in the environment. Environ Pollut 150: 7–22.

Oberdörster E (2004) Manufactured nanomaterials (fullerenes, C_{60}) induce oxidative stress in the brain of juvenile largemouth bass. Environ Health Perspect 112: 1058–1062.

Oberdörster E, Zhu SQ, Blickley TM, McClellan-Green P, Haasch ML (2006) Ecotoxicology of carbon-based engineered nanoparticles: effects of fullerene (C-60) on aquatic organisms. Carbon 44: 1112–1120.

Pal S, Tak YK, Song JM (2007) Does the antibacterial activity of silver nanoparticles depend on the shape of the nanoparticle? A study of the gram-negative bacterium *Escherichia coli*. Appl Eviron Microbiol 73: 1712–1720.

Petersen EJ, Huang Q, Weber WJ Jr. (2008) Ecological uptake and depuration of carbon nanotubes by *Lumbriculus variegatus*. Environ Health Perspect 116: 496–500.

Petersen EJ, Akkanen J, Kukkonen JVK, Weber WJJ (2009) Biological uptake and depuration of carbon nanotubes by *Dapnia magna*. Environ Sci Technol 43: 2969–2975.

Phenrat T, Long TC, Lowry GV, Veronesi B (2009) Partial oxidation ("aging") and surface modification decrease the toxicity of nanosized zerovalent iron. Environ Sci Technol 43: 195–200.

Poland CA, Duffin R, Kinloch I et al. (2008) Carbon nanotubes introduced into the abdominal cavity of mice show asbestos-like pathogenicity in a pilot study. Nat Nanotechnol 3: 423–428.

Roberts AP, Mount AS, Seda B, Souther J, Qiao R, Lin S, Ke PC, Rao AM, Klaine SJ (2007) In vivo biomodification of lipid-coated carbon nanotubes by *Daphnia magna*. Environ Sci Technol 41: 3025–3029.

Roh JY, Sim SJ, Yi J, Park K, Chung KH, Ryu DY, Choi J (2009) Ecotoxicity of silver nanoparticles on the soil nematode *Caenorhabditis elegans* using functional ecotoxicogneomics. Environ Sci Technol 3: 3933–3940.

Sayes CM, Liang F, Hudson JL et al. (2006) Functionalization density dependence of single-walled carbon nanotubes cytotoxicity in vitro. Toxicol Lett 161: 135–142.

Schmid K, Riediker M (2008) Use of nanoparticles in Swiss industry: a targeted survey. Environ Sci Technol 42: 2253–2256.

Smith CJ, Shaw BJ, Handy RD (2007) Toxicity of single walled carbon nanotubes to rainbow trout (*Oncorhynchus mykiss*): respiratory toxicity, organ pathologies, and other physiological effects. Aquat Toxicol 82: 94–109.

Stern ST, McNeil SE (2008) Nanotechnology safety concerns revisited. Toxicol Sci 101: 4–21.

Templeton RC, Ferguson PL, Washburn KM, Scrivens WA, Chandler GT (2006) Life-cycle effects of single-walled carbon nanotubes (SWNTs) on an estuarine meiobenthic copepod. Environ Sci Technol 40: 7387–7393.

Thibodeau MS, Giardina C, Knecht DA, Helble J, Hubbard AK (2004) Silica-induced apoptosis in mouse alveolar macrophages is initiated by lysosomal enzyme activity. Toxicol Sci 80: 34–48.

Usenko CY, Harper SL, Tanguay RL (2008) Fullerene C_{60} exposure elicits an oxidative stress response in embryonic zebrafish. Toxicol Appl Pharmacol 229: 44–55.

Van Hoecke K, De Schamphelaere KAC, Van der Meeren P, Lucas S, Janssen CR (2008) Ecotoxicity of silica nanoparticles to the green alga *Pseudokirchneriella subcapitata*: importance of surface area. Environ Toxicol Chem 27: 1948–1957.

Van Hoeke K, Quik JTK, et al. (2009) Fate and effects of CeO_2 nanoparticles in aquatic ecotoxicity tests. Environ Sci Technol 43: 4537–4546.

Vevers WF, Jha AN (2008) Genotoxic and cytotoxic potential of titanium dioxide (TiO_2) nanoparticles on fish cells in vitro. Ecotoxicology 17: 410–420.

Wang H, Wick RL, Xing B (2009) Toxicity of nanoparticulate and bulk ZnO, Al_2O_3 and TiO_2 to the nematode *Caenorhabditis elegans*. Environ Pollut 157: 1171–1177.

Zhang LW, Monteiro-Riviere NA (2009) Mechanisms of quantum dot nanoparticle cellular uptake. Toxicol Sci 110: 138–155.

Zhu L, Zhao Q, Li Y, Cai X, Li W (2006a) The interaction and toxicity of multi-walled carbon nanotubes with *Stylonychia mytilus*. J Nanosci Nanotechnol 6: 1357–1364.

Zhu S, Oberdörster E, Haasch ML (2006b) Toxicity of an engineered nanoparticle (fullerene, C_{60}) in two aquatic species, Daphnia and fathead minnow. Mar Environ Res 62(Suppl.): S5–S9.

Zhu X, Zhu L, Li Y, Duan Z, Chen W, Alvarez PJJ (2007) Developmental toxicity in zebrafish (*Danio rerio*) embryos after exposure to manufactured nanomaterials: buckminsterfullerene aggregates (nC_{60}) and fullerol. Environ Toxicol Chem 26: 976–979.

Zhu X, Zhu L, Duan Z, Qi R, Li Y, Lang Y (2008a) Comparative toxicity of several metal oxide nanoparticle aqueous suspensions to zebrafish (*Danio rerio*) early developmental stage. J Environ Sci Health A 43: 278–284.

Zhu X, Zhu L, Lang Y, Chen Y (2008b) Oxidative stress and growth inhibition in the freshwater fish *Carassius auratus* induced by chronic exposure to sublethal fullerene aggregates. Environ Toxicol Chem 27: 1979–1985.

Chapter 12
Standardisation

Birgit C. Gordalla

12.1 Introduction

The potential of nanoparticles as a promising novel type of materials is now undoubted, and considerable research efforts and amounts of money are being expended to investigate, develop, and exploit the options given by the unique properties of these nanoscale versions of matter. Besides research and development activities aiming at the intended applications of the nanoparticles, more and more studies are being carried out which deal with unintended effects that may arise in their life cycle – production, processing, distribution, application, disposal – in order to avoid adverse effects from their use (Biswas and Wu, 2005; Balbus et al., 2007). This includes the aspects occupational health, human health, and environmental fate. A compendium of research and development projects in the EU in the field of health and environmental impacts of nanoparticles is given in Aguar and Murcia Nicolás (2008). The efforts for risk assessment, sometimes taking place in industry venues (Lauterwasser, 2005; Verband der Chemischen Industrie e. V, 2008a, b), but often driven by governmental initiatives (BAUA, BfR and UBA, 2007; Department for Environment, Food and Rural Affairs, 2007; NanoKommission der Deutschen Bundesregierung, 2008; National Science and Technology Council, 2008; Science Policy Council and U.S. Environmental Protection Agency, 2007), have an anticipatory character and might result in regulatory measures.

One aspect of the systematic efforts for characterisation, exploitation, and risk assessment of nanoparticles is the question of whether there is a need for special standards to deal with these materials. Different kinds of standards are possible: characterisation methods that have the format of metrological standards, product standards for materials specification, standards describing minimum requirements concerning health, safety and environmental issues, or standardised examination and assessment protocols. Below, first standardisation activities in the field of

B.C. Gordalla (✉)
Engler-Bunte-Institut, Chair for Water Chemistry, Karlsruhe Institute of Technology (KIT), Engler-Bunte-Ring 1, D-76131 Karlsruhe, Germany
e-mail: birgit.gordalla@kit.edu

F.H. Frimmel, R. Niessner (eds.), *Nanoparticles in the Water Cycle*,
DOI 10.1007/978-3-642-10318-6_12, © Springer-Verlag Berlin Heidelberg 2010

nanomaterials are reviewed, with especial focus on nano-sized particles and the water cycle as an environmental compartment.

As ingredients of aquatic environment, nanoparticles are a subset of aquatic colloids and hence implicitly always have been a subject to the scientific approaches of colloid chemistry with its experimental and analytical instruments. The perception of nanoparticles in aquatic systems as a special class of aquatic colloids results from the fact that properties and especially reactivity of some nanoparticles change dramatically with decreasing size in a manner that even goes beyond the effect normalised to surface area. Practically, this perception was only made possible when suitable metrological and analytical instruments were developed. It is not long since analytical methods for reliable detection of these very small particles have become available, especially in the presence of larger sized particles and in the extremely low concentrations usual in many natural waters. An overview about the suitable methods for analysis and characterisation is given in Hassellöv et al. (2008). Considering nanoparticles in aquatic systems, the following cases can be separated (Wiggington et al., 2007): (i) aquatic nanoparticles of natural origin that are formed by natural (geo)chemical or biological processes in aquatic systems; (ii) aquatic nanoparticles of natural origin that are unintentionally formed in aquatic environments by human activities, e.g. mining; (iii) manufactured ("engineered") nanoparticles getting into the aquatic environment in the course of production, use, or disposal (e.g., industrial or domestic wastewater); (iv) nanoparticles intentionally added to aquatic systems to achieve a desired effect. Examples for the latter are Fe(0) particles for remediation of groundwater or TiO_2 as a photocatalyst in water treatment. This chapter will mainly focus on manufactured nanoparticles and standardisation activities in this field.

There may be different reasons for having a specification or a procedure in a standardised format or even as a standard delivered by a standards body:

- Documentary standards either for material specification or for description of operations facilitate the exchange of goods and services and therefore are helpful for industry and economics.
- Standards are often referred to by authorities for reasons of justifiability and comparability. In the field of aquatic environment, this may apply to monitoring programmes or quality control and to regulating practices. For chemical monitoring of the so-called priority pollutants on the basis of the European Water Framework Directive, validated and documented methods have to be used which meet well-defined performance criteria (Commission Directive 2009/90/EC). In German wastewater regulations, the calculation of sewage taxes (AbwAG, 2005) as well as compliance with the wastewater ordinance (AbwV, 2002), which gives the minimum requirements for the quality of discharged wastewater, is based on parameters measured according to official standards. According to the German drinking water ordinance (TrinkwV, 2001), only materials whose specifications are defined in a standard may be included in the list of registered chemicals allowed for drinking water treatment according to § 11 (http://www.umweltbundesamt.de/wasser/themen/trinkwasser/trinkwasseraufbereitungstoffliste.htm).

- Standardised protocols are immanent to several core parameters in water analysis, unaffected by whether they are delivered by a standards body. This applies not only to the so-called conventional – operationally defined – parameters as, e.g. dissolved organic carbon (DOC), adsorbable organic halogens (AOX), biochemical oxygen demand (BOD), or chemical oxygen demand (COD), but also for the large group of effect, toxicity, and ecotoxicity testing protocols.

Several organisations deal with developing standardised specifications or protocols on a national or super-national level for different users and interests.

12.2 Testing Protocols for Chemicals – OECD Activities Concerning Nanoparticles

For the testing of chemicals with regard to a potential risk for environment and health, the Organisation for Economic Cooperation and Development (OECD) has developed a collection of about 130 testing protocols under the aspects (i) physical–chemical properties, (ii) effects on biotic systems, (iii) degradation and accumulation, and (iv) health effects (OECD, 2008). Current regulations for existing chemicals normally do not distinguish between the nanoscale and the macroscale forms of a substance. In order to identify future needs for risk assessment within a regulatory context, the OECD council established the OECD Working Party on Manufactured Nanomaterials (WPMN) to help member countries to address the safety challenges of nanomaterials. Several projects were initiated to identify the most relevant nanomaterials, to specify properties, and to propose suitable testing strategies relevant with respect to environmental safety and human health (Working Party on Manufactured Nanomaterials, 2008).

In the project "Safety testing of a representative set of manufactured nanomaterials", a programme was developed to create an understanding of the kind of information on intrinsic properties of nanomaterials that may be relevant for exposure and effect assessment by testing. For this, a priority list of 14 representative nanomaterials now, or soon to be, in commerce was worked out (Table 12.1). The list comprises eight inorganic and six carbon-based or organic materials. They are intended as reference materials to support measurement, toxicology, and risk assessment of nanomaterials.

A list of endpoints was specified which should be taken into account when testing for human health and environmental safety. These endpoints shall be part of a dossier about each material that should comprise information on the categories (i) material information/identification, (ii) physical–chemical properties and materials characterisation, (iii) environmental fate, (iv) environmental toxicology, (v) mammalian toxicology, and (vi) material safety.

The proposed endpoints on environmental fate (see Table 12.2) comprise several tests on degradability and degradation products. In the spectrum of recommended issues for testing, the water cycle is highly represented, thus a test on ultimate degradation in surface water and a sewage simulation test are requested. As relevant

Table 12.1 List of 14 representative manufactured nanomaterials selected for testing by the OECD working party on manufactured nanomaterials for phase one of testing (Working Party on Manufactured Nanomaterials, 2008)

Fullerenes (C_{60})	Silver nanoparticles
Single-walled carbon nanotubes (SWCNTs)	Iron nanoparticles
Multi-walled carbon nanotubes (MWCNTs)	Titanium dioxide
Carbon black	Aluminium oxide
Polystyrene	Cerium oxide
Dendrimers	Zinc oxide
	Silicon dioxide
	Nanoclays

Table 12.2 List of endpoints, proposed by the OECD working party on manufactured nanomaterials for phase one of testing, concerning physical–chemical properties, environmental fate, and environmental toxicology (Working Party on Manufactured Nanomaterials, 2008)

• Physical–chemical properties and material characterisation	
Dustiness	Water solubility
Pour density	Zeta potential (surface charge)
Porosity	Surface chemistry (where appropriate)
Crystalline phase	Photocatalytic activity
Crystallite size	Redox potential
Representative TEM picture(s)	Radical formation potential
Particle size distribution	Octanol–water partition coefficient, where relevant
Specific surface area	
Agglomeration/aggregation	
• Environmental fate	
Dispersion stability in water	Biotic degradability
Hydrolysis, for surface-modified nanomaterials	Simulation testing on ultimate degradation in surface water
Adsorption–desorption	Ready biodegradability
Adsorption to soil or sediment	Soil simulation testing
Abiotic degradability and fate	Sediment simulation testing
Identification of degradation products	Sewage treatment simulation testing
Further testing of degradation products as required	Bioaccumulation potential
• Environmental toxicology	
Effects on pelagic species (short term/long term)	Effects on soil species (short term/long term)
Effects on sediment species (short term/long term)	Effects on terrestrial species
Effects on microorganisms	

aspects for toxicity testing, short-term and long-term effects on pelagic species, on sediment species, and on microorganisms are taken into account, the latter might be especially significant for wastewater treatment plants.

At present, the repertoire of testing methods for aquatic toxicity is being checked for applicability on manufactured nano-sized materials and is discussed in the literature (Hund-Rinke and Herrchen, 2007). A comprehensive review on the suitability of already existing OECD methods for ecotoxicological assessment of engineered nanoparticles is given in Crane et al. (2008). As it is stressed there, the ultimate goal of ecotoxicological tests is to predict effects on survival, growth, or reproduction. Chronic ecotoxicity data are necessary in order to look for reliable acute-to-toxic ratios and standard assessment factors. It is suggested that rapid test systems might be used to check whether nanoparticles have hazard properties similar to the macroscale substances and whether hazard data on demographic endpoints can be read across. These fast systems should cover overall toxicity (cell viability assay or microbial population growth test), e.g. with *Vibrio fischeri,* and major toxicity mechanisms like genotoxicity (abbreviated Ames test), immunotoxicity, or oxidative stress (Crane et al., 2008). A major issue for ecotoxicity testing is the problem of dosing, i.e. to present the nanoparticles to the test organisms in a way that is environmentally relevant. Several methods for homogeneous dispersion, e.g. sonication, use of detergents, or stirring, are discussed (Hund-Rinke and Herrchen, 2007; Crane et al., 2008).

It should be kept in mind that risk assessment requires the assessment of both effect (hazard) and exposure (Wiesner et al., 2006). To define meaningful metrics or endpoints based on scientific understanding, empirical data, and available measurement techniques, that allow to predict an exposure scenario, i.e. partition behaviour and actual status and morphology of nanoparticles in the water cycle, is not trivial (Hund-Rinke and Herrchen, 2007). Besides characteristics that refer to the bulk phase, in Table 12.2 several physical–chemical properties of nanomaterials are included that are supposed to be indicative of their fate and partition behaviour in water and adjacent compartments, e.g. water solubility, zeta potential, and octanol–water partition coefficient. The latter is often interpreted as an indicator for bioaccumulation potential, but recently there were concerns, whether the K_{OW} partition test (OECD, 1995) works on nanoparticles (Crane et al., 2008). Properties like photocatalytic activity, radical formation potential, and redox potential might allow conclusions concerning reactions and reactivity of the nanoparticles in the aqueous phase.

12.3 Standardisation by Standardisation Organisations

Standards in the narrower sense are delivered by national, European, or international standards bodies. The German national standards body is the DIN – *Deutsches Institut für Normung* (German Institute for Standardisation). On European level, standardisation is organised in the CEN – *Comité Européen de Normalisation*

(European Committee for Standardization), international standardisation activities are coordinated by the ISO – International Organization for Standardization. The participating countries are represented in CEN and in ISO by their national standards bodies. Adoption of European Standards into the national standards collection as DIN EN standards is mandatory, adoption of ISO International Standards as DIN ISO standards is optional and performed more and more frequently. Adoption of ISO Standards into the European Standards collection is regulated by the Vienna Agreement; accordingly, ISO standards can be adopted either by parallel vote (PV) or by unique acceptance procedure (UAP).

Standardisation work is organised in technical committees covering larger subject areas. The actual work on specific standards is done in working groups of project-related specialists (see for instance flow sheet in Fig. 12.1 on the organisation of DIN work). Standards making itself is regulated by standards, too (DIN 820-1, 2009; ISO/IEC Directives, 2009). Features of the standardisation process are publicity, participation of the so-called interested parties, and, in European and international standardisation, a time schedule to be met. In the case of standards on water quality, the interested parties are authorities, industry, and universities. Depending on the method, manufacturers of analytical instruments are also involved. As already mentioned, one important objective of standards bodies is to facilitate the exchange of goods and services. Thus, materials, instruments, or methods which are unique or on which a monopoly is held are not suitable for standardisation.

Development of a standard is initiated by a formal new work item proposal to the responsible technical committee. The documents delivered are the result of a consensus which is formed in a regulated several-stage process of elaborating, commenting, and voting (see for instance Fig. 12.2). In some technical committees, a validation step is prerequisite for consideration of a document as a standard.

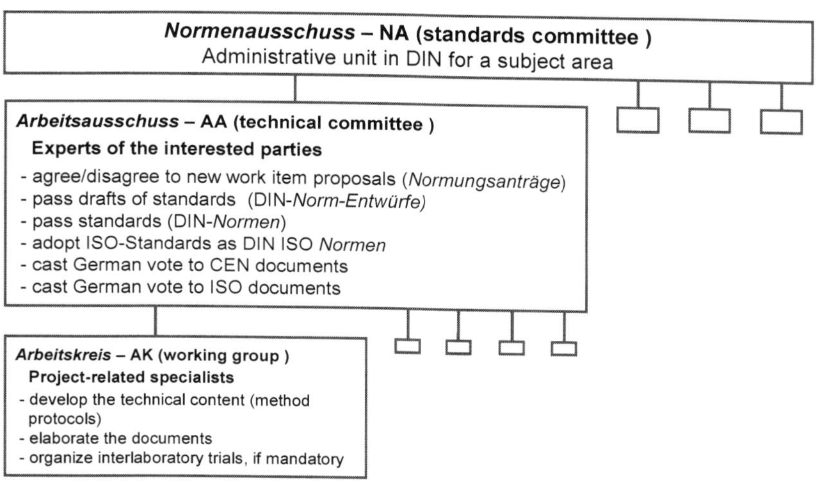

Fig. 12.1 Standardisation in the DIN – panels involved on different levels in a subject area. For further procedural details see DIN 820-1 (2009)

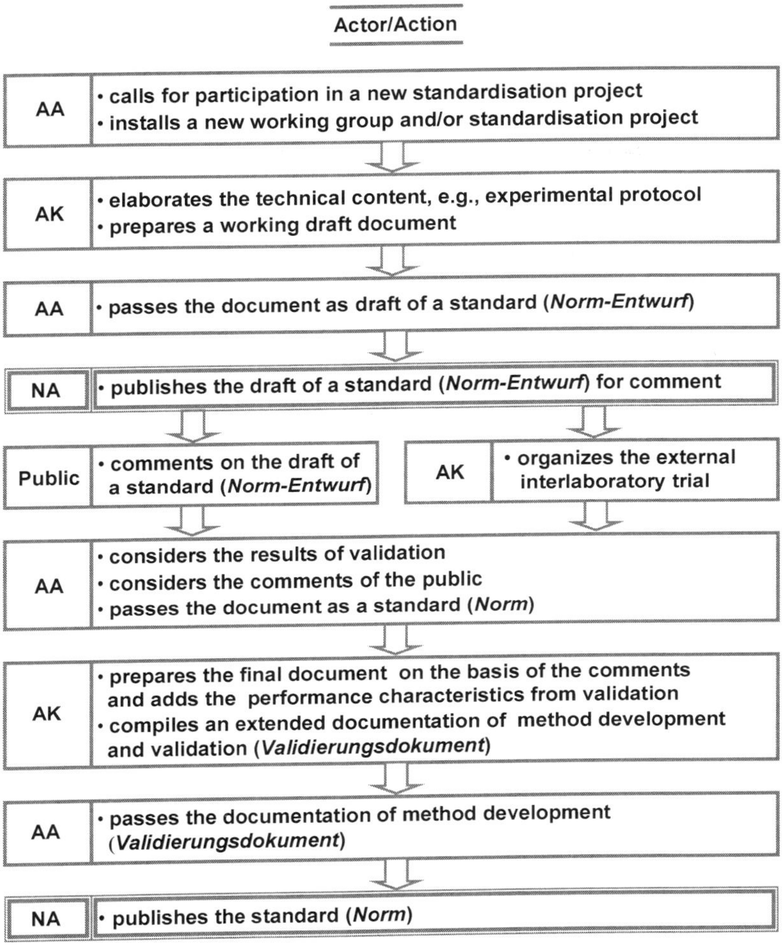

Fig. 12.2 Standardisation process – setting up a German Standard (DIN-*Norm*). "Draft of a standard" (*Norm-Entwurf*) and "standard" (*Norm*) are the two official document stages. Validation by an interlaboratory trial and supplementary documentation by a validation document are special requirements for German standards on water quality examination developed by the DIN technical committee on water examination, DIN NA 119-01-03 AA "*Wasseruntersuchung*". Abbreviations see Fig. 12.1

Validation might be based on certified reference materials or on a successful interlaboratory trial.

Besides International Standards, some ISO committees deliver documents with a normative content that are based on a lower level of consensus and require a shorter processing. This holds for the so-called Publicly Available Specification (PAS) or the Technical Specification (TS). Non-normative documents with a more informative or guideline character are published in the format of a Technical Report (TR).

12.3.1 Standardisation Activities in the Field of Water Quality

In the German standards body DIN, the technical committees that deliver standards concerning the issues of water quality assessment, water treatment, and water distribution are pooled in the Standards Committee "Water Practice Standards" (*Normenausschuss* DIN NA 119 "*Wasserwesen*"). Its standards portfolio, a list of current standardisation projects, and its annual report can be found at http://www.naw.din.de/.

12.3.1.1 Water Analysis

The DIN technical committee in charge of standardisation in the field of water analysis is the *Arbeitssausschuss* DIN NA 119-01-03 AA "*Wasseruntersuchung*" ("Water analysis"). The DIN NA 119-01-03 AA "*Wasseruntersuchung*" mirrors the committees CEN/TC 230 "Water analysis" (secr. DIN, Germany), where European standards are developed, and ISO/TC 147 "Water quality" (secr. DIN, Germany), where international standardisation takes place.

The CEN/TC 230 "Water analysis" focusses mainly on the development of biological methods which are suitable to meet the requirements of the European Water Framework Directive. Furthermore, it deals with the problem of the so-called priority pollutants. A list of standards and standardisation projects of this committee is given at http://www.cen.eu/CENORM/Sectors/TechnicalCommitteesWorkshops/CENTechnicalCommittees/.

The majority of water analytical standards valid in Germany are meanwhile developed at an international level in the ISO/TC 147 "Water quality" and adopted as European standards and further transposed into national standards. According to its scope, ISO/TC 147 is in charge of standardisation in the field of water quality, including definition of terms, sampling of waters, measurement, and reporting of water characteristics. A total of 34 countries participate in ISO/TC 147 as so-called *p*-members with the right and the obligation to vote, further 53 observing countries are informed about the activities of the committee. The standards portfolio of this committee is available at http://www.iso.org/iso/standards development/technical committees/list of iso technical committees/. About 40 active working groups are working on current standardisation projects, organised in the 5 subcommittees (SC) SC 1 "Terminology" (secr. SABS; South Africa), SC 2 "Physical, chemical and biochemical methods" (secr. DIN, Germany), SC 4 "Microbiological methods" (secr. DIN, Germany), SC 5 "Biological methods" (secr. DIN, Germany), and SC 6 "Sampling" (secr. BSI; UK). In ISO/TC 147, "International Standard" is the intended format for documents with a normative content, TS is delivered only as an exception.

Water analytical standards are set up for a defined scope concerning concentration range and matrices. For parameters that are continuously measurable, a validation of the method by an external interlaboratory trial is obligatory. This holds for all the three technical committees on water quality, i.e. at German, European, and international level. In short, the philosophy of these round robin tests is as follows: The mean value is the reference value. An identical sample has to be measured according to the overall procedure at least in duplicate, preferably in three or four replicates. The evaluation of the test is performed according to ISO 5725-2 or based on it (Deutsche Einheitsverfahren, Verfahren A 0-3, 2003). The data are checked for type 1 (Grubbs test), type 2 (Grubbs test), and type 3 (Cochran test) outliers, the percentage of relative outliers should not exceed 25%. The minimum performance to be achieved is that at least 8 valid data sets and 24 outlier-free single data should remain. The coefficient of variation of reproducibility (interlaboratory) CV_R should not exceed 30%. If a true value does exist, there will be additional requirements concerning bias, which depend on the method. For national German standards authored by the AA "*Wasseruntersuchung*" (standards series 38400), the process of method development and especially the interlaboratory trial has to be documented extensively and published as the so-called *Validierungsdokument*. These documents are accessible on the web site of the German Chemical Society (GDCh) (http://www.gdch.de/strukturen/fg/wasser/publikat/vali/vd.htm).

Standards passed by the technical committee DIN NA 119-01-03 AA "*Wasseruntersuchung*", i.e. DIN, DIN EN, DIN ISO, or DIN EN ISO standards, are included into the German loose-leaf collection of standard methods for the examination of water, wastewater, and sludge, *Deutsche Einheitsverfahren zur Wasser-, Abwasser- und Schlammuntersuchung* (Deutsche Einheitsverfahren, 2009), which comprises 10 volumes with about 250 methods covering a large field of chemical, microbiological, and biological aspects of aquatic systems. Analytical methods are provided for quantification of a large number of inorganic and organic single compounds or jointly determinable substances, but also for general measures of effects and substances ("sum parameters"). For effects testing, bio-assays and sub-animal testing methods are available.

12.3.1.2 Material Specifications Relevant for Water Treatment

Standardisation of methods for examination is only one aspect of standardisation activities in the field of water quality. The technical committee DIN NA 119-04-02 AA *Arbeitsausschuss "Wasseraufbereitung"* ("Water treatment") delivers product standards, mostly as DIN EN standards, which were developed for chemicals and substances that are intended for use in drinking water treatment. These standards are a prerequisite for having the substances included in the official list of chemicals allowed for drinking water treatment in Germany.

12.3.2 Standardisation Activities in the Field of Nanomaterials

Standardisation activities relevant for nanoparticles are pursued by different standardisation organisations. Explicitly dedicated to nanomaterials and nanotechnology at international level are the ISO committee TC 229 "Nanotechnologies" and the technical committee E56 "Nanotechnology" of the industry-oriented international standardisation organisation ASTM International, which emerged from the American Society for Testing and Materials.

12.3.2.1 ISO/TC 229 "Nanotechnologies" and CEN/TC 352 "Nanotechnologies"

The ISO committee TC 229 "Nanotechnologies" was founded in 2005 and deals with standardisation in the field of nanotechnologies that, according to its scope, includes either or both the following: (i) "understanding and control of matter and processes at the nanoscale, typically, but not exclusively, below 100 nm in one or more dimensions where the onset of size-dependent phenomena usually enables novel applications" and (ii) "utilizing the properties of nanoscale materials that differ from the properties of individual atoms, molecules, and bulk matter, to create improved materials, devices, and systems that exploit these new properties."

The secretary and convenorship of ISO/TC 229 is held by the British Standards Institution (BSI). Meanwhile, 32 countries are involved by their standards bodies as participating countries and 11 further countries are attached as observing members. The technical work is organised in the working groups WG 1 "Terminology and nomenclature" (convenorship: Canada), WG 2 "Measurement and characterization" (convenorship: Japan), WG 3 "Health, safety and environmental aspects of nanotechnologies" (convenorship: USA), and WG 4 "Material specifications" (convenorship: China).

At European level, this committee is mirrored by the CEN/TC 352 "Nanotechnologies", also held by BSI and guided by the same convenor as ISO/TC 229. In Germany, the DIN has established the technical committee DIN NA 062-08-17 AA *Arbeitssausschuss "Nanotechnologien"* within the standards committee "Materials testing" (*Normenausschuss* DIN NA 062 "*Materialprüfung*") to deal with the subjects of ISO/TC 229 and CEN/TC 352 at national level.

ISO/TC 229 is working on about 40 standardisation projects, roughly subdivided into the series "Nanotechnologies", "Nanomaterials", and "Nanotubes" (see Table 12.3). The proposals are predominantly of Asian, US, and UK origin. Aspects of materials sciences, i.e. characterisation and specifications, make up a large part of the working programme. Several projects deal with the characterisation of single-walled carbon nanotubes or their purity evaluation by different methods (scanning electron microscopy, transmission electron microscopy, energy-dispersive X-ray analysis, UV-vis-NIR absorption, near-infrared photoluminescence/fluorescence, evolved gas analysis-gas chromatography mass spectrometry). Two projects deal with multi-walled carbon nanotubes. As nanomaterials of mineral origin, calcium carbonate and titanium dioxide are given attention.

Many of the projects are at the very beginning. More than half of the documents are intended to be delivered in the format of technical specifications. A technical specification on terminology (CEN ISO/TS 27687) and a technical report dealing with occupational health and safety (ISO/TR 12885) have already been delivered as final documents. Three documents on toxicity testing are already in the advanced stage of draft (DIS).

Table 12.3 Standards and standardisation projects of ISO/TC 229 "Nanotechnologies" (see also: http://www.iso.org/iso/standards_development/technical_committees/list_of_iso_technical_committees/). Additionally, the corresponding status in CEN (see also: http://www.cen.eu/CENORM/Sectors/TechnicalCommitteesWorkshops/CENTechnicalCommittees/) and DIN (see also: http://www.nmp.din.de/) is given, if documents on those levels already exist

Intended ISO deliverable	Status	Title
• Terminology		
ISO/TS 27687	ISO/TS 27687:2008, CEN ISO/TS 27687:2008, Vornorm DIN CEN ISO/TS 27687	Nanotechnologies – Terminology and definitions for nano-objects – Nanoparticle, nanofibre, and nanoplate
ISO/TS 80004-1	AWI	Nanotechnologies –Terminology and definitions – Part 1: Core terms
ISO/TS 80004-3	CD	Nanotechnologies –Terminology and definitions – Part 3: Carbon nano-objects
ISO/TS 80004-4	AWI	Nanotechnologies –Terminology and definitions – Part 4: Nanostructured materials
ISO/TS 80004-5	AWI	Nanotechnologies –Terminology and definitions – Part 5: Bio/nano interface
ISO 80004-6	AWI	Nanotechnologies –Terminology and definitions – Part 6: Nanoscale measurement and instrumentation
ISO/TS 80004-7	AWI	Nanotechnologies –Terminology and definitions – Part 7: Medical, health and personal care applications
ISO/TS 80004-8	NP	Nanotechnologies –Terminology and definitions – Part 8: Nanomanufacturing processes
• Material specification		
ISO/TS 11931-1	AWI	Nanotechnologies – Nano-calcium carbonate – Part 1: Characteristics and measurement methods
ISO/TS 11931-2	NP	Nanotechnologies – Nano-calcium carbonate – Part 2: Specifications in selected application areas
ISO/TS 11937-1	AWI	Nanotechnologies – Nano-titanium dioxide – Part 1: Characteristics and measurement methods
ISO/TS 11937-2	NP	Nanotechnologies – Nano-titanium dioxide – Part 2: Specifications in selected application areas
ISO 12025	CD	Nanomaterials – General framework for determining nanoparticle content in nanomaterials by generation of aerosols

Table 12.3 (continued)

Intended ISO deliverable	Status	Title
ISO/TS 12805	AWI	Nanomaterials – Guidance on specifying nanomaterials
• Toxicity testing		
ISO 10801	DIS prEN ISO 10801	Nanotechnologies – Generation of metal nanoparticles with the evaporation/condensation method for inhalation toxicity testing
ISO 10808	DIS prEN ISO 10808	Nanotechnologies – Characterisation of nanoparticles in inhalation exposure chambers for inhalation toxicity testing
ISO 29701	DIS prEN ISO 29701	Nanotechnologies – Endotoxin test on nanomaterial samples for in vitro systems – Limulus amebocyte lysate (LAL) test
• Occupational safety		
ISO/TR 12885	ISO/TR 12885:2008	Nanotechnologies – Health and safety practices in occupational settings relevant to nanotechnologies
ISO/TS 12901-1	AWI	Nanotechnologies – Guidance on safe handling and disposal of manufactured nanomaterials
ISO/TS 12901-2	NP	Guidelines for occupational risk management applied to engineered nanomaterials based on "control banding approach"
ISO/TR 13121	AWI	Nanotechnologies – Nanomaterial risk evaluation framework
ISO/TR 13329	NP	Nanomaterials – Preparation of Material Safety Data Sheets (MSDS)
• Analytical methods for material testing		
ISO/TR 11808	AWI prCEN ISO/TR 11808	Nanotechnologies – Guide to nanoparticle measurement methods and their limitations
ISO/TR 11811	AWI prCEN ISO/TR 11811	Nanotechnologies – Guide to methods for nanotribology measurements
ISO/TR 13014	AWI	Nanotechnologies – Guidance on physico-chemical characterisation of engineered nanoscale materials for toxicologic assessment
ISO/TS 10797	WD	Nanotubes – Use of transmission electron microscopy (TEM) in the characterisation of single-walled carbon nanotubes (SWCNTs)

Table 12.3 (continued)

Intended ISO deliverable	Status	Title
ISO/TS 10798	CD	Nanotubes – Scanning electron microscopy (SEM) and energy dispersive X-ray analysis (EDXA) in the characterisation of single-walled carbon nanotubes (SWCNTs)
ISO/TS 11251	CD	Nanotechnologies – Use of evolved gas analysis-gas chromatography mass spectrometry (EGA-GCMS) in the characterisation of single-walled carbon nanotubes (SWCNTs)
ISO/TS 11308	CD	Nanotechnologies – Use of thermo gravimetric analysis (TGA) in the purity evaluation of single-walled carbon nanotubes (SWCNT)
ISO/TS 10812	AWI	Nanotechnologies – Use of Raman spectroscopy in the characterisation of single-walled carbon nanotubes (SWCNTs)
ISO/TS 10867	CD	Nanotubes – Use of NIR-photoluminescence (NIR-PL) spectroscopy in the characterisation of single-walled carbon nanotubes (SWCNTs)
ISO/TS 10868	CD	Nanotubes – Use of UV-vis-NIR absorption spectroscopy in the characterisation of single-walled carbon nanotubes (SWCNTs)
ISO/TR 10929	CD	Measurement methods for the characterisation of multi-walled carbon nanotubes (MWCNTs)
ISO/TS 11888	AWI	Determination of mesoscopic shape factors of multi-walled carbon nanotubes (MWCNTs)
ISO/TS 13126	NP	Artificial gratings used in nanotechnology – Description and measurement of dimensional quality parameters
ISO/TS 13278	NP	Carbon nanotubes – Determination of metal impurities in carbon nanotubes (CNTs) using inductively coupled plasma mass spectroscopy (ICP-MS)

AWI – accepted work item; CD – Committee Draft; DIS – Draft of an International Standard; NP – new project; prCEN – Draft of a CEN deliverable; prEN – Draft of a European Standard; TR – Technical Report; TS – Technical Specification; WD – Working Draft.

12.3.2.2 ASTM E56 "Nanotechnology"

This committee, which now has about 250 members (individuals and organisations) and in which 22 countries are represented, was founded in 2005. Its technical work is organised in three subcommittees, whose active standards and current standardisation projects are given in Table 12.4.

The standardisation activities obviously focus on the medical applications of nanoparticles. Several guideline papers on cytotoxic effects of nanoparticles are in the portfolio of this committee (ASTM E 2524-08; ASTM E 2525-08; ASTM E 2526-08). The second focus is characterisation methods of nanoparticles, also in the aqueous phase. Of interest for applications on aquatic systems might be the guide on the measurement of particle size distribution of nanomaterials in suspension by photon correlation spectroscopy (PCS) (ASTM E 2490-09). Ancillary data to support the bias statement of this method are intended to be taken by atomic force microscopy (AFM) and transmission electron microscopy (TEM). As reference materials, the following substances are considered: nominal 10 nm gold nanoparticles (RM 8011), nominal 30 nm gold nanoparticles (RM 8012), nominal 60 nm gold nanoparticles (RM 8013), amine-terminated G6 PAMAM (polyamidoamine) dendrimers, hydroxy-terminated G6 PAMAM (polyamidoamine)dendrimers (ISO, IEC and NIST and OECD International Workshop, 2008). A standard practice document to calculate mean sizes and standard deviations of particle size distributions is already available (ASTM E 2578-07).

A standard on terminology, developed by the Subcommittee E56.01, appeared already in 2006 (ASTM E 2456-06). Inter alia, this standard specifies the term

Table 12.4 Active standards (E) and work items (WK) of the ASTM Committee E56 "Nanotechnology" (see also http://www.astm.org/COMMIT/SUBCOMMIT/E56.htm)

Responsible Subcommittee: E56-01 "Terminology"	
E 2456-06	Standard terminology relating to nanotechnology
Responsible Subcommittee: E56-02 "Physical, chemical, and toxicological properties"	
E 2524-08	Standard test method for analysis of haemolytic properties of nanoparticles
E 2525-08	Standard test method for evaluation of the effect of nanoparticulate materials on the formation of mouse granulocyte–macrophage colonies
E 2526-08	Standard test method for evaluation of cytotoxicity of nanoparticulate materials in porcine kidney cells and human hepatocarcinoma cells
E 2490-09	Standard guide for measurement of particle size distribution of nanomaterials in suspension by photon correlation spectroscopy (PCS)
E 2578-07	Standard practice for calculation of mean sizes/diameters and standard deviations of particle size distributions
WK 9952	Standard practice for measuring length and thickness of carbon nanotubes using atomic force microscopy methods
WK 21915	Zeta potential measurement by electrophoretic mobility
Responsible Subcommittee: E56-03 "Environment, health, and safety"	
E 2535-07	Standard guide for handling unbound engineered nanoscale particles in occupational settings (ASTM E 2535-07)

"transitive nanoparticle" for nanoparticles "exhibiting a size-related intensive property that differs significantly from that observed in fine particles or bulk materials", to be distinguished from the so-called non-transitive nanoparticles, which do not exhibit size-related intensive properties.

The work of the Subcommittee E56.03, dedicated to environment, health, and safety, up to now is limited to a standardisation project on handling of unbound nanoparticles in occupational settings (ASTM E 2535-07). The aquatic environment is not an issue.

12.3.2.3 Supplementary Standardisation Work of Further Technical Committees

Some metrological committees, some of them liaised to ISO/TC 229, standardise analytical or characterisation methods not especially tailored to nanoparticles, but applicable to them. This applies to several standards and standardisation projects of the technical committees ASTM E42 "Surface analysis" and ISO/TC 201 "Surface chemical analysis", on Auger electron spectroscopy (ASTM E827-02; ASTM E983-05; ASTM E995-04; ISO 15471; ISO 17973; ISO 17974; ISO 18118; ISO/TR 18394; ISO 18516; ISO/TR 19319; ISO 20903; ISO 21270; ISO 24236), X-ray photoelectron spectroscopy (ASTM E902-05; ASTM E995-04; ISO 15470; ISO 15472; ISO 18118; ISO/TR 18392; ISO 18516; ISO 19318; ISO 20903; ISO 21270; ISO 242375; ISO/TC 201/SC 7/CD 10810), secondary ion mass spectrometry (SIMS) (ISO 14237; ISO 17560; ISO 18114; ISO 20341; ISO/DIS 23812; ISO 23830), atomic force microscopy (AFM) (ASTM E 2382-04; ASTM E 2530-06; ISO/TC 201/SC 9/AWI 11939), and scanning probe microscopy (ASTM E 1813-96, 2007; ISO/TC 201/SC 9/CD 27911; ISO/TC 201/SC 9/WD 11775; ISO/TC 201/SC 9/WD 11952). In the ISO/TC 202 "Microbeam analysis", metrological standards using electron probe microanalysis (ISO 14594; ISO 14595; ISO 16592; ISO 17470; ISO 22489), scanning electron microscopy (SEM) (ISO 16700; ISO 22493; ISO/DIS 24597), transmission electron microscopy (TEM) (ISO/TC 202/SC 3/CD 25498; ISO/TC 202/SC 3/CD 29301), and energy dispersive spectrometry (EDS) (ISO 22309) are developed. Most of these methods are to be applied to the pure nanomaterials, not to nanomaterials in the aqueous phase. Particle characterisation and sizing is an issue to the subcommittee ISO/TC 24/SC 4 "Sizing by methods other than sieving", which develops standards on light scattering and diffraction, photon correlation spectroscopy (ISO 9276-1; ISO 9276-2; ISO 13320-1; ISO 13321; ISO 13323-1; ISO/TS 13762; ISO/DIS 21501-1; ISO 21501-2; ISO 22412), and centrifugal liquid sedimentation (ISO 13318-1; ISO 13318-2; ISO 13318-3). This subcommittee considers the aqueous phase as well (ISO 14887).

12.3.2.4 Standardisation in the Field of Nanomaterials – Coordination, Gaps, and Needs

In order to be mutually informed of standardisation activities in the field of nanotechnologies, convenors and stakeholders of ASTM, ISO, NIST (National

Institute of Standards and Technology), IEC (International Electrotechnical Commission), and OECD, dealing with these issues, met for an "International Workshop on Documentary Standards for Measurement and Characterisation for Nanotechnologies" (NIST, Gaithersburg, Maryland, USA) in February 2008. An extended report of this meeting is available (ISO, IEC and NIST and OECD International Workshop, 2008). On occasion of the workshop, an inventory of already existing standards and standardisation projects was compiled. Based on this, necessities and topical issues for further work were addressed.

In human toxicology, the tendency was reported to preferably apply in vitro instead of in vivo methods. The applicability of non-testing methods (quantitative structure–activity relationship – QSAR) for assessment in this field is in discussion. In ecotoxicology testing and assessment, it might be difficult to distinguish engineered nanoparticles from the natural background. Generally, a need was stated for characteristic reference materials in order to develop or verify suitable toxicity testing protocols. The issue of reference materials for toxicity testing is often addressed in pre-normative studies (BAUA, BfR and UBA, 2007; Aitken et al., 2007). Reference materials chosen for hypothesis testing should represent distinguishable classes concerning effects and should be usable for distribution in interlaboratory trials (Aitken et al., 2007).

In order to prepare meaningful toxicity tests, the reproducible dispersion of nanoparticles in the aqueous phase, their stabilisation, and reliable measurement of their concentration and size (distribution) were stated as essential (see also Warheit (2008)). In Crane et al. (2008) for instance, a minimum set of metrological characteristics required to prepare an ecotoxicity testing is postulated that should comprise the number or surface area per volume, the zeta potential, an electron microscope image (or similar) of the material in solution, and measurement of the individual particle size in the stock solutions used for dosing the test system. This would include mean particle size $\pm$ standard deviation for replicate samples of the stock solution and measurement of the size of any agglomerated or aggregated material in the test solution (means and standard deviation). It was stressed that particle size distribution in suspensions is not a material property, but a temporary state of dynamic equilibrium between dispersing and agglomerating effects. Measurement needs were stated on the meeting for polydispersity in suspensions. For validating instrument performance, polydisperse reference materials are necessary. In this context, a standardisation project on dispersion index or indices is envisaged by the ISO/TC 24/SC 4. Standardisation projects on zeta potential measurement are intended both by the committees ASTM E56 and ISO/TC 24/SC 4.

Especially for particle characterisation, there is still a need for suitable methods rather than for standards. Guidance documents are needed for correct choice of measurement techniques. A further challenge will be to determine the morphological state of measured particles, which is considered to be crucial for bacterial toxicity (Wiesner et al., 2006).

As a result of the Gaithersburg meeting, a nanotechnologies liaison coordination group was established, which will meet at each ISO/TC 229 plenary week.

12.4 Relevance of Current Standardisation Activities with Respect to Nanoparticles in the Water Cycle and Outlook on Further Needs

The majority of current standardisation projects of the standards bodies concerning the characterisation of nanomaterials aim at the specification of the bulk material.

Standardisation projects motivated by health and safety concerns mainly focus on occupational safety, above all respiratory health. Toxicity testing of nanomaterials is primarily directed by the intended use for medical applications. In this context, nanomaterials will be subject to the testing methods and the risk management process developed by the committee ISO/TC 194 "Biological evaluation of medical devices", which is in liaison with ISO/TC 229 (see for instance the standards series ISO 10993, especially EN ISO 10993-3; EN ISO 10993-4; EN ISO 10993-5; EN ISO 10993-9; EN ISO 10993-12; EN ISO 10993-16; EN ISO 10993-17; ISO 10993-20, some under review at present, and the guidance document ISO/TS 20993). One major aspect here is leaching behaviour.

The main impetus for the hitherto existing standardisation activities on nanoparticles in the aqueous phase arises from the needs of toxicity testing. From this point of view, dispersion procedures for particles in water and measurement of dispersion stability and particle distribution are an issue for standardisation.

It is striking that separation and fractionation methods are scarcely represented in the choice of methods for standardisation, neither in the portfolio of the technical committees dedicated to nanomaterials nor amongst the metrological standards considered as applicable for nanoparticles issues at the Gaithersburg meeting (ISO, IEC and NIST and OECD International Workshop, 2008). As the only method of this kind, centrifugal liquid sedimentation is to be mentioned (ISO 13318-1; ISO 13318-2; ISO 13318-3). Methods like membrane filtration, field-flow fractionation (FFF), and chromatographic techniques like size exclusion chromatography (SEC) or gel permeation chromatography, reported in the literature as suitable techniques applicable in nanoparticle analysis of aquatic systems (Hassellöv et al., 2008; Wiggington et al., 2007), are not included in standardisation activities at present. Nanoparticles do not seem to be perceived as a component of a more complex mixture in aqueous phase.

Obviously, the water cycle as a recipient for manufactured nanomaterials is not of major concern for the standardisation committees being in charge of nanomaterials or at least is not reflected in their current standardisation activities for risk assessment. This is in line with the fact that ISO/TC 147 "Water quality" is not amongst the internal liaison committees of ISO/TC 229 "Nanotechnologies".

On the subject of water quality examination and assessment, there are no active standards explicitly dealing with nanoparticles at present. On the level of current standardisation projects, nanoparticles are not yet considered, neither as part of the water matrix nor as target analyte. It is likely that many aspects and measurement needs that might arise from nanoparticles in the aquatic environment are already covered by existing methods of the standards collection, because aquatic sciences can look back on their long experience of dealing with particle-containing matrices and developed have been designing their standards based on this situation for quite some time.

As a component, nanoparticles are included in the term "whole water phase" in several analytical standards for trace analysis with respect to the European Water Framework Directive, and in the terms "suspended matter" and "turbidity", the latter a core parameter with respect to drinking water treatment. The nanoscale fraction of humic substances is included in the sum parameter dissolved organic carbon. In order to better understand the reactions in natural waters and in order to optimise drinking water treatment processes there are ongoing efforts to resolve the world of very small particles in the water matrix, to capture the particles beyond turbidity in very low concentrations, e.g. by LIBD (Wagner et al., 2002; Kim and Walther, 2007), and to characterise the nature of the DOC (Perminova et al., 2003; Schmitt-Kopplin and Junkers, 2007; Thieme et al., 2002). Whether these investigations result in reconsidering some conventions concerning filtration, e.g. the threshold 0.45 μm marking the dissolved fraction, or lead to the entry of highly sensitive particle sizing methods into the standards portfolio of water analytical methods is still open. Methods for measuring particle size and particles size distribution are needed in any case to meet the requirements of toxicity and ecotoxicology testing (Murdock et al., 2008), where the descriptor number concentration is requested (Crane et al., 2008), and in order to estimate the surface of the particles.

Should particular nanoparticles be assessed as a new type of toxicants that require special regulatory measures, the need for an analytical standard delivered by a standards body most probably might arise in the field of wastewater survey, as, besides the atmosphere, wastewater is the most likely input source for engineered nanoparticles (BAUA, BfR and UBA, 2007). When using methods of elemental analysis (e.g. ICP-OES, ICP-MS, TOC measurement devices) in order to trace nanoparticles or to prove their absence, the recovery of those methods should be critically checked. Whether a new conventional parameter is necessary which mimics the partition behaviour of nanoparticles, e.g. test columns to represent porous media, or which is indicative for their aggregation/coagulation behaviour, e.g., a kind of jar test, is too early to decide.

If nano-sized titanium dioxide is considered as a photocatalyst for drinking water treatment, a standardised materials specification will be necessary.

A portfolio of tests for degradability, as one of the endpoints that is proposed by OECD for nanomaterials testing, does already exist in the German standards collection for water and wastewater examination (DEV) (Deutsche Einheitsverfahren, 2009). In detail these are several tests for ultimate aerobic degradability in the aqueous phase – CO_2 headspace test (DEV, L 17) (EN ISO 14593), closed-respirometer test (DEV, L 22) (EN ISO 9408), carbon dioxide evolution test (DEV, L 23) (EN ISO 9439), and closed bottle test (DEV, L 46) (EN ISO 10707) – furthermore an activated sludge simulation test (DEV, L 41) (EN ISO 11733) and a test for ultimate anaerobic biodegradability in digested sludge (DEV, L 47) (EN ISO 11734). The method L 19 (EN ISO 10634) describes dispersion methods to prepare poorly water soluble compounds for testing by these methods. A guideline for selection of suitable tests for biodegradability is given in ISO/TR 15462. A prerequisite for performing meaningful degradation tests is that the carbon content of the test solution or dispersion is quantifiable. Moreover, the substances to be tested should not be toxic or inhibiting towards the bacteria used in the test. This has to be considered, as bacterial toxicity is suspected for some carbon-based nanomaterials (Lyon et al., 2005).

Also toxicity tests for different trophic levels are already available in the standards collection, e.g. a freshwater algal growth inhibition test with unicellular green algae (DEV, L 9) (EN ISO 8692), a *Daphnia* acute toxicity test (DEV, L 40) (EN ISO 6341), and a fish egg test (DEV, T 6) (EN ISO 15088). They are primarily designed for wastewaters, but should be applicable also for waters containing nanoparticles or after having ensured dispersion of particles by special measures according to ISO 14442 or EN ISO 5667-16 (DEV, L 1). A contact toxicity test designed for solids in suspension is given by the *Arthrobacter globiformis* test (DIN 38412-48), which might be suitable for application on nanoparticles. A kinetic luminescent bacteria test to be applied for sediments and solids is in preparation (ISO/TC 147/SC 5/DIS 21338).

The question to which extent nanoparticles have to be taken in account as a class of ingredients of their own or even as novel toxicants of aquatic systems and whether they call for additional standardisation activities in the field of water quality examination and assessment is still on the level of pre-normative research. In ISO/TC 147, the Ad-hoc Group "Key toxicity testing issues" within SC 5 "Biological methods", which is dealing with upcoming issues, is observing the discussion about nanoparticles as target toxicants in the water cycle, so that ISO/TC 147 can start action as soon as it is needed.

Acknowledgements The author would like to thank Christian Kabbe, Hans-Jürgen Pluta, and Andreas Paetz for thoroughly reading the manuscript.

References

AbwAG (2005) - Gesetz über Abgaben für das Einleiten von Abwasser in Gewässer (Abwasserabgabengesetz - AbwAG) in der Fassung vom 18. Januar 2005. Bundesgesetzblatt Teil I Nr. 5. 114–119.

AbwV (2002) - Verordnung über Anforderungen an das Einleiten von Abwasser in Gewässer (Abwasserverordnung - AbwV) in der Fassung vom 15. Oktober 2002. Bundesgesetzblatt Teil I Nr. 74, 4047–4121.

Aguar P, Murcia Nicolás JJ (2008) EU nanotechnology R&D in the field of health and environmental impact of nanoparticles. Version: 28 January 2008. Unit G4 Nano and Converging Sciences and Technologies, European Commission, Research DG. ftp://ftp.cordis.europa.eu/pub/nanotechnology/docs/final-version.pdf. Accessed 22 January 2009.

Aitken RJ, Hankin SM, Tran CL, Donaldson K, Stone V, Cumpson P, Johnstone J, Chaudhry Q, Cash S (2007) REFNANO: Reference materials for engineered nanoparticle toxicology and metrology. Final report on Project CB01099, 21 August 2007. http://www.iom-world.org/pubs/REFNANOReport.pdf. Accessed 15 January 2009.

ASTM E 827-02 – Standard practice for identifying elements by the peaks in Auger electron spectroscopy.

ASTM E 902-05 – Standard practice for checking the operating characteristics of X-ray photoelectron spectrometers.
ASTM E 983-05 – Standard guide for minimizing unwanted electron beam effects in Auger electron spectroscopy.
ASTM E 995-04 – Standard guide for background subtraction techniques in Auger electron spectroscopy and X-ray photoelectron spectroscopy.
ASTM E 1813-96 (2007) – Standard practice for measuring and reporting probe tip shape in scanning probe microscopy.
ASTM E 2382-04 – Guide to scanner and tip related artifacts in scanning tunneling microscopy and atomic force microscopy.
ASTM E 2456-06 – Standard terminology relating to nanotechnology.
ASTM E 2490-09 – Standard guide for measurement of particle size distribution of nanomaterials in suspension by photon correlation spectroscopy (PCS).
ASTM E 2524-08 – Standard test method for analysis of haemolytic properties of nano particles.
ASTM E 2525-08 – Standard test method for evaluation of the effect of nanoparticulate materials on the formation of mouse granulocyte-macrophage colonies.
ASTM E 2526-08 – Standard test method for evaluation of cytotoxicity of nanoparticulate materials in porcine kidney cells and human hepatocarcinoma cells.
ASTM E 2530-06 – Standard practice for calibrating the Z-magnification of an atomic force microscope at subnanometer displacement levels using Si(111) monatomic.
ASTM E 2535-07 – Standard guide for handling unbound engineered nanoscale particles in occupational settings.
ASTM E 2578-07 – Standard practice for calculation of mean sizes/diameters and standard deviations of particle size distributions.
Balbus JM et al. (2007) Meeting report: hazard assessment for nanoparticles – report from an interdisciplinary workshop. Environ Health Perspect 115: 1654–1659.
Biswas P, Wu C-Y (2005) Nanoparticles and the environment. J Air Waste Manage Assoc 55: 708–746.
Bundesanstalt für Arbeitsschutz und Arbeitsmedizin (BAUA), Bundesanstalt für Risikobewertung (BfR), Umweltbundesamt (UBA) (2007) Nanotechnologie: Gesundheits- und Umweltrisiken von Nanomaterialien – Forschungsstrategie. http://www.umweltbundesamt.de/technik-verfahren-sicherheit/dokumente/forschungsstrategie.pdf. Accessed 2 February 2009.
CEN ISO/TS 27687:2008 – Nanotechnologies – Terminology and definitions for nano-objects – Nanoparticle, nanofibre and nanoplate (ISO/TS 27687:2008). German version: Vornorm DIN CEN ISO/TS 27687:2008.
Commission Directive 2009/90/EC of 31 July 2009 laying down, pursuant to Directive 2000/60/EC of the European Parliament and of the Council, technical specifications for chemical analysis and montoring of water status. OJ L 201 (2009) 36–38.
Crane M, Handy RD, Garrod J, Owen R (2008) Ecotoxicity test methods and environmental hazard assessment for engineered nanoparticles. Ecotoxicology 17: 421–437.
Department for Environment, Food and Rural Affairs, UK (2007) Characterising the potential risks posed by engineered nanoparticles. A second UK Government Research Report. http://www.defra.gov.uk/environment/nanotech/research/pdf/nanoparticles-riskreport07.pdf. Accessed 22 January 2009.
Deutsche Einheitsverfahren zur Wasser-, Abwasser- und Schlammuntersuchung – Verfahren A 0-3 – Strategien für die Wasseranalytik: Anleitung zur Durchführung von Ringversuchen zur Validierung von Analysenverfahren. Wiley-VCH, Weinheim, und Beuth, Berlin, 57. Lieferung 2003.
Deutsche Einheitsverfahren zur Wasser-, Abwasser- und Schlammuntersuchung. Herausgegeben von der Wasserchemischen Gesellschaft – Fachgruppe in der Gesellschaft Deutscher Chemiker in Gemeinschaft mit dem Normenausschuß Wasserwesen (NAW) im DIN Deutsches Institut für Normung e. V. Wiley-VCH, Weinheim, und Beuth, Berlin, 75. Lieferung 2009. Table of content: http://www.wiley-vch.de/contents/dev/devih.pdf

DIN 820-1:2009 – Normungsarbeit; Grundsätze. DIN 820-2:2008 – Normungsarbeit – Teil 2: Gestaltung von Dokumenten. DIN 820-4:2000 – Normungsarbeit – Teil 4: Geschäftsgang.

DIN 38412-48:2002 – Deutsche Einheitsverfahren zur Wasser-, Abwasser- und Schlammuntersuchung – Testverfahren mit Wasserorganismen (Gruppe L) – Teil 48: *Arthrobacter globiformis*-Kontakttest für kontaminierte Feststoffe (L 48).

EN ISO 5667-16:1998 – Water quality – Sampling – Part 16: Guidance on biotesting of samples (ISO 5667-16:1998). German version: DIN EN ISO 5667-16:1999.

EN ISO 6341:1996 – Water quality – Determination of the inhibition of the mobility of *Daphnia magna* Straus (Cladocera, Crustacea) – Acute toxicity test (ISO 6341:1996). German version: DIN EN ISO 6341:1996.

EN ISO 8692:2004 – Water quality – Freshwater algal growth inhibition test with unicellular green algae (ISO 8692:2004). German version: DIN EN ISO 8692:2005.

EN ISO 9408:1999 – Water quality – Evaluation of ultimate aerobic biodegradability of organic compounds in aqueous medium by determination of oxygen demand in a closed respirometer (ISO 9408:1999). German version: DIN EN ISO 9408:1999.

EN ISO 9439:2000 – Water quality – Evaluation of the ultimate aerobic biodegradability of organic compounds in aqueous medium – Carbon dioxide evolution test (ISO 9439:1999); German version: DIN EN ISO 9439:2000.

EN ISO 10634:1995 – Water quality – Guidance for the preparation and treatment of poorly water-soluble organic compounds for the subsequent evaluation of their biodegradability in an aqueous medium (ISO 10634:1995). German version: DIN EN ISO 10634:1995.

EN ISO 10707:1997 – Water quality – Evaluation in an aqueous medium of the „ultimate“ aerobic biodegradability of organic compounds – Method by analysis of biochemical oxygen demand (closed bottle test) (ISO 10707:1994). German version: DIN EN ISO 10707:1998.

EN ISO 10993-3:2004 – Biological evaluation of medical devices – Part 3: Tests for genotoxicity, carcinogenity and reproductive toxicity (ISO 10993-3:2003). German version: DIN EN ISO 10993-3:2004.

EN ISO 10993-4:2007 – Biological evaluation of medical devices – Part 4: Selection of tests for interactions with blood. (ISO 10993-4:2002+Amd.1:2006). German version: DIN EN ISO 10993-4:2007.

EN ISO 10993-5:1999 – Biological evaluation of medical devices – Part 5: Tests for in vitro cytotoxicity (ISO 10993-5:1999). German version: DIN EN ISO 10993-5:1999.

EN ISO 10993-9:1999 – Biological evaluation of medical devices – Part 9: Framework for identification and quantification of potential degradation products. (ISO 10993-9:1999). German version: DIN EN ISO 10993-9:1999.

EN ISO 10993-12:2008 – Biological evaluation of medical devices – Part 12: Sample preparation and reference materials (ISO 10993-12:2007). German version: DIN EN ISO 10993-12:2008.

EN ISO 10993-16:2008 – Biological evaluation of medical devices – Part 16: Toxicokinetic study design for degradation products and leachables. (ISO 10993-16:2008). German version: DIN EN ISO 10993-16:2008.

EN ISO 10993-17:2003 – Biological evaluation of medical devices – Part 17: Establishment of allowable limits for leachable substances. (ISO 10993-17:2002). German version: DIN EN ISO 10993-17:2003.

EN ISO 11733:2004 – Water quality – Determination of the elimination and biodegradability of organic compounds in an aqueous medium – Activated sludge simulation test (ISO 11733:2004). German version: DIN EN ISO 11733:2004.

EN ISO 11734:1998 – Water quality – Evaluation of the „ultimate" anaerobic biodegradability of organic compounds in digested sludge – Method by measurement of the biogas production (ISO 11734:1995). German version: DIN EN ISO 11734:1998.

EN ISO 14593:2005 – Water quality – Evaluation of ultimate aerobic biodegradability of organic compounds in aqueous medium – Method by analysis of inorganic carbon in sealed vessels (CO_2 headspace test) (ISO 14593:1999). German version: DIN EN ISO 14593:2005.

EN ISO 15088:2008 – Water quality – Determination of the acute toxicity of waste water to zebrafish eggs (*Danio rerio*) (ISO 15088:2007). German version: DIN EN ISO 15088:2009.

Hassellöv M, Readman JW, Ranville JF, Tiede K (2008) Nanoparticle analysis and characterization methodologies in environmental risk assessment of engineered nanoparticles. Ecotoxicology 17: 344–361. doi: 10.1007/s10646-008-0225-x.

Hund-Rinke K, Herrchen M (2007) Technisches Vorgehen bei der Testung von Nanopartikeln. Bericht zum Forschungsprojekt im Auftrag des Umweltbundesamtes FuE-Vorhaben Förderkennzeichen 206 61 203/03. Fraunhofer Institut Molekularbiologie und Angewandte Oekologie IME, Schmallenberg und Aachen. Publikationen des Umweltbundesamtes, http://www.umweltdaten.de/publikationen/fpdf-l/3484.pdf

ISO 5725-2:1994 including Technical Corrigendum 1:2002 – Accuracy (trueness and precision) of measurement methods and results – Part 2: Basic method for the determination of repeatability and reproducibility of a standards measurement method. German version: DIN ISO 5725-2:2002.

ISO 9276-1:1998 – Representation of results of particle size analysis – Part 1: Graphical representation. German version: DIN ISO 9276-1:2004.

ISO 9276-2:2001 – Representation of results of particle size analysis – Part 2: Calculation of average particle sizes/diameters and moments from particle size distributions. German version: DIN ISO 9276-2:2006.

ISO 10993-20:2006 – Biological evaluation of medical devices – Part 20: Principles and methods for immunotoxicology testing of medical devices.

ISO 13318-1:2001 – Determination of particle size distribution by centrifugal liquid sedimentation methods – Part 1: General principles and guidelines.

ISO 13318-2:2001 – Determination of particle size distribution by centrifugal liquid sedimentation methods– Part 2: Photocentrifuge method.

ISO 13318-3:2004 – Determination of particle size distribution by centrifugal liquid sedimentation methods – Part 3: Centrifugal X-ray method.

ISO 13320-1:1999 – Particle size analysis – Laser diffraction methods – Part 1: General principles.

ISO 13321:1996 – Particle size analysis – Photon correlation spectroscopy. German version: DIN ISO 13321:2004.

ISO 13323-1:2000 – Determination of particle size distribution – Single-particle light interaction methods – Part 1: Light interaction considerations.

ISO 14237:2000 – Surface chemical analysis – Secondary-ion mass spectrometry – Determination of boron atomic concentration in silicon using uniformly doped materials.

ISO 14442:2006 – Water quality – Guidelines for algal growth inhibition tests with poorly soluble materials, volatile compounds, metals and waste water.

ISO 14594:2003 – Microbeam analysis – Electron probe microanalysis – Guidelines for the determination of experimental parameters for wavelength dispersive spectroscopy.

ISO 14595:2003 – Microbeam analysis – Electron probe microanalysis – Guidelines for the specification of certified reference materials (CRMs).

ISO 14887:2000 – Sample preparation - Dispersing procedures for powders in liquids.

ISO 15470:2004 – Surface chemical analysis – X-ray photoelectron spectroscopy – Description of selected instrumental performance parameters.

ISO 15471:2004 – Surface chemical analysis – Auger electron spectroscopy – Description of selected instrumental performance parameters.

ISO 15472:2001 – Surface chemical analysis – X-ray photoelectron spectrometers – Calibration of energy scales.

ISO 16592:2006 – Microbeam analysis – Electron probe microanalysis – Guidelines for determining the carbon content of steels using a calibration curve method.

ISO 16700:2004 – Microbeam analysis – Scanning electron microscopy – Guidelines for calibrating image magnification.

ISO 17470:2004 – Microbeam analysis – Electron probe microanalysis – Guidelines for qualitative point analysis by wavelength dispersive X-ray spectrometry.

ISO 17560:2002 – Surface chemical analysis – Secondary-ion mass spectrometry – Method for depth profiling of boron in silicon.
ISO 17973:2002 – Surface chemical analysis – Medium-resolution Auger electron spectrometers – Calibration of energy scales for elemental analysis.
ISO 17974:2002 – Surface chemical analysis – High-resolution Auger electron spectrometers – Calibration of energy scales for elemental and chemical-state analysis.
ISO 18114:2003 – Surface chemical analysis – Secondary-ion mass spectrometry – Determination of relative sensitivity factors from ion-implanted reference materials.
ISO 18118:2004 Surface chemical analysis – Auger electron spectroscopy and X-ray photoelectron spectroscopy – Guide to the use of experimentally determined relative sensitivity factors for the quantitative analysis of homogeneous materials.
ISO 18516:2006 – Surface chemical analysis – Auger electron spectroscopy and X-ray photoelectron spectroscopy – Determination of lateral resolution.
ISO 19318:2004 – Surface chemical analysis – X-ray photoelectron spectroscopy – Reporting of methods used for charge control and charge correction.
ISO 20341:2003 – Surface chemical analysis – Secondary-ion mass spectrometry – Method for estimating depth resolution parameters with multiple delta-layer reference materials.
ISO 20903:2006 – Surface chemical analysis – Auger electron spectroscopy and X-ray photoelectron spectroscopy – Methods used to determine peak intensities and information required when reporting results.
ISO 21270:2004 Surface chemical analysis – X-ray photoelectron and Auger electron spectrometers – Linearity of intensity scale.
ISO 21501-2:2007 – Determination of particle size distribution – Single particle light interaction methods – Part 2: Light scattering liquid-borne particle counter.
ISO 22309:2006 – Microbeam analysis – Quantitative analysis using energy-dispersive spectrometry (EDS).
ISO 22412:2008 – Particle size analysis – Dynamic light scattering (DLS).
ISO 22489:2006 – Microbeam analysis – Electron probe microanalysis – Quantitative point analysis for bulk specimens using wavelength-dispersive X-ray spectroscopy.
ISO 22493:2008 – Microbeam analysis – Scanning electron microscopy – Vocabulary.
ISO 23830:2008 Surface chemical analysis – Secondary-ion mass spectrometry – Repeatability and constancy of the relative-intensity scale in static secondary-ion mass spectrometry.
ISO 24236:2005 – Surface chemical analysis – Auger electron spectroscopy – Repeatability and constancy of intensity scale.
ISO 24237:2005 – Surface chemical analysis – X-ray photoelectron spectroscopy – Repeatability and constancy of intensity scale.
ISO/DIS 21338:2009 – Water quality – Kinetic determination of the inhibitory effects of sediment, other solids and coloured samples on the light emission of *Vibrio fischeri* (kinetic luminescent bacteria test).
ISO/DIS 21501-1:2008 – Determination of particle size distribution - Single particle light interaction methods – Part 1: Light scattering aerosol spectrometer.
ISO/DIS 23812:2008 – Surface chemical analysis – Secondary-ion mass spectrometry – Method for depth calibration for silicon using multiple delta-layer reference materials.
ISO/DIS 24597:2008 – Microbeam analysis – Scanning electron microscopy – Measurement methods of image sharpness.
ISO, IEC, NIST and OECD International workshop on documentary standards for measurement and characterization for nanotechnologies NIST, Gaithersburg, Maryland, USA, 26 – 28 February 2008. Final Report, http://www.standardsinfo.net/info/livelink/fetch/2000/148478/7746082/assets/final_report.pdf. Accessed 25 January 2009.
ISO/IEC Directives, Part 1 – Procedures for the technical work. 7th edition, 2009. ISO/IEC Directives, Part 2 – Rules for the structure and drafting of International Standards. Fifth edition, 2004.
ISO/TC 201/SC 7/CD 10810 Surface chemical analysis – X-ray photoelectron spectroscopy – Guide to analysis.

ISO/TC 201/SC 9/AWI 11939 Standards on the measurement of angle between an AFM tip and surface and its certified reference material.
ISO/TC 201/SC 9/CD 27911 – Surface chemical analysis - Scanning probe microscopy- Definition and calibration of lateral resolution of a near-field optical microscope.
ISO/TC 201/SC 9/WD 11775 – Surface chemical analysis – Scanning probe microscopy – Determination of cantilever normal springconstants.
ISO/TC 201/SC 9/WD 11952 – Guideline for the determination of geometrical quantities using scanning probe microscopes - Calibration of measuring systems.
ISO/TC 202/SC 3/CD 25498 – Method of selected area electron diffraction for transmission electron microscopy.
ISO/TC 202/SC 3/CD 29301 – Microbeam analysis – Analytical transmission electron microscopy – Methods for calibrating image magnification by using periodic pattern in layered structure.
ISO/TR 12885:2008 Nanotechnologies – Health and safety practices in occupational settings relevant to nanotechnologies.
ISO/TR 15462:2006 – Water quality – Selection of tests for biodegradability.
ISO/TR 18392:2005 Surface chemical analysis – X-ray photoelectron spectroscopy – Procedures for determining backgrounds.
ISO/TR 18394:2006 – Surface chemical analysis – Auger electron spectroscopy – Derivation of chemical information.
ISO/TR 19319:2003 – Surface chemical analysis – Auger electron spectroscopy and X-ray photoelectron spectroscopy – Determination of lateral resolution, analysis area, and sample area viewed by the analyser.
ISO/TS 13762:2001 Particle size analysis - Small angle X-ray scattering method.
ISO/TS 20993:2006 – Biological evaluation of medical devices – Guidance on a risk-management process.
Kim J, Walther C (2007) Laser-induced breakdown detection. In: Wilkinson KJ, Lead, JR (eds.) Environmental Colloids and Particles. Behaviour, Separation and Characterisation. John Wiley & Sons, Chichester.
Lauterwasser C (ed.) (2005) Small sizes that matter: Opportunities and risks of nanotechnologies. Report in co-operation with the OECD Internal Futures Programme. http://www.oecd.org/dataoecd/37/19/37770473.pdf. Acessed 2 February 2009.
Lyon DY et al. (2005) Bacterial cell association and antimicrobial activity of a C_{60} water suspension. Environ Toxicol Chem 4: 2757–2762; Tsao N, et al. (2001) Inhibition of group A *Streptococcus* infection by carboxyfullerene. Antimicrob Agents Chemother 45(6):1788–1793; Tsao N, et al. (1999) Inhibition of *Escherichia Coli*-induced meningitis by carboxyfullerene. Antimicrob Agents Chemother 43:2273–2277. Cited in Crane et al., 2008.
Murdock RC, Braydich-Stolle L, Schrand AM, Schlager J, Hussain S (2008) Characterization of nanomaterial dispersion in solution prior to in vitro exposure using dynamic light scattering technique. Toxicol Sci 101(2): 239–253.
NanoKommission der Deutschen Bundesregierung (2008) Verantwortlicher Umgang mit Nanotechnologien. http://www.nanotruck.de/fileadmin/nanoTruck/redaktion/download/Druckschriften/Nanokommission_Abschlussbericht_2008.pdf. Accessed 2 February 2009.
National Science and Technology Council (2008) Strategy for nanotechnology-related environmental, health, and safety research. http://www.nano.gov/NNI_EHS_Research_Strategy.pdf. Accessed 2 February 2009.
OECD Guidelines for the Testing of Chemicals – Test No. 107: Partition Coefficient (n-octanol/water): Shake Flask Method. Adopted 25 July 1995.
OECD – Organisation for Economic Cooperation and Development (2008) OECD Guidelines for the testing of chemicals. PDF Edition, ISSN 1607-310X, http://oberon.sourceoecd.org
Perminova IV, Frimmel FH, Kudryavtsev AV, Kulikova NA, Abbt-Braun G, Hesse S, Petrosyan VS (2003) Molecular weight characteristics of humic substances from different environments

as determined by size exclusion chromatography and their statistical evaluation. Environ Sci Technol 37: 2477–2485.
Schmitt-Kopplin P, Junkers J (2007) Modern electrophoretic techniques for the characterisation of natural organic matter. In: Wilkinson KJ, Lead, JR (eds.) Environmental colloids and particles. Behaviour, separation and characterisation. John Wiley & Sons, Chichester.
Science Policy Council, U.S. Environmental Protection Agency (2007) U.S. Environmental Protection Agency Nanotechnology White Paper. EPA 100/B-07/001. http://es.epa.gov/ncer/nano/publications/whitepaper12022005.pdf. Accessed 2 February 2009.
Thieme J, Schmidt C, Abbt-Braun G, Specht C, Frimmel FH (2002) X-ray microscopy studies of refractory organic substances. In: Frimmel FH, Abbt-Braun G, Heumann KG, Hock B, Lüdemann H-D, Spiteller M (eds.) Refractory Organic Substances in the Environment. Wiley-VCH, Weinheim.
TrinkwV - Verordnung über die Qualität von Wasser für den menschlichen Gebrauch vom 21. Mai 2001(Trinkwasserverordnung - TrinkwV 2001). Bundesgesetzblatt 2001 I Nr. 24 S. 959 ff.
Verband der Chemischen Industrie e. V. (2008a) Leitfaden zur abgestuften Sammlung von Gefährdungsinformationen zur Risikobeurteilung von Nanomaterialien. http://www.vci.de/template_downloads/tmp_VCIInternet/122417 Gefaehrdungsinformatione_Risikobeurteilung_Nanomaterialien_2008_02_28~DokNr~122417~p~101.pdf. Accessed 2 February 2009.
Verband der Chemischen Industrie e. V. (2008b) Responsible production and use of nanomaterials http://www.vci.de/default~cmd~shd~docnr~122306~lastDokNr~116417.htm. Accessed 2 February 2009.
Wagner T, Bundschuh T, Schick R, Schwartz T, Köster R (2002) Investigation of colloidal water content with laser-induced breakdown detection during drinking water purification. Acta Hydrochim Hydrobiol 30: 266–274.
Warheit DB (2008) How meaningful are the results of nanotoxicity studies in the absence of adequate material characterization. Toxicol Sci 101(2): 183–185.
Wiesner MR, Lowry GV, Alvarez P, Dionysiou D, Biswas P (2006) Assessing the risks of manufactured nanomaterials. Environ Sci Technol 40: 4336–4345.
Wiggington NS, Haus KL, Hochella MF Jr. (2007) Aquatic environmental nanoparticles. J Environ Monit 9: 1306–1316.
Working Party on Manufactured Nanomaterials (2008) List of manufactured nanomaterials and list of endpoints for phase one of the OECD testing programme. OECD Environment, Health and Safety Publications, Series on the safety of manufactured nanomaterials, No. 6. ENV/JM/MONO(2008)13/REV, http://www.oecd.org/dataoecd/38/11/41040311.pdf. Accessed 29 January 2009.

Index

F.H. Frimmel, R. Niessner (eds.), *Nanoparticles in the Water Cycle*,
DOI 10.1007/978-3-642-10318-6, © Springer-Verlag Berlin Heidelberg 2010